T0140629

Studies in Computational Intelligence

Volume 1083

Series Editor

Janusz Kacprzyk, Polish Academy of Sciences, Warsaw, Poland

The series "Studies in Computational Intelligence" (SCI) publishes new developments and advances in the various areas of computational intelligence—quickly and with a high quality. The intent is to cover the theory, applications, and design methods of computational intelligence, as embedded in the fields of engineering, computer science, physics and life sciences, as well as the methodologies behind them. The series contains monographs, lecture notes and edited volumes in computational intelligence spanning the areas of neural networks, connectionist systems, genetic algorithms, evolutionary computation, artificial intelligence, cellular automata, self-organizing systems, soft computing, fuzzy systems, and hybrid intelligent systems. Of particular value to both the contributors and the readership are the short publication timeframe and the world-wide distribution, which enable both wide and rapid dissemination of research output.

Indexed by SCOPUS, DBLP, WTI Frankfurt eG, zbMATH, SCImago.

All books published in the series are submitted for consideration in Web of Science.

More information about this series at https://link.springer.com/bookseries/7092

Theodor Borangiu · Damien Trentesaux ·
Paulo Leitão

Editors

Service Oriented, Holonic and Multi-Agent Manufacturing Systems for Industry of the Future

Proceedings of SOHOMA 2022

 Springer

Editors
Theodor Borangiu
Faculty of Automatic Control
and Computers
University Politehnica of Bucharest
Bucharest, Romania

Damien Trentesaux
LAMIH UMR CNRS 8201
Université Polytechnique Hauts
de France UPHF
Valenciennes Cedex 9, France

Paulo Leitão
Research Centre in Digitalization and
Intelligent Robotics (CeDRI)
Instituto Politecnico de Bragança
Bragança, Portugal

ISSN 1860-949X   ISSN 1860-9503   (electronic)
Studies in Computational Intelligence
ISBN 978-3-031-27434-3   ISBN 978-3-031-24291-5   (eBook)
https://doi.org/10.1007/978-3-031-24291-5

© The Editor(s) (if applicable) and The Author(s), under exclusive license
to Springer Nature Switzerland AG 2023
This work is subject to copyright. All rights are solely and exclusively licensed by the Publisher, whether
the whole or part of the material is concerned, specifically the rights of translation, reprinting, reuse of
illustrations, recitation, broadcasting, reproduction on microfilms or in any other physical way, and
transmission or information storage and retrieval, electronic adaptation, computer software, or by similar
or dissimilar methodology now known or hereafter developed.
The use of general descriptive names, registered names, trademarks, service marks, etc. in this
publication does not imply, even in the absence of a specific statement, that such names are exempt from
the relevant protective laws and regulations and therefore free for general use.
The publisher, the authors, and the editors are safe to assume that the advice and information in this
book are believed to be true and accurate at the date of publication. Neither the publisher nor the
authors or the editors give a warranty, expressed or implied, with respect to the material contained
herein or for any errors or omissions that may have been made. The publisher remains neutral with regard
to jurisdictional claims in published maps and institutional affiliations.

This Springer imprint is published by the registered company Springer Nature Switzerland AG
The registered company address is: Gewerbestrasse 11, 6330 Cham, Switzerland

Preface

This volume gathers the peer-reviewed papers presented at the 12th edition of the International Workshop on Service-oriented, Holonic and Multi-Agent Manufacturing Systems for the Industry of the Future, SOHOMA'22, organized on 22–23 September 2022 by the CIMR Research Centre in Computer Integrated Manufacturing and Robotics of University Politehnica of Bucharest in collaboration with Polytechnic University Hauts-de-France (the LAMIH Laboratory of Industrial and Human Automation Control, Mechanical Engineering and Computer Science) and Polytechnic Institute of Bragança (the CeDRI Research Centre in Digitalization and Intelligent Robotics).

The main objective of SOHOMA workshops is to foster innovation in smart and sustainable manufacturing and logistics systems and in this context to promote concepts, methods and solutions for the digital transformation of manufacturing through service orientation and agent-based control with distributed intelligence.

The theme of the SOHOMA'22 Workshop is '**Virtualization—a multifaceted key enabler of Industry 4.0 from holonic to cloud manufacturing**'.

Virtualization in industrial control with computing and information-based system support signifies creating virtual versions of computer hardware platforms, computer network resources, industrial equipment or software control systems. There are two virtualization technologies currently available that allow creating virtual instances of various physical devices on different layers of digital environments including CPUs and process controllers, sensors, storage, networking, communication protocols and interfaces: (1) *digital twins* and (2) *agentification* of reality-reflecting class entities such as activities/processes, outcomes/products, facilities/resources and procedures/orders. The latter is either directly associated with the distribution of intelligence in heterarchical control systems or used as implementing framework of *holonic* orchestration.

The distribution of intelligence within industrial control systems and the need for collaborative decisions of strongly coupled shop floor entities—specific for Cyber-Physical Production Systems (CPPS)—led to the adoption of a new modelling approach for optimized and robust orchestration of batch processes: the *holonic control* paradigm. This approach is based on the virtualization of a set of abstract entities: *products* (reflecting the client's needs and value propositions), *resources* (technology, humans—reflecting the producer's profile, capabilities, skills) and *orders* (reflecting business solutions) that are modelled by autonomous holons collaborating in holarchies by means of their information counterparts— **intelligent agents** that are organized in dynamic clusters to reach a common, production-related goal.

At present, the *digital twin* concept evolved to a highly advanced modelling and simulation technology used for enhanced product and process design, event-driven resource maintenance and health monitoring, anomaly detection and intelligent manufacturing control with prediction of behaviours, production costs, reality-awareness and optimization. Modelling reality is done with two classes of digital twins (DT): *data driven* (with physical counterpart) that relies on distributed edge computing and storage of shop floor data in Industrial IoT frameworks and *model driven* (without physical twin)—a digital simulation for design and analysis. The multi-purpose character of digital twin virtualization in manufacturing derives from its main application areas:

- *Simulation with software in the loop*: a model-driven DT with control software operating in simulation exactly as it will be deployed; the virtual model is used to design equipment, tune parameters, validate layouts and compare performances in special conditions.
- *Synchronizing the virtual twin with its physical twin*: this type of DT is involved in all activities that imply the physical twin (e.g., shop floor resource, product with embedded intelligence) and provide control with situation awareness such as health monitoring or diagnostic.
- *Embedded simulation faster than real time*: a combination of a model-driven DT with physical, non-material twins of activity instances (e.g., individual product execution orders) allowing the latter to virtually execute much faster than real time (at computer speed) the intentions of the decision-makers (e.g., operation scheduling and allocation to resources, cost optimization, control tuning) on the twins of the resource instances, offering predictive situation awareness.

In cyber-physical systems for manufacturing, virtualization is performed on two layers: (1) shop floor entities (resources, products) and (2) MES (information system):

1. The set of workloads are directly connected to the *physical entities* that are virtualized by agents representing a resource or product; these are either *local individual workloads* of a physical device, e.g., resource monitoring, product routing, dispatching and tracking a production order, or *local collaborative workloads* of the shop floor, e.g., assigning jobs to resources for products in simultaneous execution.
2. The set of *global workloads* connected to *MES computing entities* are virtualized by expert or advisor agents acting as predictor of resource team utilization cost, optimization engine, system scheduler or anomaly detector; these are

workloads relative to the full production life cycle at the farthest horizon—the batch of ordered products.

Virtualization also represents the main enabling technology for *cloud computing applied to the industrial domain* regardless of the delivery method considered. The IT model of cloud computing (CC) is extended to services that orchestrate operational technology hardware and software elements (industrial controllers, application software) that monitor and control factory resources, products, processes and events. This dual cloud control and computing model (CCoC) is the real-time partition of the *cloud manufacturing* (CMfg) enterprise model mapped on its technical layer, and which:

- Transposes pools of factory resources (robots, machines, controllers, proprietary software) into on-demand making services.
- Enables pervasive, on-demand network access to a shared pool of configurable high-performance computing (HPC) resources—servers, storage and applications that can be rapidly provisioned and released as services to global MES workloads with minimal management.

The research of the SOHOMA scientific community is aligned to these technological developments that build up actual trends and development priorities for cyber-physical systems in the manufacturing, supply chain and logistics industries:

1. Sustainable Holonic Architectures (HCAs), able to optimize the production costs with reality-awareness, adaptability at environment changes and robustness at technical disturbances that ensure fault tolerance of the entire production system.
2. Reconfigurable Manufacturing Systems (RMS), designed for a) rapid adjustment of production capacity and functionality, in response to new circumstances, by rearrangement or change of their components: machines, robots and conveyors for the entire production system, new sensors and new controller algorithms; b) quick reaction to changing product demand, producing a new product on an existing system or integrating new process technology into existing manufacturing systems.
3. Mapping the factory data streams and global MES functions to specific workloads in the cloud, defined in terms of activity scheduling, resource assignment and behaviour forecast; the latter incorporate AI and ML capabilities. The industrial sector is interested in deploying autonomous workloads to achieve higher productivity and better operational safety.
4. Developing the four key drivers to sustain the paradigm Logistics 4.0: data automation and transparency (end-to-end visibility over the supply chain, logistics control tower, optimization software); new production methods (robotized palletizing, stereovision-based picking); new methods of physical transport (driverless vehicles, autonomous pickers, drones); digital platforms (shared warehouse and transport capacity, cross-border platforms).

5. Manufacturing as a Service (MaaS)—new models of service-oriented, knowledge-based manufacturing systems, virtualizing and encapsulating shop floor and MES workloads into cloud networked services—will also address product design for 'open manufacturing' and the knowledge and infrastructure sharing in cloud collaborative manufacturing enterprises.
6. Developing Industry 4.0 models and frameworks using advanced technologies: robotics, additive manufacturing, augmented reality, extended digital modelling and simulation through digital twins, horizontal/vertical enterprise integration, cloud, cybersecurity, big data and analytics and the industrial Internet.

This integrated approach derives from the research performed in the last years by members of the scientific community SOHOMA, based on recently developed key digital and cyber-technologies: cloud and fog computing, digital twins, edge computing, optimization, cognitive robotics, intelligent image processing, machine vision, additive machining, artificial intelligence and machine learning, and ethical and societal models in future industrial systems: human–machine cooperation, human integration in cyber-physical systems, ethics of the artificial, humans in Industry 4.0 (I4.0), low-cost digital solutions for manufacturing:

- Data mining and analysis of data collected during the utilization phase to design new product-service systems.
- CPS-enabled reconfiguration of automated manufacturing systems: (1) deployment of legacy production equipment and systems; (2) increasing autonomy and intelligence of existing machinery and robots; (3) adaptation through context awareness and reasoning aiming at making machinery and robots aware of their surroundings; (4) developing multi-layered, decentralized control architectures in which resources can take autonomous decisions.
- Intelligent decision-making in cloud manufacturing through big data streaming and machine learning; combining data-driven digital twins for predictive situation-awareness with model-driven digital twins simulating the reality of interest faster than real time with software in the loop.
- Sharing of data/information from all the supply chain's elements to support continuous monitoring and automatic control of all the production phases while preserving security and confidentiality of data shared along the supply network.
- The adoption of IoT and CPS as implementing frameworks for digital supervision and control of manufacturing.
- Digital manufacturing on a shoestring—low-cost digital solutions for SMEs.
- Service manufacturing which includes design for open manufacturing, optimization, maintenance, supply and distribution activities, all of them being offered in service-oriented architectures.
- Fostering the open and universal manufacturing enterprise—responsive to the X-as-a service model, where X covers design, manufacturing, supply and distribution and supports resource sharing and networking in the cloud.

Following the workshop's technical programme, this book is structured into seven parts that group chapters reporting research results in topics related to service-oriented, holonic and multi-agent manufacturing systems for the industry of the future: *Part 1*: Applications of multi-agent and holonic systems in smart manufacturing; *Part 2*: Digital twins in industrial systems; *Part 3*: Factory—product life cycle value stream for Industry 4.0; *Part 4*: Education for Industry 4.0; *Part 5*: Performance, ethics and operations management in internal logistics 4.0; *Part 6*: Industry 4.0 technologies and low-cost digitization; and *Part 7*: Reconfigurable Manufacturing Systems.

Three main categories of changes in manufacturing foreshadowed by the transformations defined by Industry 4.0 are approached in the book: (1) The evolution of the manufacturing paradigm to mass customization, agility to frequent changes in product types and characteristics, and reconfigurability in response to unexpected environment and business conditions; the organizational perspective of this paradigm—with important implications in logistics—results from the relationship between products and manufacturing that will expand in future years from the 'many-to-one' production model (represented by integrated manufacturing) to the 'many-to-many' model with highly distributed production facilities (represented by open and universal manufacturing); (2) The use of new digital information, communication and control technologies such as holonic and multi-agent organization, digital twins, product intelligence, edge and cloud computing, machine learning integrated in IIoT and CPS architectures; (3) Human–system integration: Industry 4.0 developments integrate physical-, human- and software components leading to human CPS which is a system of interconnected systems (computers, devices, people) that interact in real time, working together to achieve the goals of production systems at environmental changes.

A number of workshop papers interpret and implement the special characteristics of the reference architecture model of the Industry 4.0 framework (RAMI4.0) reflecting the combination of life cycle and value stream with a hierarchically structured approach for the definition of I4.0 components. These papers highlight the interaction between four I4.0 structuring aspects: horizontal integration through value networks; vertical integration within a factory or shop floor; life cycle management and end-to-end engineering; human beings orchestrating the value stream. Proposals for I4.0 component implementation are made relative to the layers that define the structure of their virtual representation, the position and stage in the dual factory-product life cycles and the location of functionalities within the enterprise's hierarchy of information and control systems.

Cyber-physical systems are seen here as the backbone implementing infrastructure to realize Industry 4.0 compliant solutions that rely on a network of manufacturing components with cyber and physical counterparts, which have the ability of performing decentralized and autonomous decisions. CPS are combined with emergent digital technologies like Internet of Things, digital twin (a novel approach associated with the virtualization of production models, embedded in the design, virtual commissioning and optimization of processes and machines), big data, cloud control and computing, machine learning and cobots and are applied in the book's papers to different sectors, such as manufacturing, energy, logistics, construction and health care.

For the presented developments, RAMI4.0 serves as reference for engineering I4.0 systems, allowing designers to develop Industry 4.0 compliant solutions, based on the CPS approach, covering the hierarchy of industrial infrastructures and the life cycle/value stream of different production assets.

All these aspects are presented in this book, which we hope you will find useful reading.

October 2022   Theodor Borangiu
                                                    Damien Trentesaux
                                                       Paulo Leitão

Contents

Reconfigurable Manufacturing Systems

Applications of Multi-Agent and Holonic Systems in Smart Manufacturing

A Review of Multi-agent Systems Used in Industrial Applications

Silviu Răileanu[(⊠)] and Theodor Borangiu

Department of Automation and Applied Informatics, University Politehnica of Bucharest,
Bucharest, Romania
{silviu.raileanu,theodor.borangiu}@upb.ro

Abstract. The purpose of this review is to present the different facets of multi-agent systems (MAS) with an accent on their two main areas: MAS for developing decentralized systems and MAS for modelling and simulations. In this respect a brief state of the art is presented with emphasis on the principal applications of MAS, mainly in manufacturing. Some current platforms are briefly analyzed in order to show their potential in developing MAS applications and simulations.

Keywords: Multi-agent system · Agent-oriented programming · Agent-based modelling and simulation · Multi-agent platform

1 Introduction

As stated in the literature [40], a multi-agent system is a system composed by multiple computing entities identified as agents. Originating from computer science, the agents have two main capabilities: i) operating in an autonomous manner which is exhibited through local intelligence, and ii) interacting with other agents by communication. Nowadays, mainly due to the advances in the electronics field, the autonomy has been greatly enhanced by allowing software agents [3] to be executed on mobile, dedicated, powerful (in terms of processing capabilities) and miniaturized platforms with low energy consumption rates. This allows increasing their mobility (and portability), while at the same time increasing the way they perceive the environment through a wide range of sensors and network connections. The observations mentioned above reflect the evolution of embedded devices which will most likely govern our future both in what concerns daily life (e.g., people interacting with dedicated software such as personal assistant), and more importantly in what concerns the industrial domain, as is the case for decentralized and heterogenous systems. Thus, an interconnected environment which is characterized by the decentralization of the computation power has risen, an aspect that goes hand in hand with the principles of MAS. Therefore, the MAS formalism has been increasingly adopted for the realization of decentralized control systems of distributed processes and services.

With its origins rooted in the artificial intelligence (AI) domain [64], the multi-agent formalism has gained more attention over the last years as can be seen from the number of publications on this topic (Fig. 1). Considering its particular name, which does

© The Author(s), under exclusive license to Springer Nature Switzerland AG 2023
T. Borangiu et al. (Eds.): SOHOMA 2022, SCI 1083, pp. 3–22, 2023.
https://doi.org/10.1007/978-3-031-24291-5_1

not have overlays with other non-relevant domains, a simple search using the "multi-agent" keyword on a standard bibliographic database (www.sciencedirect.com) containing information from academic journals, conference proceedings, and other documents in various academic disciplines, reveals an increase interest in this domain (Fig. 1). Even though the multi-agent formalism is used to solve a wide variety of problems, two main usage areas can be identified by thoroughly analysing literature. These areas are: i) *research/theoretical*, for modelling, simulating, and optimizing systems which are inherently decentralized (such as control systems for different and geographically separated resources, supply chains, distributed utilities - gas, water, energy, or even animal populations), and ii) *practical* to support the effective development of decentralized (control) systems using agent-based technologies such as frameworks [36], toolkits [42, 43], and even methodologies since some programming languages that are used to implement distributed systems [25] are not inherently labelled as multi-agent.

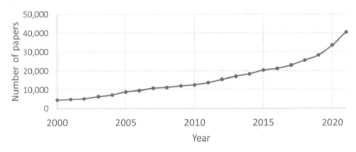

Fig. 1. Evolution of the number of articles on the "multi-agent" topic in the last 20 years

In this respect, the objectives of this article are to review relevant work making use of the MAS formalism (the considered work consists of scientific articles, projects/implementations, and platforms), synthetise the main domains that use MAS and focus on the manufacturing domain - both as a standalone element and also as a part of the value added chain. The reminder of the article is structured as follows: Sect. 2 presents the applications of multi-agent systems for modelling, simulation and automation of processes, Sect. 3 focuses on the usage of multi-agent in manufacturing control and Sect. 4 synthetizes the conclusions of the considered applications while also dealing with future research directions and usages of the MAS formalism.

2 Multi-agent Systems for Process Automation, Modelling and Simulation

Process automation and simulation have been recognized by academia and industry as vital areas, with many research programs and industrial projects initiated. The current state of such projects is characterized by an increase in complexity due to the large number of entities involved and, consequently, of the interactions between them. This is why the MAS formalism has been employed for such big systems. A thorough analysis of the current research [1] has differentiated between: a) Multi-Agent Systems (identified simply as MAS) - a concept used To for split work between multiple cooperating and decision-making nodes, traditionally used to implement decentralized control and monitoring systems, and b) Agent Based Modelling and Simulation (ABMS) [12] - a concept focusing on the analysis of individual and collective behaviours of agents in a simulated environment. Depending on its usage, the MAS formalism can be applied to a wide variety of domains such as factory automation [10], transport and distribution of goods in logistics, energy systems [6], building automation [10, 11], business processes [8], networking [9]. These domains are managed by systems engineering concepts and methods that use MAS frameworks and software tools to develop decentralized control systems, while ABMS is mostly used to analyse emergent behaviours, permit self-organization, and evaluate MAS-related architectures and interaction protocols in domains such as chemistry [13], biology[14], finance [15, 16], transport [17], and also for process optimization (through iterative simulations) [18].

The reminder of this section presents representative applications using the MAS formalism and associated frameworks to solve complex control and optimization problems in the domains previously presented.

2.1 Manufacturing Control and Factory Automation

Decentralized by definition, MAS, especially MAS frameworks, are employed in systems engineering to implement the control part of technical systems whose models resemble complex networks [1]. One of the main domains where good results have been obtained is factory automation, as reported in [4, 30]. These results have put in evidence what the authors have called *industrial agents* [11]; they propose a solution to integrate software agents with physical hardware devices and a set of interfacing practices. In this respect, it is differentiated between two decision-making levels (Fig. 2):

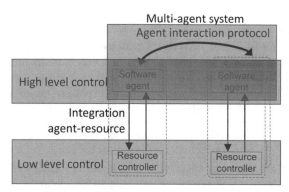

Fig. 2. Agentified resources [55] and interaction between software agents and automation devices (control units) [19]

1. Low level control associated with real-time control and physical (usually embedded) devices such as PLC (programmable logic controllers), COM (computer-on-module) and industrial computers, with direct electrical connection to the controlled process,
2. High level control implementing strategic decision-making tasks such as optimisation, integration/communication, configuration, a.o.

Figure 2 illustrates the agentification of control entities [55]: each control entity has an associated software agent in charge with the decision-making process and with the communication network through a standard medium. These agents are seen as automation objects - abstractions of mechanical devices with encapsulated intelligence [7], thus allowing component reusability.

Usually, this high level control is implemented using MAS platforms; however, due to its highly intensive computation and to the fact that there are limitations on what low level control devices can execute (e.g., a PLC cannot execute a JAVA agent such as JADE), the high level control layer cannot operate in real-time. Recommendations are thus computed on the MAS layer, and are synchronized with the low control level in different ways as depicted in Fig. 3.

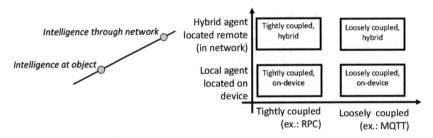

Fig. 3. Different locations of agent decision-making and interaction modes [10, 19, 56]

This interface concept has evolved from software agents associated to intelligent products [56] to industrial agents associated to all entities involved in the manufacturing process. In this respect the initial 'local and remote' axis (vertical in Fig. 3) describing the location of intelligence (agent) has now been completed with another axis (horizontal in Fig. 3) describing a more technological aspect on how the interaction is done (tightly, e.g., RPC call and loosely coupled).

2.2 Power and Energy Systems

The current green trend in energy generation which focuses on the transition from fossil fuel to renewable energy is characterized by decentralization and intelligent decision-making both on the consumer and on the provider side. In this respect, both MAS formalism and practical MAS frameworks are used to model energy distribution systems and consequently implement them. Such an example is reported in [6] where the authors present the problem with its MAS/decentralized aspects, the important requirements of the energy market, and finally a real implementation using a common MAS framework – JADE. Like for the manufacturing domain, energy systems split the control problem in two layers: a) a *cyber layer* modelling a virtual power plant (VPP) responsible with gathering offers and requests from agentified providers and consumers implemented through a VPP agent, and b) a *physical layer* responsible with the implementation of the agents onto runtime environments, labelled here as technical units (TU). These TUs are split into b1) *automation devices* such as PLCs or System-on-a-Chip boards with Ethernet interfaces that implement gateway functions to communicate with modern and legacy field bus systems capable of direct control over the electrical infrastructure and b2) *hardware devices* such as industrial PC or IoT gateways running the low-level TU agents. In this regard the proposed architecture looks like the FIPA interaction protocols [57], where VPP agents represent the initiators while TU agents act as responders.

The article presents a solution which closes the gap between MAS research and its application in industry, power, and energy systems in this case. The implementation also validates the idea that, by using a dedicated MAS framework (e.g., JADE), standard interaction protocols (such as FIPA) and dedicated hardware (PLCs, IoT gateways, industrial PCs) and software (OPC, webservices, MODBUS TCP, NodeRed) infrastructures, it is possible to meet market demands such as real-time capabilities (1 s for decision-making processes), fault-tolerance/redundancy, media break between Internet from the Internet Protocol, and security concerns by using private networks (Fig. 4).

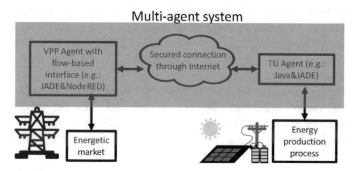

Fig. 4. Implementation example of a MAS-inspired energy control system [6]

2.3 Building Automation

An intelligent building is a building equipped with embedded devices to support automatic adaptation to the changing environment conditions, and to provide comfortable living conditions for its occupants. The main objectives of a building automation system are: reducing energy consumption, decreasing operating costs (e.g., automating access control), security, remote access control and operation, and improved equipment usage. Due to the increased complexity of such systems, MAS is nowadays employed to assure their decentralized operation that was traditionally performed in a centralized mode. As in the case of the previous article, MAS is employed here to automate energy- related tasks such as minimization of energy consumption.

Related to this problem, the authors of [20] propose a multi-agent system whose scope is to combine the control of comfort conditions with an energy saving strategy.

The resulting MAS is composed of four types of agents, each implementing a specific aspect of the building automation system: *occupant agent* is a human-representing agent that monitors and adapts to the user's activities, learns the user's styles and preferences; *zone agent* is an artefact-representing agent in charge with the control of a particular zone considering sensor data and acting as a negotiator between itself and the occupant agent; *manager agent* directly interfaces to the building management system; *environmental control agent* monitors and controls different environmental parameters in each zone.

This system has been designed for two types of scenarios: single occupant, where the decision is made through reconciliation between occupant and environment agents, and multi-occupant scenario, where the decision is made through negotiation between occupant agents using a blackboard system.

2.4 Complex Networks

Nowadays research has three major challenges: i) long distances (e.g., reaching other planets), ii) small distances (e.g., manipulate microorganisms) and iii) complexity (e.g., dealing with interconnected systems, controlling their emergent behaviour and assuring coherence). By modelling these interconnected systems as complex networks, the authors of [1] show how the MAS formalism can be applied to control and optimize such systems. A large majority of distributed systems can be modelled as graphs where the vertices represent the composing entities, and the edges represent the relations/communications/interactions between them. In practice, these models take non-regular topologies to better represent the real-world systems. This type of model is called a complex network.

The computer networks, the Internet, the social networks [9], the utility networks [6, 7], the transportation networks [40, 65], the chemical, bio and bio-chemical systems [14] are some examples of real-world systems which can be modelled as complex networks and the MAS formalism can be applied to analyse different scenarios. In this respect the authors separate MAS frameworks, used to implement decentralized systems, from agent-based modelling and simulation (ABMS), used as a tool to analyse sets of individual entities modelled as agents.

The research article [1] presents a set of representative applications of complex networks and agent-based controls used to implement important processes in different Systems Engineering domains such as: *manufacturing* and *supply chains* (scheduling, simultaneous processing of orders, quality assurance, real-time customization and context-aware servicing and maintenance), *electricity power grids* (MAS using smart meters implement distributed control for the power systems without a hierarchical or central supervisor thus increasing the system's capacity, reliability and economic efficiency), *transportation systems* (cyber-physical systems [5] composed of agents associated to vehicles and transportation nodes are disposed in a hierarchical control architecture where strategic decisions are taken at network layer and operational decision, influenced by the upper layer, are implemented at vehicle layer), *utilities distribution* (agent-based systems suit well at dealing with the nowadays ubiquity of sensors used to operate individual gas/water source and demand points).

Based on the analysis of these decentralized systems a set of partial conclusions is drawn: a) MAS can represent more realistic examples of real cases, b) MAS will play an essential role providing complex networks with intelligence for flow control, evolution and security, c) the validity of agent-based models is given by the real data acquired through a wide network of sensors, d) artificial intelligence through machine learning will be employed to develop new agent models and behaviours, e) the term MAS is evolving into Cyber Physical Systems due to the closer link between the informational and physical layers, f) MAS can be employed in the blockchain technology.

2.5   Networking Analysis and Security

Employed in the development of decentralized systems, MAS always operate in a digital environment implemented as a local or wide network of computers or, more commonly embedded devices. With application covering a multitude of domains - some of them with critical requirements such as electrical power distribution [6] - security and traffic analysis become key factors in the development of such decentralized control systems [27] that should be reliable and resilient in face of disturbances and cyber events. In this respect, the authors of [21] propose a distribution automation system (DAS) implemented using a MAS, the latter using detection and mitigation algorithms that can identify anomalies, abnormal activities, and unusual system operations of the DAS. Thus, operations are divided between a set of layered agents responsible with central operation, circuit breakers and terminal units. The MAS formalism in this report is used to model interactions between composing agent which monitor themselves to detect Denial-of-Service (DoS) attacks.

In another study [22], the MAS formalism is used to implement an Intrusion Detection System (IDS) in which software agents implement security monitoring functions directly at the host element. The presented architecture is divided into modules implemented as software agents: reflex agents (which sense the environment and additionally make operation logs and generate alarms), analysis agents (in charge with attack analysis based on the data captured by the reflex agents), and decision agents (which act as goal-based agents to make the appropriate decisions). In addition to the fixed IDS architectures, the authors study the usage of mobile agents (agents travelling along different hosts in the network to be monitored) to detect events that may relate to intrusions.

The main advantages offered by mobile agents are:

a)   Mobility (by relocating the analysis and decisional agents closer to the source lowers the response latency),
b)   Robustness, and
c)   Fault tolerance (due to the replication of the agents).

In todays interconnected digital ecosystem which represents the operating environment for MAS the emergence of cyber-attacks is an important concern for system process operation. From a process control point of view cyber-attacks typically interfere with the Supervisory Control and Data Acquisition (SCADA) [1], affecting the software agents connected directly to the Internet.

2.6 Business Processes Automation (RPA Agent)

Through implementation of a networked solution, MAS can be employed at the level of its composing entities - the agents, which are characterized by intelligence and bidirectional interaction with the environment (perception and action). In this respect, software agents are employed to automate business tasks (usually document-related) such as automatic fill in of electronic documents, digitalization of written documents using OCR, a.o. The aforementioned processes are realized by software agents while being subordinated to a centralized system. This is how the term Robotic Process Automation (RPA) has appeared: the use of software that incorporates technologies such as artificial intelligence (AI) and machine learning (ML) to automate routine, high-volume tasks that are sensitive to human error [8].

The agents represent software applications which capture the computer environment (presentation layer as seen by the human operator) and operate in an accelerated time allowing the processing of high volumes of data. By integrating and usage of proprietary systems, RPA agents can operate in a near-automated way. A classical architecture (Fig. 5) is composed of several autonomous RPA agents interconnected through a common database with other peer RPA agents and with higher level agents (such as MES in the case of manufacturing). RPA agents are subject to the same monitoring, analytics and reporting scrutiny as their human counterparts. Although designed for business automation, RPA agents are employed in manufacturing [23] for automatic status check, equipment setup, material tracking, alarm detection and data collection, especially in the situations where total integration is not possible, and the developer makes use of the presentation layer. If total integration is possible RPA agents are perceived as industrial agents [11] and interact through standard communication protocols with the low-level control (e.g., OPC, MODBUS, a.o.).

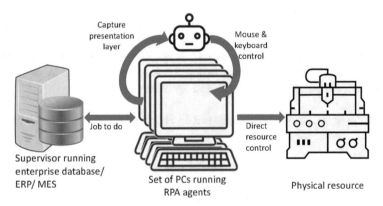

Fig. 5. RPA agent operation

2.7 ABMS Applications

More numerous, the applications in which the ABMS concept is used are suited generally for cognitive sciences, complex social interaction [1] and study of biologic or chemical systems where individual components (microbes, molecules, bacteria, atoms) are modelled as agents. In this respect, a MAS results where agents can interact in an accelerated time (to obtain results faster than doing the physical experiment); it is also possible to model and simulate the evolution of theoretical, hard to replicate systems, such as bioreactors [14]) or potentially dangerous ones (microbial systems [13, 14]). As stated in the beginning of this article, the fundamentals of this branch of multi-agent systems, namely agent-based modelling and simulation (ABMS), differ from the standard MAS formalism. In this regard, various agent-based modelling and simulation platforms have been developed that deal with such complex interactions, such as NetLogo [41, 42], MaDKit [43, 44], and Repast [47, 48].

In literature there are two approaches to model a living system [14]: a) *population-based*, a top-down approach that relies on mathematical laws and modelling, and b) *individual-based*, a discrete, bottom-up approach which relies on modelling individual entities and their interactions; these interactions are both with peer entities and with the environment. The two approaches do not exclude each other, but they are rather complementary in the sense that mathematical modelling (characteristic to the population-based approach) can be used to model individual interactions. Thus, by applying the MAS formalism, realistic computational models result which can be used to accurately simulate processes such as: bioreactors, chemical plants or biological processes (e.g., prey-predator model). An example of a model for a biological process is given in Fig. 6 where animals (wolf and sheep) are modelled as mobile agents while the environment is modelled as a stationary agent.

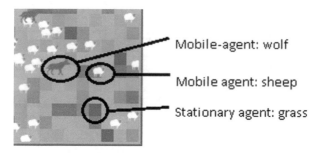

Fig. 6. Translation of a biological process into an individual-based model [41]

3   Agent Platforms

An *agent platform* (or framework) is a software system used by the agents and by the environment to execute operations. In this software ecosystem agents operate to achieve their goals [2, 26]. The agent platform may additionally support the development of agents and agent-based applications or simulations. On this point, there are two types of agent platforms: a) used for agent-oriented programming (AOP) [40], and b) used for agent-based modelling (ABM) [40]. From a historic point of view AOP is an evolution of the OOP concept where the basic unit of the programming paradigm is the agent.

The AOP models and implements a project as a collection of components called agents that are characterized by autonomy, proactivity, and the ability to communicate. In the case of AOP it is correct to talk about projects, not applications, because a MAS is composed of several software agents, each of them representing a different application usually running on different machine. A broad distinction in this field can be made between middleware- and reasoning-oriented systems [28]. The middleware-oriented systems focus on the social intelligence of agents (communication) and how this can be standardized both at message level (structure, grammar) and at interaction protocol level (sequence of messages). The most cited reference for interoperability, which is used in the development of the majority of AOP platforms is the Foundation of Intelligent and Physical Agents (FIPA [57]). The reasoning-oriented systems focus on local intelligence and decision making, mainly using the BDI (Belief Desire Intent) software development model.

Throughout all the references cited so far, the JADE framework [36] is the most used in agent-related applications. Technically any cross-platform programming language that supports network communication can be used to develop decentralized projects that comply with MAS requests (autonomy, ability to communicate and decision-making). Based on a thorough study of available AOP platforms [29, 51] throughout the years, it has been noticed that most of them tend to become obsolete very soon as is the case with most products from the IT domain [26]. As a consequence, in this paper the authors have chosen to mention only the most used platforms as depicted in Table 1. An exhaustive comparison of different alternatives (24 such platforms) can be found in [29] with some additions in [2013].

Table 1. Comparison of different AOP platforms

AOP platform	Cross platform	Latest release	Programming language	Comments
JADE [36, 37]	Yes	JADE 4.5, 2017	Java	Open source platform for peer-to-peer agent-based applications
WADE [34, 35]	Yes	WADE 3.6.0, 2017	Java	Open source platform based on JADE for workflows and agent-based applications
JADEX [32, 33]	Yes	Jadex 4, 2021	Java and XML	Implementation of a BDI-infrastructure for JADE agents

(*continued*)

Table 1. (*continued*)

AOP platform	Cross platform	Latest release	Programming language	Comments
JACK [38, 39]	Yes	JACK 5.6, 2015	Java extension	Commercial agent platform with proprietary plan language and GUI that uses the BDI software model
Erlang [25, 31]	Yes	Erlang/OTP25, 2022	Erlang	Distributed applications with built-in fault tolerance, able to run on a multitude of systems; is suitable for industrial implementations

Unlike an AOP platform, which is used to develop decentralized systems and whose main advantage over a standard programming language resides in the facility of multi-platform execution, an ABMS toolkit is a modelling software where agents (representing both mobile entities and environment) can interact based on an explicit set of rules [24]. It is an alternative to the mathematical modelling of a system or to performing a real-world experiment. The reasons why ABMS gain popularity among disciplines not related to software development are its simplicity and flexibility in terms of defining a distributed system, providing at the same time its natural description while also capturing its emergent behaviour [40]. Based on a detailed study of the available literature the ABMS toolkits [29] are a little bit more numerous and tend to be more up to date, mainly because they are employed by a wider range of researchers. Below is an open list of the most cited ABMS toolkits with some observations associated to each of them.

Table 2. Comparison of different ABMS platforms

ABMS toolkit	Latest release	Programming language	Observations
Netlogo [41, 42]	Version 6, 2021	NetLogo, need JVM	Free and easy to learn MAS programmable and modelling environment which enables exploration of emergent phenomena

(*continued*)

Table 2. (*continued*)

ABMS toolkit	Latest release	Programming language	Observations
MaDKit [43, 44]	Version 5, 2021	Java	Lightweight Java library for MAS design and simulation
Mesa [45, 46]	2022, GitHub	Python	ABM framework developed for Python with the features: modular components, visualization using browser, embedded analysis tool
Repast [47, 48]	Version 2.9.1, 2022	JAVA, C++, Python	Free and open-source ABM under continuous development. The Repast Suite can be used both at workstation level and on clustered infrastructures and supercomputers
AnyLogic [18, 49]	Version 8.7.8, 2022	Graphical modelling language and Java	ABM platform, available both free and commercial, used for MAS and event-driven simulations with applications in domains related to supply chains

4   Multi-agent Formalism in Industrial Control

This section deals with integrating intelligent software agents with low-level automation devices, a concept encountered in MAS-related literature as *industrial agents*. Relatively a new concept [11], industrial agents refer to the link between the informational and physical layers, which provides low level control devices such as PLCs with high level functionalities such as optimization algorithms, Artificial Intelligence (AI), interaction protocols, a.o.

The evolution to the physical-informational link [50] and its materialization into industrial agents started from the extension of the software agents influence towards the physical resources which resulted in the niche term *holon* [53, 54], traditionally encountered in manufacturing control [52], and the more generic term cyber-physical system (CPS) [51]. Consequently, this section will present applications of the MAS formalism in manufacturing control and its extension to supply chain management, monitoring, and traceability.

4.1 Optimization: Operation Scheduling on Resources

Until the 4th Industrial Revolution manufacturing control systems were influenced by organizational structures resulting in top-down control architectures. However, in today's hard to predict and fast evolving marketplace, a more dynamic scheduling method is preferred to deal with frequent changes in a more independent way from upper layers. In this respect the MAS formalism is adopted by associating agents to manufacturing entities. These agents perform local tasks based on upper-layer recommendations [66], one of them being decentralized allocation of operations to resources. A standard MAS interaction protocol based on market rules (sellers and buyers/participants and initiators [57]) is used to realize the optimization process - Contract Net Protocol (CNP [58]).

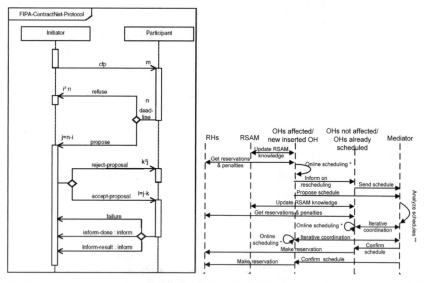

Fig. 7. Comparison between the original CNP [57, 58] and its adaptation for operation scheduling on resources in a manufacturing system [59]

Initially designed to allocate tasks to resources (Fig. 7 left), the protocol has been adapted to consider some particular aspects of the manufacturing systems such as: a) differentiation of operations into processing and transportation, b) the fact that in order to execute a final product several processing operations are needed, and between these operations precedencies might appear so the process has to be reiterated, c) responses from resources and operation execution time are not instantaneous (Fig. 7 right).

4.2 Configuring Groups of Resources for Production

Another mechanism used along with decentralized operation scheduling is the selection of the needed resources to execute a batch of products. This problem has been dealt with by the CoBASA (Coalition Based Approach for Shop floor Agility) architecture [60]. Agents are associated to resources extending their physical operations with software capacities such as negotiation, contracting, and servicing, able to participate in coalitions/consortia. The result of this process is the Manufacturing Resource Agent (MRA). Multiple MRAs can associate, forming a coalition based on production requirements and under the guidance of a Coordinator agent (CA) to generate aggregated functionalities beyond the individual capabilities of each composing agent. This process is described in Fig. 8.

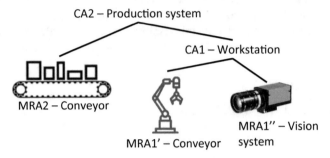

Fig. 8. Formation of an agent-based manufacturing coalition (inspired from [60])

4.3 Dynamic Routing of Work in Process

An important task of a manufacturing control system is routing products towards the workstations scheduled to perform operations.

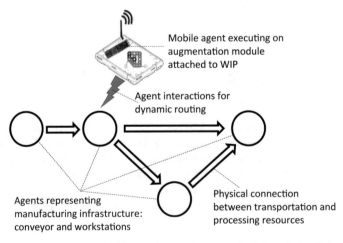

Fig. 9. Agent-based solution for intelligent routing of products [61]

Currently, due to advances in electronics, it is possible to associate embedded devices with decisional capabilities directly to the entities composing the flow of Work in Process (WIP). Thus, the concept of active product [61] or *intelligent product* [62, 63], appeared. An intelligent product is composed of the passive product with an associated augmentation module, executing an agent in charge of computing local decisions. Authors of article [61] present a system where MAS is used to measure and update online transportation times; this information is used for adaptive routing: mobile agents representing products measure and share information in order to dynamically choose a path which minimizes the transportation time. A graphical description of this process along with the localization of agents (tightly coupled and on-device according to [11]) is given in Fig. 9.

4.4   Work in Process Traceability. Implementing the Intelligent Product Concept

A current trend in manufacturing is the Make-to-Order (MTO) concept [67], which is characterized by the fact that the customer is pulling the orders, and can also personalize and track them. In this respect, the MAS technologies are employed here in order to track orders both in the manufacturing stage [56] and in the supply chain stage [63]. From a practical point of view current technologies such as auto-identification, wireless communications, miniaturized sensors and actuators allow virtualizing physical events and hence the WIP and final products, while also allowing the customer to dynamically influence the way the order is produced, stored, or transported. The key concept here is the "intelligent product" which represents an upgrade of the classical physical product, see Fig. 10.

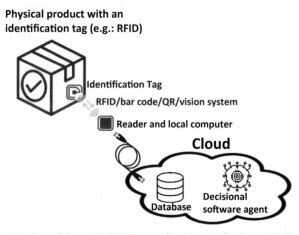

Fig. 10. Intelligent product with remote intelligence implemented using agent-based technology (adapted from [63])

This intelligence can be understood in two ways: a product with intelligent information (that is virtualized and processed at fixed points by agents representing manufacturing or supply chain entities such as processing or handling resources) or a product with intelligence. In the second case the intelligence is represented by an agent located locally or remote [56, 62, 63]. The advantage of an associated agent is that it can produce events of interest for the lifecycle of the associated product based both on its physical location (determined, for example, from an auto-identification – RFID system) and on its history and internal state stored in a database. A graphical description of such a system is given in Fig. 10.

5 Discussion and Perspectives

As stated in the introduction MAS (Fig. 1) gains more and more ground in the design, analysis, simulation, and development of decentralised projects. Nevertheless, a comparison between the different domains of MAS, namely AOP (Table 1) and ABMS (Table 2) emphasizes that projects involving simulation and consequently ABMS tools are more numerous, more up to date and more used. In brief, in what concerns industrial agents, there is the JADE platform with its extensions (WADE and JADEX), and some scarce projects involving JACK/Erlang, while in what concerns MAS modelling and simulation toolkits these are generally standalone applications (e.g., NetLogo, Swarm, a.o., see Table 2).

In this respect, our belief is that the power of MAS stands in its capacity to model a decentralized system both from structural point of view (agents with associated roles and locations) and from dynamic point of view (interaction protocols and the standardization of exchanged messages, see FIPA), thus making MAS looking as a formalism rather than a tool. This is emphasized by the fact that decentralized projects can be implemented without the agent term. These projects contain just software applications running on separated resources and applications developed in dedicated programming languages, the only requirement being that these applications understand each other (e.g., standardization of communication protocols, which makes it possible to communicate between PLCs and embedded devices). Another advantage of MAS AOP platforms is that they offer the possibility to easily implement control strategies based on states by using the concept of behaviour (e.g., sequential, parallel, FSM behaviours – JADE). The above observation is reinforced by the fact that all projects presented here use AOP platforms as a decisional layer which is completed/extended by dedicated applications which interconnect the informational and physical layers (e.g., the agentification of an industrial robot implies the association of a dedicated PC/embedded system which executes the industrial agent due to the practical fact that developers do not have access to the controller's operating system or in some cases - such as PLCs it is not even possible to run the AOP platform on that particular system). All authors of the papers cited in this article, with applications in manufacturing and electrical energy, state the importance of associating the software agents with industrial protocols to close the gap between the informational and physical layers.

Despite the gap between AOP and ABMS our belief is that: a) the MAS approach will gain more terrain in industrial applications mainly due to miniaturization of embedded devices allowing thus the distribution of intelligence and decision making, b) by executing software agents directly on the control devices their operating time will decrease (approach real-time), c) the MAS formalism is a solution for implementing cyber-physical systems, and d) it will be possible to measure agents' performance.

References

1. Herrera, M., Perez Hernandez, M., Parlikad, A.K., Izquierdo, J.: A Review on Control and Optimisation of Multi-Agent Systems and Complex Networks for Systems Engineering (2020). https://doi.org/10.20944/preprints202001.0282.v1
2. Leszczyna, R.: Evaluation of Agent Platforms (ver. 2.0). EUR 23508 EN. Luxembourg (Luxembourg): European Commission; JRC47224 (2008)
3. Bhamra, G.S., Verma, A.K., Patel, R.B.: Intelligent software agent technology: an overview. Int. J. Comput. Appl. **89**, 19–31 (2014). https://doi.org/10.5120/15474-4160
4. Latsou, C., Farsi, M., Erkoyuncu, J., Morris, G.: Digital twin integration in multi-agent cyber physical manufacturing systems. IFAC-PapersOnLine **54**, 811–816 (2021). https://doi.org/10.1016/j.ifacol.2021.08.096
5. Lyu, G., Fazlirad, A., Brennan, R.: Multi-agent modeling of cyber-physical systems for IEC 61499 based distributed automation. Procedia Manufact. **51**, 1200–1206 (2020). https://doi.org/10.1016/j.promfg.2020.10.168
6. Woltmann, S., Kittel, J., Stomberg, M., Coordes, A.: Using multi-agent systems for demand response aggregators: a technical implementation. In: 25th IEEE International Conference on Emerging Technologies and Factory Automation (ETFA), pp. 911–918 (2020). https://doi.org/10.1109/ETFA46521.2020.9212168
7. Obitko, M., Mařík, V.: Ontologies for multi-agent systems in manufacturing domain. In: DEXA Workshop, pp. 597–602 (2002). https://doi.org/10.1109/DEXA.2002.1045963
8. Jamison, N.: Robotic Process Automation: A New Era of Agent Engagement, A Frost & Sullivan White Paper (2017). www.frost.com
9. Costa-Montenegro, E., et al.: Multi-agent system model of a BitTorrent network. In: 9th ACIS International Conference on Software Engineering, Artificial Intelligence, Networking, and Parallel/Distributed Computing, pp. 586–591 (2008). https://doi.org/10.1109/SNPD.2008.168
10. Leitão, P., Karnouskos, S., Ribeiro, L., Moutis, P., Barbosa, J., Strasser, T.: Common practices for integrating industrial agents and low level automation functions. In: 43rd Annual Conference of the IEEE Industrial Electronics Society (IECON), pp. 6665–6670 (2017). https://doi.org/10.1109/IECON.2017.8217164
11. IEEE Recommended Practice for Industrial Agents: Integration of Software Agents and Low-Level Automation Functions, Developed by the Standards Committee of the IEEE Industrial Electronics Society, Approved 24 September 2020, IEEE SA Standards Board
12. Lane, J.: Method, theory, and multi-agent artificial intelligence: creating computer models of complex social interaction. J. Cogn. Sci. Relig. **1**, 161–180 (2014). https://doi.org/10.1558/jcsr.v1i2.161
13. Taherian, M., Mousavi, S., Chamani, H.: An agent-based simulation with NetLogo platform to evaluate forward osmosis process (PRO Mode). Chin. J. Chem. Eng. **26**, 2487–2494 (2018). https://doi.org/10.1016/j.cjche.2018.01.032
14. Ginovart, M., Prats, C.: A bacterial individual-based virtual bioreactor to test handling protocols in a Netlogo platform. IFAC Proc. Vol. **45**(2), 647–652 (2012). ISSN 1474-6670, ISBN 9783902823236
15. Damaceanu, R.-C.: An agent-based computational study of wealth distribution in function of resource growth interval using NetLogo. Appl. Math. Comput. **201**(1–2), 371–377 (2008). ISSN 0096-3003
16. Souissi, M., Bensaid, K., Rachid, E.: Multi-agent modeling and simulation of a stock market. Invest. Manag. Financ. Innov. **15**, 123–134 (2018). https://doi.org/10.21511/imfi.15(4).2018.10

17. He, B.Y., et al.: A validated multi-agent simulation test bed to evaluate congestion pricing policies on population segments by time-of-day in New York City. Transp. Policy **101**, 145–161 (2021). ISSN 0967-070X

18. Muravev, D.F., Hu, H., Rakhmangulov, A., Mishkurov, P.: Multi-agent optimization of the intermodal terminal main parameters by using AnyLogic simulation platform: case study on the Ningbo-Zhoushan Port. Int. J. Inf. Manag. **57**, 102133 (2021). ISSN 0268-4012

19. Leitão, P., Karnouskos, S., Ribeiro, L., Lee, J., Strasser, T., Colombo, A.W.: Smart agents in industrial cyber–physical systems. Proc. of IEEE **104**(5), 1086–1101 (2016)

20. Duangsuwan, J., Liu, K.: A multi-agent system for intelligent building control - norm approach. In: 2nd International Conference on Agents and Artificial Intelligence ICAART, vol. 2, pp. 22–29 (2010)

21. Choi, I.-S., Hong, J., Kim, T.-E.W.: Multi-agent based cyber attack detection and mitigation for distribution automation system. IEEE Access **8**, 183495–183504 (2020). https://doi.org/10.1109/ACCESS.2020.3029765

22. Herrero, Á., Corchado, E.: Multiagent systems for network intrusion detection: a review. In: Herrero, Á., Gastaldo, P., Zunino, R., Corchado, E. (eds.) Advances in Intelligent and Soft Computing, vol. 63, pp. 143–154. Springer, Heidelberg (2009). https://doi.org/10.1007/978-3-642-04091-7_18

23. George, A., Ali, M., Papakostas, N.: Utilising robotic process automation technologies for streamlining the additive manufacturing design workflow. CIRP Ann. **70**, 119–122 (2021). https://doi.org/10.1016/j.cirp.2021.04.017

24. Wilensky, U., Rand, W.: Introduction to Agent-Based Modeling: Modeling Natural, Social and Engineered Complex Systems with NetLogo. MIT Press, Cambridge (2015). ISBN 978-0262731898

25. Krzywicki, D., Turek, W., Byrski, A., Kisiel-Dorohinicki, M.: Massively-concurrent agent-based evolutionary computing. J. Comput. Sci. **11**, 153–162 (2015). https://doi.org/10.1016/j.jocs.2015.07.003

26. Leszczyna, R.: Evaluation of Agent Platforms, ver. 2.0, EUR 23508 EN, Luxembourg: European Commission, JRC47224 (2008)

27. Leszczyna, R.: Architecture supporting security of agent systems. Ph.D. thesis, Gdansk University of Technology, Gdansk, Poland (2006)

28. Braubach, L., Pokahr, A., Lamersdorf, W.: Jadex: a BDI-agent system combining middleware and reasoning. In: Unland, R., Calisti, M., Klusch, M. (eds.) Software Agent-Based Applications Platforms and Development Kits, pp. 143–168. Birkhäuser Basel, Basel (2005). https://doi.org/10.1007/3-7643-7348-2_7

29. Kravari, K., Bassiliades, N.: A survey of agent platforms. J. Artif. Soc. Soc. Simul. **18**, 11 (2015). https://doi.org/10.18564/jasss.2661

30. Leitão, P., Mařík, V., Vrba, P.: Past, present, and future of industrial agent applications. IEEE Trans. Industr. Inform. **9**, 2360–2372 (2013). https://doi.org/10.1109/TII.2012.2222034

31. Kruger, K., Basson, A.: Evaluation of JADE multi-agent system and Erlang holonic control implementations for a manufacturing cell. Int. J. Comput. Integr. Manuf. **32**, 1–16 (2019). https://doi.org/10.1080/0951192X.2019.1571231

32. Braubach, L., Pokahr, A., Moldt, D., Lamersdorf, W.: Goal representation for BDI agent systems. In: Bordini, R.H., Dastani, M., Dix, J., El Fallah Seghrouchni, A. (eds.) ProMAS 2004. LNCS (LNAI), vol. 3346, pp. 44–65. Springer, Heidelberg (2005). https://doi.org/10.1007/978-3-540-32260-3_3

33. Active Components JADEX. https://www.activecomponents.org/#/docs/overview. Accessed June 2022

34. Bergenti, F., Caire, G., Gotta, D.: Agent-based social gaming with AMUSE. Procedia Comput. Sci. **32**, 914–919 (2014). https://doi.org/10.1016/j.procs.2014.05.511

35. Wade. Workflows and Agents Development Environment. https://jade.tilab.com/wadepr oject/. Accessed June 2022
36. Bellifemine, F., Carie, G., Greenwood, D.: Developing Multi-Agent Systems with JADE. Wiley, Hoboken (2007). ISBN 978-0-470-05747-6
37. JAVA Agent Development Framework. https://jade.tilab.com/. Accessed June 2022
38. Winikoff, M.: Jack™ intelligent agents: an industrial strength platform. In: Bordini, R.H., Dastani, M., Dix, J., El Fallah Seghrouchni, A. (eds.) Multi-Agent Programming. MSASSO, vol. 15, pp. 175–193. Springer, Boston, MA (2005). https://doi.org/10.1007/0-387-26350-0_7
39. JACK autonomous software. https://aosgrp.com/products/jack/. Accessed June 2022
40. Odell, J.: Agent Technology - An Overview, paper/booklet (2011). http://www.jamesodell.com/Agent_Technology-An_Overview.pdf. Accessed June 2022
41. NetLogo. https://ccl.northwestern.edu/netlogo/. Accessed June 2022
42. Marcon, E., Chaabane, S., Sallez, Y., Bonte, T., Trentesaux, D.: A multi-agent system based on reactive decision rules for solving the caregiver routing problem in home health care. Simul. Model. Pract. Theory **74**, 134–151 (2017) ISSN 1569-190X
43. MaDKit, The Multiagent Development Kit. https://www.madkit.net/madkit/. Accessed June 2022
44. Gutknecht, O., Ferber, J.: Madkit: a generic multi-agent platform. In: Autonomous Agents, AGENTS 2000, Barcelona, pp. 78–79. ACM Press (2000). https://doi.org/10.1145/336595.337048
45. Mesa: Agent-based modeling in Python 3+. https://mesa.readthedocs.io/en/latest/. Accessed June 2022
46. Simoiu, M., Fagarasan, I., Ploix, S., Calofir, V., Iliescu, S.: Towards energy communities: a multi-agent case study. In: IEEE International Conference on Automation, Quality and Testing, Robotics (AQTR), pp. 1–6 (2022). https://doi.org/10.1109/AQTR55203.2022.980 2060
47. The Repast Suite. https://repast.github.io/. Accessed June 2022
48. North, M.J., et al.: Complex adaptive systems modeling with repast symphony. Complex Adapt. Syst. Model. **1**(1), 1–26 (2013). https://doi.org/10.1186/2194-3206-1-3
49. AnyLogic Simulation Software. https://www.anylogic.com/. Accessed June 2022
50. Răileanu, S.: Proposition of a generic model for the control of a guided flow system, Application of the holonic concepts in intelligent transportation (FMS/PRT), Ph.D. thesis, Univ. of Valenciennes, France (2011)
51. Leitão, P., Colombo, A.W., Karnouskos, S.: Industrial automation based on cyber-physical systems technologies: Prototype implementations and challenges. Comput. Ind. **81**, 11–25 (2016). https://doi.org/10.1016/j.compind.2015.08.004
52. Cardin, O., Trentesaux, D., Thomas, A., Castagna, P., Berger, T., Bril, H.: Coupling predictive scheduling and reactive control in manufacturing: state of the art and future challenges. In: Borangiu, T., Thomas, A., Trentesaux, D. (eds.) Service Orientation in Holonic and Multi-agent Manufacturing. Studies in Computational Intelligence, vol. 594, pp. 29–37. Springer, Cham (2015). https://doi.org/10.1007/978-3-319-15159-5_3
53. Derigent, W., Cardin, O., Trentesaux, D.: Industry 4.0: contributions of holonic manufacturing control architectures and future challenges. J. Intell. Manuf. **32**(7), 1797–1818 (2020). https://doi.org/10.1007/s10845-020-01532-x
54. Valckenaers, P., van Brussel, H.: Design for the Unexpected. From Holonic Manufacturing Systems towards a Humane Mechatronics Society. Butterworth-Heinemann, Elsevier (2015) ISBN 978-0-12-803662-4
55. Răileanu, S., Borangiu, T., Rădulescu, S.: Towards an ontology for distributed manufacturing control. In: Borangiu, T., Trentesaux, D., Thomas, A. (eds.) Service Orientation in Holonic and Multi-Agent Manufacturing and Robotics. Studies in Computational Intelligence, vol. 544, pp. 97–109. Springer, Cham (2014). https://doi.org/10.1007/978-3-319-04735-5-7

56. Meyer, G., Främling, K., Holmström, J.: Intelligent products: a survey. Comput. Ind. **60**, 137–148 (2009). https://doi.org/10.1016/j.compind.2008.12.005
57. The Foundation for Intelligent Agents. http://www.fipa.org/. Accessed June 2022
58. Smith, R.G.: The contract net protocol: high-level communication and control in a distributed problem solver. IEEE Trans. Comput. **29**(12), 1104–1113 (1980). https://doi.org/10.1109/TC.1980.1675516
59. Borangiu, T., et al.: Product-driven automation in a service oriented manufacturing cell. In: International Conference on Industrial Engineering and Systems Management (IESM), Metz, France (2011)
60. Barata, J., Camarinha-Matos, L.: Coalitions of manufacturing components for shop floor agility - the CoBASA architecture. Int. J. Netw. Virtual Organ. **2**(1), 50–77 (2003). https://doi.org/10.1504/IJNVO.2003.003518
61. Sallez, Y., Berger, T., Trentesaux, D.: Management du cycle de vie d'un produit actif: Concept d'agent d'augmentation, 8ème Congrès international de Génie Industriel (2009). file:///C:/Users/BT/Downloads/Congres_GI_2009_paper179.pdf
62. McFarlane, D., Vaggelis, G., Wong, A., Harrison, M.: Product intelligence in industrial control: theory and practice. Annu. Rev. Control. **37**, 69–88 (2013). https://doi.org/10.1016/j.arcontrol.2013.03.003
63. Wong, C., McFarlane, D., Zaharudin, A., Agarwal, V.: The intelligent product driven supply chain. In: 2002 IEEE International Conference on Systems, Man and Cybernetics, vol. 4, p. 6. IEEE (2002)
64. Wooldridge, M.: An introduction to multi-agent systems. J. Artif. Soc. Soc. Simul. **7** (2004). https://doi.org/10.1007/978-3-642-01904-3_2
65. Lu, L., Wang, G.: A study on multi-agent supply chain framework based on network economy. Comput. Ind. Eng. **54**(2), 288–300 (2008). https://doi.org/10.1016/j.cie.2007.07.010
66. Pach, C., Berger, T., Sallez, Y., Trentesaux, D.: Instantiation of the open-control concept in FMS based on potential fields. In: Proceedings Industrial Electronics Conference, IECON 2012 (2012). https://doi.org/10.1109/IECON.2012.6389486
67. Roehrich, J.K., Parry, G., Graves, A.: Implementing build-to-order strategies: enablers and barriers in the European automotive industry. Int. J. Autom. Technol. Manag. **11**(3), 221–235 (2011). https://doi.org/10.1504/IJATM.2011.040869

Alignment of Digital Twin Systems with the RAMI 4.0 Model Using Multi-agent Systems

Victória Melo[1(✉)], Fernando de la Prieta[2], and Paulo Leitão[1]

[1] Research Centre in Digitalization and Intelligent Robotics (CeDRI), Instituto
Politécnico de Bragança, Campus de Santa Apolónia, 5300-253 Bragança, Portugal
{victoria,pleitao}@ipb.pt
[2] BISITE Digital Innovation Hub, University of Salamanca, Edificio Multiusos
I+D+i, 37007 Salamanca, Spain
fer@usal.es

Abstract. The Digital Twin approach has emerged in the Industry 4.0
context to perform the design, virtual commissioning and optimization
of processes and machines, considering the simulation model feed with
real-time data, and articulated with AI data-driven algorithms to extract
value and knowledge. The Digital Twin can be seen as a way to proceed
with the digitalization of production assets, covering the layers dimension
defined by the RAMI 4.0 model. However, a pertinent question is related
to understand the relationship between the Digital Twin concept and the
other two dimensions of the RAMI 4.0 model, and particularly the way to
address the distributed nature exhibited by the hierarchy levels and the
life-cycle management defined by the life-cycle value stream dimension.
Having this in mind, this paper aims to analyse the coverage of the
Digital Twin concept according to the three dimensions of the RAMI 4.0
model, particularly considering the inherent capabilities of Multi-agent
Systems to support the development of more distributed and pro-active
Digital Twin ecosystems that cover the life-cycle management, facing the
demands established by Industry 4.0 compliant solutions.

Keywords: Digital Twin · RAMI4.0 · Multi-agent systems

1 Introduction

The digital transformation, fueled by the 4th industrial revolution, is re-shaping
the way persons, machines, processes, systems and companies operate, with the
availability of data and the use of Artificial Intelligence (AI) techniques being the
main drivers. The manufacturing industry, that continues to have a significant
importance in the world economy (e.g., Deloitte projections anticipate a GDP
growth in US manufacturing of 4.1% for 2022 [4]), is also facing this change,
with the Industry 4.0 initiative defining a set of design principles that allows the
transformation of existing production systems into more efficient, flexible and

© The Author(s), under exclusive license to Springer Nature Switzerland AG 2023
T. Borangiu et al. (Eds.): SOHOMA 2022, SCI 1083, pp. 23–35, 2023.
https://doi.org/10.1007/978-3-031-24291-5_2

responsive ones, with the ability to learn and make autonomous decisions [9]. In this context, Cyber-physical Systems (CPS) have a pivotal role, and can be seen as the backbone infrastructure to realize Industry 4.0 compliant solutions [7] that rely on a network of manufacturing components with cyber and physical counterparts, which have the ability of performing decentralised and autonomous decisions [8]. CPS are combined with several emergent digital technologies like Internet of Things (IoT), AI, Big data, cloud computing and cobots, applied to different sectors, such as energy, logistics, health and manufacturing.

RAMI 4.0, recognized as DIN 91345 standard, is a three-dimensional architecture established by the Industrie 4.0 platform to serve as reference for engineering Industry 4.0 systems [1], guiding the enterprises through a digital transformation roadmap [5]. This model allows to develop Industry 4.0 compliant solutions (based on the CPS approach), covering the hierarchy of an industrial infrastructure and the life-cycle/value stream of different production assets. For this purpose, it is structured according to three axes: Hierarchy Levels, Layers and Life Cycle Value Stream. The Hierarchy levels dimension represents the roles and responsibilities/functionalities of the HW/SW assets within the factory or plant, following the levels specified by IEC 62264 and IEC 61512. The Layers dimension represents the IT perspectives for the digitalization of the production assets, namely communication protocols, data models and functional descriptions, described in a structured way layer by layer. Finally, the Life-Cycle Value Stream dimension represents the life cycle of plants and products based on the IEC 62890 standard, which considers a distinction between type and instance: a "type" becomes an "instance" when the development and prototype production is completed and the actual product is manufactured.

As referred, Industry 4.0 establishes several key design principles, serving as guidelines for its implementation. Virtualization is one of these principles, being defined as the capability to create a virtual copy of the physical system, providing a connection between these two counterparts in order to fed the simulation model with the real-time data. The Digital Twin is a novel approach pushed by the Industry 4.0 context as an evolution of the virtualization concept, that can be defined as *a digital copy of a physical object or system that is connected and shares functional and operational data, allowing the historical and real-time data collected to be used to assess the conditions of the physical asset and support the execution of simulation and data analysis techniques to allow the optimization and improvement of the physical object* [14, 15]. Intense research work is being developed in this field, particularly focusing on design, virtual commissioning and optimization tasks. Basically, it considers the simulation of the virtual model fed by the real-time data that are collected by IoT technologies, and articulated with AI-based data-driven algorithms to extract value and knowledge aiming at the implementation of the referred functionalities.

The Digital Twin can be seen as a way to proceed with the digitalization of assets, covering the Layers dimension defined by RAMI 4.0. However, a pertinent question is related to understand the relationship between the Digital Twin concept and the other two dimensions of the RAMI4.0 model, expressed in the following research questions:

- RQ1: In which way the Digital Twin covers the Hierarchy levels dimension that comprises assets belonging to OT and IT layers, that interact each other in a distributed manner? In particular, it is convenient to consider one single Digital Twin to represent the entire process (traditional approach) or to consider a distributed approach with one Digital Twin for each individual asset?
- RQ2: In which way the Digital Twin covers the Life-cycle and Value stream dimension of one production asset? In particular, is it convenient to consider one single Digital Twin to represent the entire life-cycle of one asset or to consider one Digital Twin for each phase of the asset's life-cycle?

Having this in mind, the paper discusses these two research questions, presenting guidelines and models that can be used to develop more efficient Digital Twin-based systems that cover the diversity of assets belonging to a production process and along their life-cycle. Based on this analysis, the main challenges to structure and build these pro-active and distributed systems are defined, mainly based on the holonic and Multi-agent Systems (MAS) principles.

The rest of the paper is organized as follows: Sect. 2 analyses the alignment of the Digital Twin concept with the Layers dimension, and particularly with the Asset Administration Shell (AAS). Sections 3 and 4 answer to the research questions, discussing the alignment of the Digital Twin concept with the Hierarchy levels and Life-cycle and Value stream dimensions of RAMI 4.0. Based on this analysis, Sect. 5 presents guidelines to develop Digital Twin solutions that cover the distribution of Hierarchy levels and the asset Life-cycle management, and discusses the main challenges for their deployment. Finally, Sect. 6 rounds up the paper with the conclusions and points out the future work.

2   Digital Twin Alignment with the Layers Dimension

Analyzing the RAMI 4.0 model, the 2D surface composed by the Hierarchy levels and the Life-cycle & Value stream dimensions, illustrated in Fig. 1, contains 28 cells, each one comprising at least an asset in a specific life-cycle phase. As can be observed, an asset, e.g., a manipulator robot in the figure, can be positioned in different cells according to its life-cycle phase and its role in the industrial infrastructure. Particularly, in cell #1, the robot is a product type in the design phase (product element), but in cell #2 the robot is already an instance that is being produced. The same robot in cell #3 is not anymore a product but instead is a control device installed in a production line. Each one of these different perspectives for the same asset present different requirements, roles and functionalities, needing to be managed in a different manner.

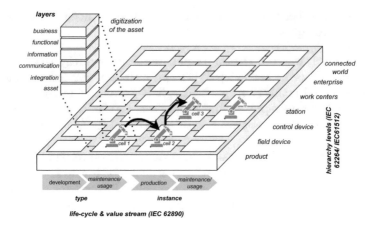

Fig. 1. 2D digitalization matrix in the RAMI 4.0 model.

The digitalization of each asset placed in these cells follows the vertical dimension, i.e. the Layers dimension of RAMI 4.0. Briefly, the asset layer includes physical and nonphysical objects, e.g., machines, documents or services that can be digitalized. These assets are made available to the other layers via the integration layer, and the data gathered from the assets is made available to be processed on higher levels. The communication layer establishes the standard communication protocols and methods to ensure the proper exchange of data, the information layer describes the information shared between services and components structuring the data properly, and the functional layer comprises the features and services that the asset can offer. The business layer provides the business perspective on the functionalities offered by the digitalized asset.

The concrete implementation of this vertical digital dimension can be performed by using the AAS [2,3]. The AAS is an important concept defined by the RAMI 4.0 model that transforms an ordinary object into an Industry 4.0 component, providing the following main functionalities: stores all data and information about the asset, serves as the standardized communication interface to interconnect the physical asset to the Industry 4.0 environment, and integrates passive assets. Basically, each physical object needs its own administration shell to support its integration into the Industry 4.0 ecosystem, with the AAS forming the digital part and the object forming the real part.

As pointed out by [6,16], it is nowadays recognized that AAS is the basis of building Digital Twins since it provides the required structure and composition, particularly *sub-models* based on the digitalized data and information associated to different structural and functional specifications of the asset (see Fig. 2). In this context, the Digital Twin and the AAS implement the business, functional, information and communication layers defined by the Layers dimension established by the RAMI 4.0 model.

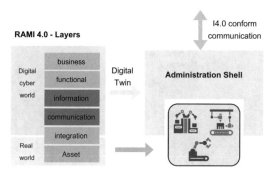

Fig. 2. Relationship of Digital Twins and the layers dimension of the RAMI 4.0 model.

3   Digital Twin Alignment with the Hierarchy Levels

Traditionally, the Digital Twin focuses on an entire process through a virtual model that can be fed with real-time data and simulated under different scenarios to derive optimized system operation configurations. The use of this monolithic structure to represent the entire process can turn the model large and complex, making it difficult to be manipulated and time consuming to be simulated and analyzed. Additionally, this approach presents weak response to condition changes and system reconfiguration, since the modification of the process requires the complete reformulation of the system model which is time-constrained.

Taking inspiration from distributed systems and particularly from MAS, alternative solutions to this centralized approach are related to the principle of 'divide to conquer', i.e. splitting the entire system model into several modular and simple sub-systems or assets, each one represented by a Digital Twin as illustrated in Fig. 3. This approach is aligned with the new structure and principles defined by the hierarchy levels of the RAMI 4.0 model, that introduces the following main innovations related to the precursor IEC 62264 (ANSI/ISA 95, also known as the automation pyramid): i) the rigidity of layers are not anymore present, each asset being able to communicate with other assets (not restricting the communication between neighbouring layers), ii) products are now considered as assets and are presented in the automation structure (not restricting the focus of the industrial infrastructure on the process), and iii) the world connected is introduced through the use of IoT technologies.

Modularization and distribution of assets, reflected in the virtual models of individual Digital Twins, can follow different architectures, e.g., a more hierarchical or decentralized structure, each one providing different levels of optimization, flexibility and reconfiguration, and supporting the aggregation of Digital Twins.

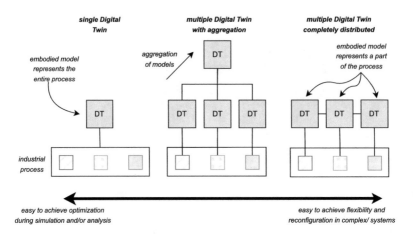

Fig. 3. Single Digital Twin vs multiple Digital Twins to represent a production process.

4   Digital Twin Alignment with the Life-Cycle and Value Stream Dimension

In terms of the life-cycle, traditionally, the Digital Twin covers the operational model of a specific asset. However, considering the asset's life-cycle as defined in the Life-cycle & Value stream dimension defined by the RAMI 4.0 model, it is crucial that the different phases and perspectives be considered, as well as the interaction among them. If the entire life-cycle of the asset/process is represented by an unique Digital Twin, the data, knowledge and behaviour related to a particular type of product are centralized, not being possible to decouple different instances of the same product type. For example, consider a robot manipulator that holds the design information about the production process and the CAD design files of the robot, and the thousands of manipulator robots from that robot type produced in the shop floor: if the manipulator robot asset is represented by a single Digital Twin, it is not possible to decouple the info and data for the design phase associated to the robot type from the production/operation data concerning each produced manipulator robot (i.e. instances of the robot type).

In this way, it is important to consider several Digital Twins to represent the different phases along the assets' life-cycle, as illustrated in Fig. 4, each one managing different types of data provided along the digital thread of the physical product life-cycle. These Digital Twins should be collaborative by interacting each other in a forward and backward manner.

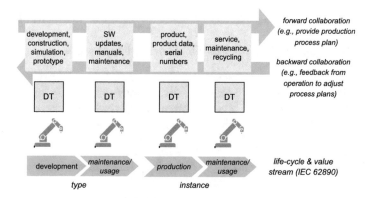

Fig. 4. Single Digital Twin vs multiple Digital Twins to represent an asset's life-cycle.

The forward collaboration is related to the interaction among Digital Twins to exchange information that is passed from the Digital Twin associated to an asset (product or machine) placed in the development phase of the asset's type to the child Digital Twins representing the associated assets placed in production and maintenance/usage phases of the asset's instance (see Fig. 4). The backward collaboration is related to the feedback that the Digital Twins associated to the instances (assets placed in production and maintenance/usage life-cycle phases), based on their individual behaviour and performance, can sent to the Digital Twin associated to the asset placed in the development phase, which is able to refine and adjust the design and production plans of such asset. This feature is crucial to ensure the continuous improvement of the production process.

Another interesting concept that emerges from this perspective is the intelligent product [13], which are related to products that have knowledge about their characteristics and are connected to each other to share information in real-time about their state or environment or to communicate with other cooperative objects [12]. These intelligent products have the ability to acquire life-cycle data, which is used to support the implementation of monitoring, traceability and decision-making functions [12,19]. This brings significant benefits, namely improvement of the next generation of the product, improvement of the entire life-cycle of the product, establishment of a product-driven production approach and improvement of the product quality through the application of self-learning, self-diagnosis, self-adaptation and self-optimization methods [12].

5   Developing Collaborative Agent-Based Digital Twins

5.1   Agent-Based Principles to Create Digital Twins

As pointed out in the previous sections, the need to relate Digital Twins with the Hierarchy levels and the Life-cycle dimensions defined by the RAMI 4.0 model imposes a reflection on how to design the Digital Twin systems that are virtualizing a particular production process/asset. Particularly:

- Creating a Digital Twin that represents the entire process may turn the model very large and complex, making it difficult to be manipulated and time consuming to be simulated and analyzed.
- Adapting to condition changes and system reconfiguration, e.g., modifying the layout or production process by adding/removing resources, can be difficult and time consuming.
- Creating a Digital Twin that represents the entire life-cycle of the process/asset may turn difficult to decouple the data, knowledge and behaviour of the assets' types and instances, i.e. the design of the asset's type from the production and operation of asset's instances.

As result, and as conclusion for the two established research questions, it is required to consider a network of interconnected Digital Twins covering the different production assets along their life-cycle, with individual Digital Twins organized in a kind of federated structure. This allows reaching high decision capabilities in two directions: i) autonomy with self-decision processes (local view on single models), and ii) collaboration with decision processes based on negotiation schemes (global view based on the interaction among single models).

For this purpose, holonic principles and MAS [11,18] can be used to support the design and deployment of such systems, as illustrated in Fig. 5 that shows the use of agents to support the development of such systems, with particular focus to the alignment of their functional modules with the digital layers dimension defined by the RAMI 4.0 model.

The implementation of this distributed Digital Twin architecture requires that individual Digital Twins interact following a collaboration scheme, with the overall behaviour emerging from the non-linear interactions. MAS is suitable to support the challenge of implementing the collaboration models among these Digital Twins with the scope to realize the overall functionality. In fact, MAS fits well with this distributed nature of Digital Twins, contributing with its inherent characteristics of autonomy, cooperation and intelligence to build a Digital Twin collaborative ecosystem that focuses on the life-cycle of its individual assets. Note that since the real-time requirements are ensured by the low-level control of the "real world", agents are focused to provide the collaboration and adaptation features. Particularly, the use of MAS technology as infrastructure to accommodate the distributed Digital Twins system transforms the typical passive and reactive approaches followed by AAS and Digital Twins into a proactive approach by ensuring:

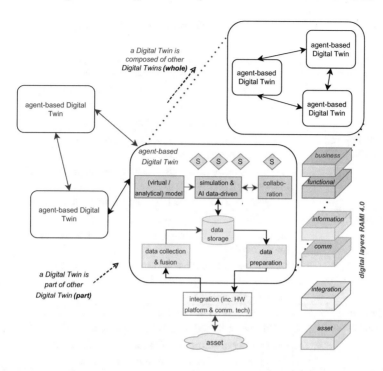

Fig. 5. MAS to support the collaborative Digital Twins systems and their alignment with the digital Layers dimension.

- Distribution of the data collection and intelligence capabilities among several Digital Twins.
- Execution of collaboration models that allow the symbiotic operation of a swarm of Digital Twins through the implementation of cooperation models.
- Evolution and reconfiguration of the system based on emergent and self-organization processes, that use a plug-and-play strategy on the fly (i.e. without the need to stop, reprogram and restart the individual components).

Embodying holonic principles in MAS systems allows to simplify the design of this kind of systems taking advantage of their inherent characteristics, namely the Janus effect and recursivity [10]. In fact, being based on the holon concept that represents simultaneously the part and the whole [10], it is possible to have a Digital Twin that simultaneously represents the entire system (e.g., a robotic manipulator) and a part of the system (e.g., an automotive assembly line).

5.2   Main Research Challenges

This agent-based Digital Twin ecosystem covering the assets' life-cycle introduces significant benefits, but the engineering of such systems also poses important research challenges, which are summarized in Table 1.

Table 1. Main challenges to develop distributed, agent-based Digital Twins systems.

Key aspects	Challenges
Design and engineering	Engineering agent-based system of systems
	Methodologies to define granularity and model distribution
	Security and data privacy
	User interaction and acceptance
Compatibility and interoperability	Servicification to encapsulate functionalities
	Data models and semantics
	Communication protocols for data exchanging (virtual - virtual interaction)
	Integration with physical assets (virtual - physical integration)
Decentralization	Modelling and simulation aggregation by collaboration
	Forward and backward collaborative models for the life-cycle management
	Distribution of intelligence among edge-cloud (management of global and local data analysis)
	Manage the myopia
Emergence and self-organization	Controlling the emergent behavior in complex contexts
	Definition of boundaries for autonomic and self-* features
	Control nervousness in self-organization processes
	Definition of self-reconfigurable mechanisms
Simulation and data analytics	Optimization and AI-based data driven systems
	MAS based what-if simulation
	Combining simulation and AI-based data-driven systems
Infrastructures and platforms	Robust and industry-oriented agent-based frameworks
	Industry testbeds
	Standards compliance

In terms of engineering, these agent-based systems, that form a system of systems it is crucial to define a proper decomposition of the process/asset model into a set of individual and modular models that are interrelated. The definition of the granularity level (e.g., if the model is established for an assembly line or for a workstation or even for each individual automation resource) assumes crucial importance to setup the distributed community of agent-based Digital Twins. Of particular relevance is the use of agents to boost the AAS functionalities through the implementation of AI algorithms and collaboration models [17]. Because these systems are based on the interchange of data between individual entities to achieve the global objectives, security and data privacy are important aspects to be considered during the design of such systems. The user interaction is also crucial to be considered for the development of sustainable and human-integrated solutions (aligned with Industry 5.0), e.g., supporting the dynamic model validation considering trust models.

Digital Twins should be interoperable in order to work together, requiring the use of standardized data models and semantics to represent the shared knowledge, and services to encapsulate functionalities they provide. For this purpose, the microservices technology can be used to increase the system interoperability. Communication technologies and protocols based on IoT technologies are crucial to ensure the proper communication between the agent-based Digital Twins (virtual - virtual interaction), as well as to properly integrate the assets (virtual - physical interaction).

In such systems, decentralization is a main pillar, being required the aggregation of simulation models by implementing collaboration models. In particular, forward and backward collaborative models are of crucial importance, since in this approach there is an explosion of Digital Twins along the life-cycle axis making evident the need of this type of interaction to support the exchange of data among Digital Twins representing the same asset at different phases of their life-cycle. Balancing the distribution of intelligence among edge to cloud computational platforms is important to manage the global and local data analysis.

Associated to decentralization, these systems are based on emergence and self-organization, which allows to pull these systems into their limits, by exploring the chaos theory, but maintaining the system under control. For this purpose, mechanisms that ensure the dynamic evolution of the Digital Twin system should be implemented, taking inspiration from nature and biology (adapting and not copying). During self-organization processes it is mandatory to control the nervousness and define the boundaries for autonomic and self-* features that will allow the emergence of new properties and behaviours (allowing the emergence of "good" behaviours and restricting the "bad" ones).

Simulation is one of the main functionalities usually associated to the Digital Twin, being present under different forms, e.g., finite element analysis (FEA) and computational fluid dynamics. This functionalities are used e.g., in manufacturing to determine the thermal, structural and material effects of processes and products. Discrete event simulation is used to determine e.g., the best layout or process before physical changes are made in the production facility. These simulation tools can assist the root cause analysis in production processes, and the use of what-if functionalities powered by the use of MAS can speed-up the exploration, analysis and assessment of alternative solutions. The use of AI-based data driven systems to analyse in real-time the collected data is also crucial to execute monitoring, diagnosis and optimization tasks, being useful to be combined with simulation to achieve more powerful and accurate systems.

Finally, in terms of infrastructures and platforms, the challenges are centred to more robust and industry-oriented frameworks to develop agent-based systems, as well as industry testbeds that allow proper testing and assessment.

6   Conclusions and Future Work

Digital Twin is a novel concept that has emerged in the Industry 4.0 context associated to the virtualization of production models, aiming to perform the design, virtual commissioning and optimization of processes and machines. Its benefits are clearly illustrated with the significant on-going research work, being advocated that it may represent AAS, defined by the RAMI 4.0 model and corresponding to the implementation of the Layers dimension of RAMI 4.0.

This paper analyses the coverage of the Digital Twin concept by the other two dimensions defined by the RAMI 4.0 model, to answer the initially established research questions about the need to develop Digital Twin ecosystems that aim to simplify the model complexity, to ensure the fast and easy reconfiguration, and the coverage of the assets' life-cycle. For this purpose, MAS is a suitable approach to empower the creation of Digital Twins by introducing distributed intelligence and collaboration models. The paper also discusses the main challenges to develop such systems, that are present in design and engineering: compatibility and interoperability, decentralization, emergence and self-organization, simulation and data analytics, and infrastructures and platforms.

Future work will be devoted to detail the proposed approach, particularly the collaboration mechanisms that use MAS and AI to support this dynamic and decentralized organization, and proceed with experimental tests that illustrate the benefits and applicability of this approach.

Acknowledgements. This work has been supported by FCT- Fundação para a Ciência e Tecnologia within the Project Scope: UIDB/05757/2020.

References

1. DIN SPEC 91345: Reference architecture model Industrie 4.0 (RAMI4.0) (2016)
2. Details of the asset administration shell - Part 2. Tech. rep., Platform Industrie 4.0 (2021)
3. Details of the asset administration shell - Part 1. Tech. rep., Platform Industrie 4.0 (2022)
4. Bachman, D.: Deloitte's US economic forecast Q3 2021. Tech. rep., Deloitte (2021)
5. Bastos, A., De Andrade, M.L.S.C., Yoshino, R.T., Santos, M.M.D.: Industry 4.0 readiness assessment method based on RAMI 4.0 standards. IEEE Access **9**, 119778–119799 (2021)
6. Boss, B., et al.: Digital twin and asset administration shell concepts and application in the industrial internet and Industrie 4.0. Plattform Industrie **4**, 12 (2020). https://www.iiconsortium.org/pdf/Digital-Twin-and-Asset-Administration-Shell-Concepts-and-Application-Joint-Whitepaper.pdf
7. Colombo, A.W., Karnouskos, S., Kaynak, O., Shi, Y., Yin, S.: Industrial cyberphysical systems: a backbone of the fourth industrial revolution. IEEE Ind. Electron. Mag. **11**(1), 6–16 (2017)
8. Colombo, A.W., Veltink, G.J., Roa, J., Caliusco, M.L.: Learning industrial cyberphysical systems and Industry 4.0-compliant solutions. In: Proceedings of the IEEE Conference on Industrial Cyberphysical Systems (ICPS 2020), vol. 1, pp. 384–390 (2020)
9. Kagermann, H., Wahlster, W., Helbig, J.: Securing the future of german manufacturing industry: recommendations for implementing the strategic initiative INDUSTRIE 4.0. Tech. rep., ACATECH (2013)

10. Koestler, A.: The Ghost in the Machine. Arkana Books, London (1969)
11. Leitão, P., Karnouskos, S. (eds.): Industrial Agents. Elsevier (2015)
12. Leitão, P., Rodrigues, N., Barbosa, J., Turrin, C., Pagani, A.: Intelligent products: the grace experience. Control. Eng. Pract. **42**, 95–105 (2015)
13. Meyer, G.G., Främling, K., Holmström, J.: Intelligent products: a survey. Comput. Ind. **60**(3), 137–148 (2009)
14. Pires, F., Cachada, A., Barbosa, J., Moreira, A.P., Leitão, P.: Digital twin in industry 4.0: technologies, applications and challenges. In: 2019 IEEE 17th International Conference on Industrial Informatics (INDIN 2019), pp. 721–726 (2019)
15. Pires, F., Melo, V., Almeida, J., Leitão, P.: Digital twin experiments focusing virtualisation, connectivity and real-time monitoring. In: Proceedings of the IEEE Conference on Industrial Cyberphysical Systems (ICPS 2020), pp. 309–314 (2020)
16. Sakurada, L., Leitao, P., De la Prieta, F.: Towards the digitization using asset administration shells. In: Proceedings of the 47th Annual Conference of the IEEE Industrial Electronics Society (IECON 2021), pp. 1–6 (2021)
17. Sakurada, L., Leitao, P., De la Prieta, F.: Agent-based asset administration shell approach for digitizing industrial assets. In: Proceedings of the 14th IFAC Workshop on Intelligent Manufacturing Systems (IMS 2022), vol. 55, pp. 193–198 (2022)
18. Wooldridge, M.: An Introduction to Multi-Agent Systems. Wiley, Chichester (2002)
19. Yang, X., Moore, P., Chong, S.K.: Intelligent products: from lifecycle data acquisition to enabling product-related services. Comput. Ind. **60**(3), 184–194 (2009)

A Design Process for Holonic Cyber-Physical Systems Using the ARTI and BASE Architectures

T. W. Defty, Karel Kruger[✉], and A. H. Basson

Department of Mechanical and Mechatronic Engineering, Stellenbosch University,
Stellenbosch 7600, South Africa
kkruger@sun.ac.za

Abstract. The manufacturing research domain has experienced the introduction of many holonic manufacturing system architectures over the past three decades. Application examples of these holonic architectures have revealed the benefit of such a control philosophy for flexible and reconfigurable production systems. The majority of these holonic architectures have remained conceptual with limited information on their application and implementation, which results in a barrier to the adoption of such architectures. This paper presents a design process for holonic systems to support their adoption and application in Cyber-Physical Systems. Specifically, this paper uses the Activity-Resource-Type-Instance (ARTI) and the Biography-Attributes-Schedule-Execution (BASE) architectures for structuring the design process.

Keywords: Holonic manufacturing system · Cyber-physical system · Human cyber-physical system

1 Introduction

Holonic architectures, such as the Product-Resource-Order-Staff Architecture (PROSA) [1], Activity-Resource-Type-Instance (ARTI) architecture [2], and Adaptive Holonic Control Architecture (ADACOR) [3], have been developed to improve the control of reconfigurable manufacturing systems, driving the paradigm of Holonic Manufacturing Systems (HMSs). HMSs was a manufacturing initiative started as part of the international manufacturing system program [4]. HMSs attempt to realize the concepts of *holons* and *holonic systems*, as developed by Arthur Koestler when studying biological and social systems. The *holonic systems* concept can also be coupled to similar concepts such as multi-agent systems, which attempt to encapsulate control and functionality in software entities called agents. Applications of HMS architectures have shown great promise in creating more flexible and adaptable control systems to guide manufacturing activities.

HMSs and multi-agent architectures have since extended beyond the manufacturing domain and have been applied to other environments and operations. Such applications typically aim to provide solutions for logistics, reactive task allocation, monitoring and control. Most of these architectures have been applied to complex Cyber-Physical

© The Author(s), under exclusive license to Springer Nature Switzerland AG 2023
T. Borangiu et al. (Eds.): SOHOMA 2022, SCI 1083, pp. 36–47, 2023.
https://doi.org/10.1007/978-3-031-24291-5_3

System (CPS) environments. CPS environments are described by the interconnectedness of systems between physical assets and computational capabilities [5]. Human Cyber-Physical Systems (HCPSs), building upon the CPS concept, aim to improve the capabilities of humans and their ability to interact with cyber and physical worlds [6]. The Biography-Attributes-Schedule-Execution (BASE) architecture presents the application of an holonic architecture to HCPS environments, to aid in delegation and communication, human interfacing and digital processing [7].

Several holonic architectures have been presented conceptually, but with little detail on the holonic system and software design process. This paper presents a holonic system design process using the ARTI and BASE architecture for (H)CPSs. The design process is formulated in response to the research gap noticed between conceptual architectures and their implementation in case study applications.

The paper begins with a review of the ARTI and BASE architectures in Sect. 2. The design process is presented in Sect. 3 and an application thereof in a healthcare case study is discussed in Sect. 4. The paper concludes in Sect. 5 with a summary of the key contributions of the paper and a perspective on future work.

2 Related Work

This section briefly reviews holonic manufacturing systems and their key philosophies. Furthermore, two relevant architectures, namely the ARTI and BASE architectures, are reviewed to aid the reader's understanding of the design process.

2.1 Holonic Systems Fundamentals

The *holon* was presented by Arthur Koestler in the book *The Ghost in the Machine* [8], noting the self-organization within biological and social systems. A holon is an entity that can communicate and make decisions, and which can have a set of sub-level holons while also being part of a network of other holons [9]. A holon thus represents the fundamental building block of complex systems. In 1992, intelligent manufacturing system research led to the development of the HMS concept, which utilised Holonic Control Architectures (HCAs) that comprise of various holons in a *holarchy* (i.e., a system of holons). HMSs and the related HCAs challenged the classic hierarchical and centralized form of manufacturing execution systems and resulted in heterarchical systems. Key architecture requirements for a HMS architecture are [10]:

- *Disturbance handling, availability, robustness* – enable system elements to have self- and cooperative planning, scheduling, and diagnosis capabilities.
- *Human integration* – provide humans with intelligence assistance and provide intuitive, responsive and customizable human interfaces
- *Flexibility* – provide improved human control over the system and allow for system self-reconfiguration.

2.2   ARTI Architecture

The ARTI architecture aims to generalize the PROSA architecture for adoption in domains outside of manufacturing [2]. *Activity instance* holons guide the processes within the system and are required to search for and utilize resources to fulfil their processes, taking on a supervision role. *Activity type* holons contain the information related to the process steps, while activity instances search for and utilize resource instances to achieve their purposes. *Resource instance* holons are representations of the physical and cyber assets which act in the system to perform certain functions and tasks. *Resource type* holons maintain the resource-related knowledge.

2.3   BASE Architecture

The BASE architecture can be viewed as a holonic implementation architecture and, as such, specifies four core components for implementing a holon's software representation [7]. The Biography stores the historical data for the holon (i.e., what it did, how it did it, why it did it, etc.). The Attributes contain data related to the holon's properties (such as age, weight, and material). The Schedule stores the information related to the planned actions of the holon. The Execution stores information about the holon's current state and/or action. The BASE architecture further specifies four types of *plugins* – Schedule plugins, Execution plugins, Reflection plugins and Analysis plugins – that encapsulate the context– or application-specific logic that determines the holon's behaviour.

3   Design Process

3.1   Overview

The design process is described through an *input-process-output* flow diagram, as seen in Fig. 1, Fig. 2 and Fig. 3. Each process step must consider specific input information (e.g., list of requirements or descriptions), which results in new output information. Although the design process is presented sequentially, the complex nature of operational CPS environments and reconfigurability of holonic systems warrant a flexible design approach. Therefore, iteration loops exist throughout the design process to cater for design refinements and the addition/elimination of system elements. Furthermore, frequent revision of the overall system design is necessary to ensure a complete holonic software system. In this paper, the overall design process is presented in a higher and a lower abstraction layer to aid the discussion.

The higher abstraction layer, seen in Fig. 1, has four process steps. Step 1, and the overall design process, begins with concrete knowledge of the user needs for the holonic software system in the form of *user requirements*. These requirements may be qualitative or quantitative and presented through written or verbal sources. During step 1, the operational environment should be considered conceptually (e.g., CAD designs, simulations, etc.) and/or observed in the real environment (e.g., operational production floor) to note any additional system requirements, processes, tasks and any interactions between system resources. Step 1 should result in the refinement of specific functional and non-functional requirements for the desired holonic system. Functional requirements

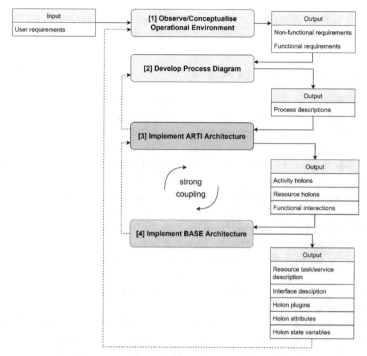

Fig. 1. Flow diagram of the higher abstraction layer of the holonic system design process.

"define what the system must be able to perform, while non-functional requirements describe attributes or characteristics that are desired in the system" [11].

Considering all the requirements and observations for the desired holonic system, a process diagram should be developed in step 2 that indicates all:

- *Workers* involved in the system operations.
- *Machines/equipment* involved in the system operations.
- *Tasks/processes* to be performed (by one or more machines/workers).
- *Physical interactions* between machines/workers.

The complexity of the interactions and sub-system processes will affect the level of detail required in the process diagram(s). The output of process descriptions should provide enough information to aid in formulating a holistic understanding of the processes and interactions in the desired system. The process diagram and descriptions will guide the application of the ARTI architecture in step 3, which maps the various processes, workers and machines in the system to activity and resource holons.

The ARTI architecture provides a framework to structure the overall system holarchy and interactions. Each holon may contain an internal holarchy, according to the *Janus* principle presented by Arthur Koestler, as holons act as both 'wholes' and 'parts'. Complementary to the ARTI architecture, the BASE architecture guides in step 4 the internal functionality and development of the ARTI holons. A strong coupling exists between step 3 and step 4, as changes in either step might lead to necessary changes in the other. This will be illustrated in the application case study in Sect. 4.

3.2 ARTI Architecture Design Process

The ARTI architecture aids in mapping the various processes, interactions, workers and machines to activity and resource holons. The lower abstraction layer flow diagram for the implementation of the ARTI architecture is presented in Fig. 2. The various processes must be mapped to activity holons in step 3.1. The system designer must generalize processes and encapsulate their functionality within different activity holons. The activity type holon, similar to a product holon in the PROSA architecture, encapsulates the process steps and logic. The activity instance should encapsulate the functionality related to the control of coordinating other holons for a single process instance. All processes identified in step 2 should be represented through the functions of various activity holons established in step 3.1. Similarly, the resource holons must be identified in step 3.2, considering the desired system requirements and observations. Resource holons are various actors in the system which perform functions required by different activity holons. Therefore, the different resources in the system and their specific functionality should be identified.

Activity holons are "free to search and utilize suitable resources in an appropriate sequence", while a resource holon should be able "to service any activity that may desire it" [12]. This design requirement leads to step 3.3, where the functional interactions between different holons need to be established and described. If an activity holon requires a specific resource holon to perform a function within its process logic, this interaction should be described. The list of activity and resource holons and their respective functional interactions will guide the implementation of the BASE architecture in step 4.

3.3 BASE Architecture Design Process

The ARTI architecture guides the identification of various activity and resource holons required for the holonic software system, whereas the BASE architecture provides an implementation architecture for intra-holon structure and functional interaction (i.e., communication) between holons. The lower abstraction layer flow diagram for the implementation of the BASE architecture (step 4) is presented in Fig. 3.

The BASE architecture can be implemented with various software languages (C++, Java, Erlang) and MAS platforms (JADE, SPADE). The implementation language is not prescribed in this design process in order to provide a generic design process.

The holon interactions from step 3.3 must be generalized into *tasks* and *services* when considering the BASE architecture in step 4.1. Tasks are functions related to the internal processes for a holon, while services are any functionality where there is an interaction between two or more holons. This classification will affect the development of the holons' BASE architecture plugins to provide the functionality identified in step 3.3.

Resource holons require a mechanism of data exchange with their physical asset counterpart. These mechanisms of interfacing (e.g., MQTT, OPC UA, web interfaces, tablets etc.) should be developed in step 4.2. Interfacing is important for ensuring that

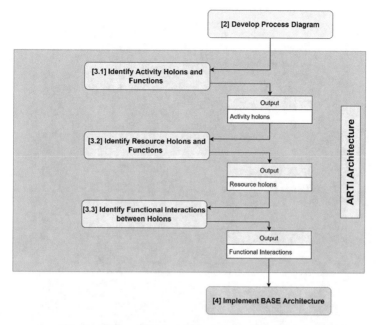

Fig. 2. ARTI architecture design process flow diagram.

the holonic software system maintains an accurate reflection of reality and allows for the effective control of all physical resources.

The first stage of plugin development only considers the schedule and execution plugins required for each holon. For tasks, the scheduled plugin should encapsulate the functionality and decision-making related to deciding on an appropriate time to begin the task. The execution plugin should encapsulate the functionality related to deciding on whether the task should begin at the scheduled time and the functionality related to performing and terminating the tasks. For services, the scheduled plugin for the service provider should encompass functionality to handle negotiations with (potential) clients and handle the outcomes of such negotiations (e.g., booking the service, rejecting offers). Schedule plugins for the clients should encapsulate functionality to initiate negotiations with service providers and decision-making. The execution plugins for other service providers and clients should encapsulate the decision-making functionality for handling the schedule serviced and the control logic and termination of the service.

The schedule and execution plugins may require the consideration of different holon attributes and state variables for decision-making, which must be identified in step 4.4. The ideal holonic software system would reflect the whole reality of its real counterparts, but such a venture would likely result in redundant data and be limited by computational resources. Therefore, step 4.4 attempts to solve that problem, by ensuring that the selection of which attributes and variables are required is based on the user needs of the system and the functions related to the holon tasks and services.

The reflection and analysis plugins are developed in step 4.5. Reflection plugins for both tasks and services should encapsulate the functionality related to selecting what

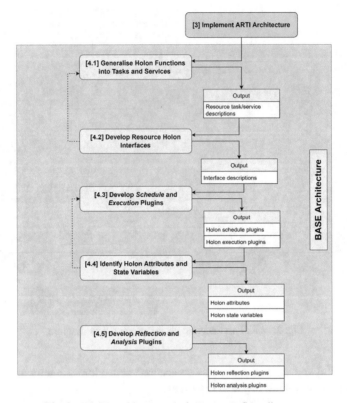

Fig. 3. BASE architecture design process flow diagram.

information of the executed service or task to store in the biography of the holon. This functionality is strongly coupled with the development of the analysis plugins, which should encapsulate any functionality related to the analysis of historic data of the holon to update the holon attributes and trigger different tasks or services.

4 Holonic Design Process Application: Healthcare Case Study

The holonic system design approach applies to different domains, such as manufacturing, agriculture and healthcare. As a part of a Master's research study, the presented design process was applied to the design and development of a holonic software system for improving the integration of humans and cyber systems in an outpatient clinic scenario, which was validated experimentally. The objectives of the case study implementation were to alleviate the resource and workload strain, improve workflow efficiencies and aid healthcare practices for healthcare facilities.

The typically observed out-patient operations are presented in Fig. 4, involving a taking of patient history, examinations, investigations (e.g., x-rays, blood tests, etc.) and diagnosis with related care. The user requirements for the new system were to:

- Reduce patient waiting times;

- Aid clinical decision-making (diagnosis);
- Automate scheduling of clinic visits and investigations; and
- Automate and digitize transport, handling, formatting and storage of patient data.

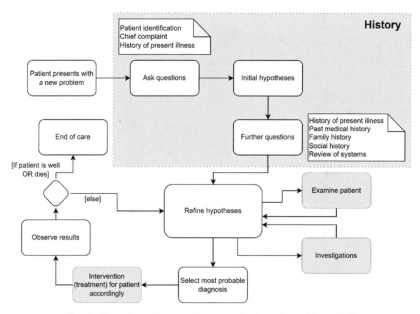

Fig. 4. Overview of out-patient consultation adapted from [13].

These user requirements and consideration of the clinic operations were used to identify the different processes, stakeholders and machinery involved in the process. This case study scenario included processes which cover the full cycle of the out-patient consultation as described in Fig. 4 and discussed in [13]. These processes and the interactions between patients, doctors, technicians and equipment had to remain flexible due to the unpredictable nature and variability of consultations due to various patient conditions. The following processes were identified within the scenario:

- The patient registers his arrival at the hospital.
- The patient coming for a consultation with a doctor at the respiratory clinic.
- The respiratory clinic doctor refers the patient to the spirometry clinic, to investigate the patient's lung function capability.
- The patient has a consultation with the respiratory doctor for diagnosis after considering all patient information and related investigations.

Figure 5 shows the complete ARTI-guided identification of activity and resource holons. The two primary activity holons were *consultation* and *spirometry* activities. The consultation activity encapsulated the functionality for handling the patient registration, doctor examination, doctor diagnosis, and discharge of the patient. An activity instance would be required for each new patient in the system. The *spirometry* activity holon encapsulated the functionality for handling the booking of the spirometry test clinic, and a spirometry test machine for the patient being investigated.

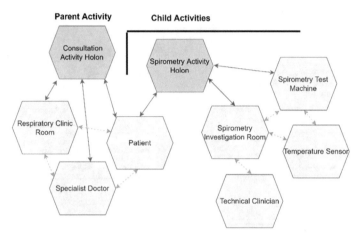

Fig. 5. ARTI Development for Healthcare Implementation.

The resources involved in the overall process included:

- Doctor – taking history, performing the examination, deciding on diagnosis, scheduling investigations.
- Respiratory clinic – respiratory facility to manage the allocation of consulting rooms.
- Patient.
- Spirometry investigation room – investigation facility to manage the allocation of testing rooms, spirometry machines and technicians.
- Technician – perform investigations.
- Spirometry machine – test the lung function of patients.
- Temperature sensor – gather atmospheric temperature of the room for the spirometry machine calibration.

The functional relationship between the consultation activity (parent) and the spirometry activity (child) is one of a parent-child. This was decided because an investigation activity will be created during a consultation process and, therefore, by the consultation activity. A child (spirometry activity) can only have one parent (consultation activity), which defines the functional relationship between the two activities. The functional relationship between the activities and resources is indicated by the solid arrows in Fig. 5, while the functional relationships between resources are indicated by the dashed lines. While the detail of this functional relationship is too broad to discuss in this paper, the identification of the functional relationships between holons aims to show that this step should provide the system designer with a more holistic understanding of the holonic system.

The BASE architecture then guided the implementation of the designed system once the activity and resource holons including their interactions were established. The encapsulated holon functions were categorized into tasks and services. Where interactions exist between holons, these were categorized as services. As an example, the specialist doctor resource holon offered an examination service (as a service provider). The consultation activity (as service client) could then search for resources that offer an examination service and the doctor's resources would be found. Another service was ambient temperature provided by the temperature sensor. The spirometry test machine (service client) could search for resources that could provide the ambient temperature as a service. Tasks were functions internal to the holons, not shown in Fig. 5. The activity holon tasks to guide the control logic of the activities were implemented as finite state machines. The patient holon has tasks for analyzing the state and attributes of the patient holon, to monitor the health and condition of the patient.

Interfaces were then developed for all the resource holons to connect to their physical counterparts. Web browsing capable devices such as tablets, laptops and mobile phones were used for all the human interfaces in the system. Web applications were developed to connect the user to their holons.

Schedule and execution plugins were developed for all the holons. The implementation software used was Erlang, due to its many benefits and previous usage in BASE architecture implementations [14, 15]. Schedule plugins were created to support the contract-net-protocol negotiations between the consultation activity, respiratory clinic, doctor and patient. The consultation activity sent out a *call for proposals* (cfp) message to all clinics that could offer consultation room services, then the clinic to doctors who could offer an examination service and so on, as seen in Fig. 6.

Attributes and state variables were selected for all the holons. Patient holons had attributes such as height, weight, and age – useful for scheduling tasks and analyzing the

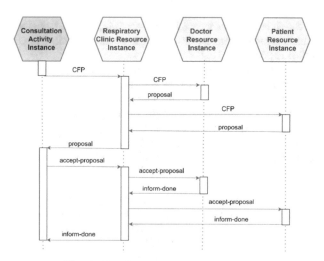

Fig. 6. CNP interactions between holons.

patient's health. Patient state variables are recorded, e.g., the status of the patient's location in the hospital. Patient holons had reflection plugins which stored the results of the spirometry test as historical data in their Biographies, and an analysis plugin constantly monitored these results to update the health status of the patient's lung function.

The design framework guided the development of the holonic system ensuring that all processes and physical components are accommodated. The iterative design steps encourage incremental improvements in the system design which reduced development time and complexity. This design process, though challenging to evaluate its benefit, has also aided research members in their recent case-specific developments. Further applications of the design framework would aid in the evaluation.

5   Conclusion and Future Work

This paper highlighted the lack of design methods for holonic systems while emphasizing the importance of such methods for the benefit of the research community. A design process for holonic systems is presented, utilizing ARTI as the holonic reference architecture and the BASE architecture for the software implementation. The design process was applied in a healthcare case study to show its application. Future work should involve the use and refinement of the design process in other holonic system applications, to lead to a well-developed design process for holonic systems.

Acknowledgements. This work is based on the research supported in part by the National Research Foundation of South Africa (grant number: 129360).

References

1. Van Brussel, H., Wyns, J., Valckenaers, P., Bongaerts, L., Peeters, P.: Reference architecture for holonic manufacturing systems: PROSA. Comput. Ind. **37**, 255–274 (1998). https://doi.org/10.1016/S0166-3615(98)00102-X
2. Valckenaers, P.: Perspective on holonic manufacturing systems: PROSA becomes ARTI. Comput. Ind. **120**, 103226 (2020). https://doi.org/10.1016/j.compind.2020.103226
3. Leitão, P., Restivo, F.: ADACOR: a holonic architecture for agile and adaptive manufacturing control. Comput. Ind. **57**, 121–130 (2006). https://doi.org/10.1016/j.compind.2005.05.005
4. Leitão, P., Mařík, V., Vrba, P.: Past, present, and future of industrial agent applications. IEEE Trans. Ind. Inform. **9**, 2360–2372 (2013). https://doi.org/10.1109/TII.2012.2222034
5. Lee, J., Bagheri, B., Kao, H.A.: A cyber-physical systems architecture for industry 4.0-based manufacturing systems. Manuf. Lett. **3**, 18–23 (2015). https://doi.org/10.1016/j.mfglet.2014.12.001
6. Romero, D., Bernus, P., Noran, O., Stahre, J., Fast-Berglund, Å.: The operator 4.0: human cyber-physical systems & adaptive automation towards human-automation symbiosis work systems. In: Nääs, I., et al. (eds.) APMS 2016. IAICT, vol. 488, pp. 677–686. Springer, Cham (2016). https://doi.org/10.1007/978-3-319-51133-7_80
7. Sparrow, D.E., Kruger, K., Basson, A.H.: An architecture to facilitate the integration of human workers in Industry 4.0 environments. Int. J. Prod. Res. (2021). https://doi.org/10.1080/00207543.2021.1937747
8. Koestler, A.: The Ghost in the Machine. Penguin Group (1967)

9. Derigent, W., Cardin, O., Trentesaux, D.: Industry 4.0: contributions of holonic manufacturing control architectures and future challenges. J. Intell. Manuf. **32**(7), 1797–1818 (2020). https://doi.org/10.1007/s10845-020-01532-x
10. Christensen, J.: Holonic Manufacturing Systems: Initial Architecture and Standards Directions. In: Holonic Manufacturing Systems, pp. 1–20 (1994)
11. SEBoK. Non-Functional Requirements (2012). https://www.sebokwiki.org/wiki/Non-Functional_Requirements_(glossary)
12. Van Brussel, H., Wyns, J., Valckenaers, P., Bongaerts, L., Peeters, P.: ARTI reference architecture - PROSA revisited. Comput. Ind. **37**, 255–274 (1998). https://doi.org/10.1016/S0166-3615(98)00102-X
13. Shortliffe, E.H., Cimino, J.J.: Biomedical Informatics: Computer Applications in Health Care and Biomedicine, 4th edn. Springer, London (2014). https://doi.org/10.1007/978-1-4471-4474-8
14. Wasserman, A.: ARTI-based holonic control implementation for a manufacturing system using the BASE architecture. Master's thesis, Stellenbosch University (2022)
15. Van Niekerk, D.J.: Extending the BASE architecture for complex and reconfigurable cyber-physical systems using holonic principles. Master's thesis, Stellenbosch University (2021)

A Recommendation Strategy Proposal for an Energy Community Modeled as a Multi-agent System

Mircea Ștefan Simoiu[1,2(✉)], Ioana Făgărășan[1], Stephane Ploix[2], Vasile Calofir[1], and Sergiu Stelian Iliescu[1]

[1] University Politehnica of Bucharest, Bucharest, Romania
{mircea_stefan.simoiu,ioana.fagarasan,vasile.calofir,
stelian.iliescu}@upb.ro
[2] Grenoble INP, Grenoble, France
stephane.ploix@grenoble-inp.fr

Abstract. Flexibility for the energy grid is expected to become a crucial aspect as the variety of both energy sources and appliances increases in diversity. Moreover, since automation becomes even more present in people's daily activities and interaction, more focus will be required from humans when dealing with the climate changes. Fortunately, energy communities have the potential to build upon the idea of cooperation between humans and energy systems. In the framework of an energy community where people are guided on a daily basis by an intelligent recommendation system, the paper proposes a recommendation strategy implemented in a multi-agent model of a community that focuses on sending straightforward signals, requiring people to modify their consumption in order to achieve their collective objective. The strategy includes a comfort threshold parameter, defined specifically to allow the implementation of several demanding/relaxing variations. Comparing to our previous work presented in [14], we focus on developing a simplistic recommendation strategy, with hourly goals that are easy to understand by the community members. Moreover, we take a step further in investigating the issue of comfort by proposing a no-alert threshold that could be adapted, depending on the community's aim for increased performance or increased effort. A case-study with several multi-agent simulation scenarios is included, emphasising the impact of the previously mentioned parameter on the collective performances, quantified through the net energy exchanged with the grid (NEEG).

Keywords: Energy community · Energy grid · Recommendation strategy · Multi-agent system

© The Author(s), under exclusive license to Springer Nature Switzerland AG 2023
T. Borangiu et al. (Eds.): SOHOMA 2022, SCI 1083, pp. 48–58, 2023.
https://doi.org/10.1007/978-3-031-24291-5_4

1   Introduction

In an increased effort to promote a more active participation of citizens in the energy sector, the European governing institutions have introduced the concept of *energy communities* as a novel organisational means for citizens [1,4]. Thus, energy communities represent a type of organisation where people or small businesses can gather voluntarily and form an entity that is focused on providing a form of environmental, social or economical benefit through collective action. The objective is to develop cooperation mechanisms between people, while also encouraging them to be increasingly informed about energy related issues and also benefit collectively in several ways.

Since energy communities represent a fairly open subject and many systems may be conceptualized around this topic, simulation instruments would represent an important step from a scientific point of view in order to provide a quantifiable perspective of the potential impact of such an organisation. Furthermore, taking into account the behaviour of people in emerging energy management strategies represent another challenge, since energy communities are focused on the contribution of people (and the potential cooperation between human and systems).

This paper proposes a multi-agent simulation model for a residential energy community in which an intelligent recommendation system provides guidance to community members and helps them to achieve their goals. The main focus of the paper is to describe through a case-study the performances of the system at global level, considering a simulation in which community members react to different simplified signals and modify their energy consumption according to the community needs. The signals are dependent on a *no-alert threshold*, a parameter that allows the strategy to be either more demanding in terms of performances, or to provide more comfort if the community requests it.

2   State-of-the-Art Review

The paradigm related to the involvement of people in energy management strategies is relatively new, and several research works have already addressed this interesting topic. For example, in [3], the authors propose a characterization of the potential flexibility that may be provided by citizens in the residential energy sector. In [19], the authors provide a comparison between direct and indirect flexibility types that may have an important impact on self-consumption at collective level. Nevertheless, more research is needed in addressing challenges such as encouraging people to provide flexibility to the grid at critical moments, without inferring a significant discomfort, possibly through excessive direct control.

Regarding energy communities, starting from the previously mentioned framework proposed by the European institutions, several works have already addressed interesting research topics for simulating the collective impact of such an organisation [6]. For example, in [5] the authors propose a multi-objective optimisation design model that involves different types of storage systems and assess the best solutions through Pareto analysis. Regarding economical profit maximisation, the paper presented in [12] describes an optimisation model applied in a case-study of an university campus, a potential energy community. Performances in such models are usually evaluated over self-consumption and self-sufficiency [9]. However, since an energy community indirectly refers to several uncertainty factors, other research papers [18] focus on quantifying this uncertainties in order to feasibly asses system performances.

Aside from the aforementioned works, multi-agent models [16] may represent a suitable solution for assessing the collective impact of an energy community as a distributed system. Specifically, the idea of community member profiles is addressed in [7], where a thermal energy-based energy community is investigated, with member profiles related to the Social Value Orientation theory. In [10], authors propose a multi-agent model to maximise the self-sufficiency of a community, with bi-directional analysis on both comfort and system performances.

This paper aims to continue our previous research related to multi-agent simulation models of energy communities [14]. Specifically, in the context of a residential energy community guided by an intelligent recommendation system, we propose a new recommendation strategy that aims to simplify the communication between agents and the community manager. The proposed strategy aims to investigate the reaction of community members (modelled as agents) to different signals sent by the community manager (another agent), signals that reflect the collective needs in terms of required flexibility.

3   The Multi-agent Simulation Environment

As it was previously mentioned, our method focuses on a multi-agent model of an energy community, with two principal agent types: the community member and the community manager. In this case, multi-agent systems are useful in the sense of providing a platform in which different behaviours may be modelled into agents, allowing a variety of scenarios to be implemented. Moreover, a certain level of intelligence can be developed in each agent to react to various external signals, offering a much more realistic view of an energy community. This could be an important research aspect, considering the sometimes unpredictable behaviour of people.

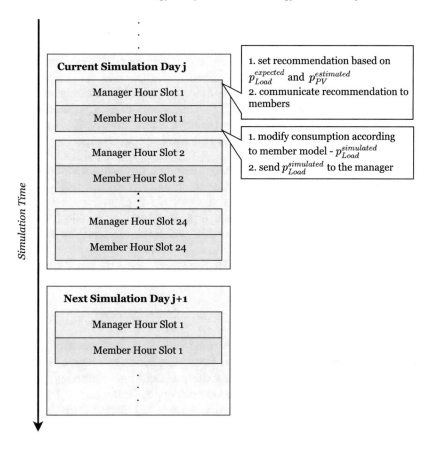

Fig. 1. The multi-agent simulation schedule

From our point of view, the community may have several members and a unique manager that implements a recommendation system. Thus, during a simulation experiment, the manager sends recommendation signals to the members, and after that each member reacts in a particular way while also allowing the manager to record his reaction.

Figure 1 depicts the schedule of a simulation, emphasising the sequence of simulation days (indexed by j), as well as the time slots in each day. As it can be noticed, the time slots come out in pairs: a *Manager Hour Slot* that is always followed by a *Member Hour Slot*. The sample time for the simulation framework is one hour. Consequently, during each hour there is a Manager Hour Slot followed by a Member Hour Slot. The communication between members and the community manager is developed based on our previous research upon the requirements of an energy community simulation model [15]:

- Firstly, the manager develops a recommendation based on the expected level of consumption $p_{Load}^{expected}$ of the community. This expected level shows an approximation of the energy needs of the community. The recommendation is further communicated to each member.
- Secondly, each member, one after the other, modifies his consumption according to a specific model. Thus, $p_{Load}^{simulated}$ is generated, as a personalised consumption profile of each community member. Consequently, this information regarding the consumption profile is sent back to the manager.

From a different perspective, we may imagine a physical implementation of this system. In this implementation the community members receive the recommendation signal through a physical device (tablet, phone) and then react to this recommendation, inferring a change in the consumption profile. This change is further measured by smart energy meters, which further send the information to the community manager.

It is assumed that the community in discussion uses a shared PV plant. The PV production model is represented by Eq. 1 (see [8]).

$$p_{PV} = n \cdot \eta \cdot p_{module} \cdot \frac{G}{G_{STC}} \quad [W] \tag{1}$$

where n represents the number of PV modules, p_{module} represents the nominal power of a module, η represents a correction factor to cover the converter efficiency, partial shadowing and other external phenomena affecting the irradiance, G represents the incident radiation (developed based on [17]) and G_{STC} represents the irradiation at standard temperature conditions ($1kW/m^2$).

The PV model is used by the manager along with the expected consumption profile to develop the recommendations.

3.1   The Recommendation Strategy

Going back to the simulation schedule, the *Manager Hour Slot*, the manager sets the recommendation for the community according to the expected consumption profile $p_{Load}^{expected}$ and the estimated production profile $p_{PV}^{estimated}$. The objective of the community is to minimise the net energy exchanged with the grid:

$$NEEG = \sum_i |p_{Load,i}^{expected} - p_{PV,i}^{estimated}| \Delta t \tag{2}$$

where time is indexed with i. The recommendation strategy proposed in this paper focuses on $\Delta_{Load}^{requested}$ as the requested modification in consumption for the community at a time step i:

$$\Delta_{Load,i}^{requested} = p_{Load,i}^{expected} - p_{PV,i}^{estimated} \tag{3}$$

The possible recommendations that may be given appear in the form of signals:

red signal recommendation to decrease consumption
green signal recommendation to increase consumption
white signal recommendation to consume as expected

This framework is chosen to provide recommendations in order to emphasise the necessity for simplicity. In a real implementation of such a solution, we may conceptualize that community members could have an indicator light device mounted at home that could show during each hour of the day the corresponding necessities of the community, as inferred by the community manager. So, during each *Manager Time Slot*, the manager computes a recommendation and sends it to the members. To compute the recommendation, the proposed strategy is developed based on a simple rule-based approach: if $|\Delta^{requested}_{Load,i}| > \tau$, then the manager provides a recommendation to either increase or decrease consumption. τ is used as a threshold for not providing a recommendation in order to lower the potential discomfort caused to the members by providing recommendations for every small $\Delta^{requested}_{Load,i}$. The whole recommendation strategy to determine the signal x at the moment i can be analysed in Eq. 4.

$$x_i = \begin{cases} \text{red} & \text{if } \Delta^{requested}_{Load} > \tau \\ \text{green} & \text{if } -\Delta^{requested}_{Load} < -\tau \\ \text{white} & \text{otherwise} \end{cases} \tag{4}$$

Thus, the proposed method allows the implementation of several strategy variations. Specifically, strategies with a large value for τ may be used to lower the required community effort in terms of consumption modification, while a relatively small value for τ would result in a demanding strategy.

3.2 Members' Reaction to Received Signals

In the multi-agent framework, after a member has received the signal during a *Manager Time Slot*, he will react to it during the *Member Time Slot*. The modification in consumption will be reflected in the $p^{simulated}_{Load}$, which is defined as the actual consumption of community members after they receive the recommendation. Thus, we propose the following model for the reaction of people:

$$p^{simulated}_{Load,i} = \begin{cases} \mathscr{U}(0.5, 1.2) \cdot p^{estimated}_{Load} & \text{if } x_i = \text{red} \\ \mathscr{U}(0.8, 1.5) \cdot p^{estimated}_{Load} & \text{if } x_i = \text{green} \\ \mathscr{U}(0.8, 1.2) \cdot p^{estimated}_{Load} & \text{if } x_i = \text{white} \end{cases} \tag{5}$$

By choosing these stochastic variables, it is assumed that a reaction to the red signal would infer a 50% decrease in consumption under normal circumstances, while the green signal would infer a 50% increase in consumption. However, due to the unpredictable nature of the human behaviour, a normal distribution is used, adjusted with 20%. In this way, it is taken into account the case where members could react in a different way when receiving the recommendation. Similarly, the reaction for the white signal is adjusted by %20 in both ways. As a consequence of the assumption that people would cooperate voluntarily during the day, we also propose to send recommendation signals only between 07:00 and 23:00.

3.3   Performance Evaluation

To evaluate the performance of the community under the influence of the recommendation system, the net energy exchanged with the grid is used (Eq. 2) as a primary metric, emphasising that the closer the value of NEEG is to zero, the lower is the dependency of the community to the grid. This aspect underlines that the community capitalizes on the available PV production (high self-consumption index) while also covering as much load as possible (high self-sufficiency index) [13]. Furthermore, it is important to compare in terms of performances the simulated behaviour of the community relative to the expected behaviour. In this way, the impact of the recommendation system is emphasised, as well as the modification in the consumption profiles.

4   Case Study

The model has been implemented in Python, using the multi-agent framework MESA [2]. To test the proposed method, a case study was conducted on an energy community of five houses in France. The houses' consumption data has been obtained from the REMODECE project [11] and spans from 16.02.2015 to 01.01.2016, with a sample time of 1h. Irradiance data has been obtained from *OpenWeatherMap.org* for the respective location and has been further used to generate the production for the community according to the model presented in Eq. 1. Consequently, the PV plant was configured to have $n = 150$, $\eta = 0.75$ and the power of a module to be $p_{module} = 500Wp$.

The simulation involves a community of five houses with a shared PV plant, as depicted in Fig. 2. The simulation period spans according to the available consumption data. The proposed case-study investigates the impact of the variation of the *no-alert threshold* τ over the system performances. The results from the simulation experiments can be investigated in Table 1, where performances from several scenarios are shown. Moreover, we have compared the daily expected NEEG value for each scenario with the simulated counterpart are illustrated in Figs. 3 three investigated scenarios, specifically Scenario 1,3 and 5 from Table 1. The images show a power profile comparison for two day selected from June, along with the corresponding signals given during the respective days.

As it can be noticed from Table 1, the recommendation mechanism manages to reduce the daily NEEG; however this aspect is affected by the choice of τ. For example, in Scenario 1 the best performances are achieved in terms of NEEG, which is significantly different from Scenario 6 where τ has a larger

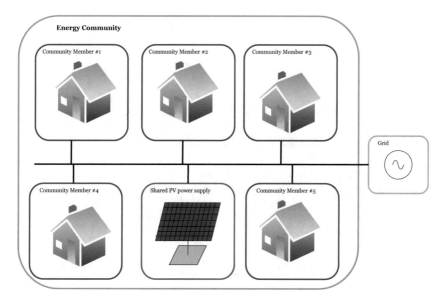

Fig. 2. Energy community configuration for the proposed case-study

value. These aspects can also be visualised in Fig. 3: as the value of τ is reduced, the number of recommendations increase, and so do the performances of the community at the expense of the member's effort. Consequently, we may consider this result important since in developing future recommendation strategies with an increased complexity, it could be relevant to achieve a certain balance between aiming for very good performances and not causing discomfort to community members.

Table 1. Simulation results comparison

Scenario	τ [W]	NEEG^{expected} [kWh/day]	NEEG^{simulated} [kWh/day]
1	100	36.84	32.4
2	500	36.84	32.57
3	1000	36.84	33.15
4	1500	36.84	33.94
5	2000	36.84	34.51
6	3500	36.84	36.03

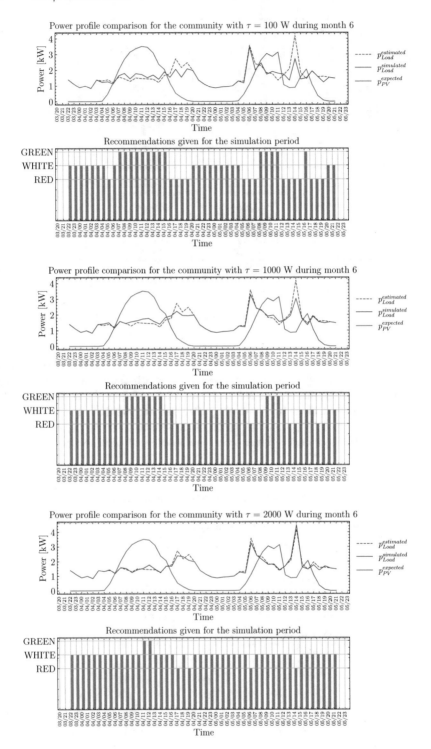

Fig. 3. Comparison of simulation scenarios, focusing on two days during the month of June

5   Conclusion

The paper presents a multi-agent simulation model that describes a residential energy community that is guided daily through recommendations by a community manager. The recommendations are given hourly in the form of simple indicator signals (red, green, white) and aim to decrease the NEEG of the community.

The main contribution of this paper is related to the recommendation strategy proposed and the related investigation into the *no-alert threshold*, a parameter that can be adjusted in order to globally offer more demanding recommendations or more relaxed ones, depending on the community choices. As it can be noticed in the results, the NEEG obtained after a simulation with the proposed recommendation strategy adapted with a very low no-alert threshold infers a significant increase compared to the expected scenario, while a relatively high no-alert threshold results in a performance decrease, but also infers increased comfort for the community members.

Future research work will address other recommendation strategies based on time correlation, as well as other indicators that may be used to assess the comfort and performance of an energy community. Moreover, future research will also focus on "coaching" recommendation strategies that will aim to increase the motivation of community members to become more active and contribute to the community.

Acknowledgement. The results presented in this article has been funded by the Ministry of Investments and European Projects through the Human Capital Sectorial Operational Program 2014–2020, Contract no. 62461/03.06.2022, SMIS code 153735.

References

1. DIRECTIVE (EU) 2018/2001 of the European Parliament and of the Council of 11 December 2018 on the promotion of the use of energy from renewable sources (recast) (Text with EEA relevance) (2018). https://eur-lex.europa.eu/legal-content/EN/TXT/?uri=uriserv:OJ.L_.2018.328.01.0082.01.ENG&toc=OJ:L:2018:328:TOC
2. MESA: agent-based modeling in Python 3+ - Mesa. 1 documentation. https://mesa.readthedocs.io/en/latest/
3. Amayri, M., Silva, C.S., Pombeiro, H., Ploix, S.: Flexibility characterization of residential electricity consumption: a machine learning approach. Sustain. Energy Grids Netw. **32**, 100801 (2022). https://doi.org/10.1016/j.segan.2022.100801
4. European Parliament, Council of the European Union: Directive (EU) 2019/944 of the European Parliament and of the Council - of 5 June 2019 - on common rules for the internal market for electricity and amending Directive 2012/27/EU (2019). https://eur-lex.europa.eu/legal-content/EN/TXT/PDF/?uri=CELEX:32019L0944&from=EN
5. Fan, G., et al.: Energy management strategies and multi-objective optimization of a near-zero energy community energy supply system combined with hybrid energy storage. Sustain. Cities Soc. **83**, 103970 (2022). https://doi.org/10.1016/j.scs.2022.103970

6. Fina, B., Monsberger, C., Auer, H.: Simulation or estimation?—Two approaches to calculate financial benefits of energy communities. J. Clean. Prod. **330**, 129733 (2022). https://doi.org/10.1016/j.jclepro.2021.129733

7. Fouladvand, J., Ghorbani, A., Sarı, Y., Hoppe, T., Kunneke, R., Herder, P.: Energy security in community energy systems: an agent-based modelling approach. J. Clean. Prod. **366**, 132765 (2022). https://doi.org/10.1016/j.jclepro.2022.132765

8. HOMER: HOMER PV energy model. https://www.homerenergy.com/products/pro/docs/latest/how_homer_calculates_the_pv_array_power_output.html

9. Lopes, R.A., Martins, J., Aelenei, D., Lima, C.P.: A cooperative net zero energy community to improve load matching. Renew. Energy **93**, 1–13 (2016)

10. Reis, I.F., Gonçalves, I., Lopes, M.A., Antunes, C.H.: A multi-agent system approach to exploit demand-side flexibility in an energy community. Util. Policy **67**, 101114 (2020). https://doi.org/10.1016/j.jup.2020.101114

11. REMODECE: REMODECE project. https://remodece.isr.uc.pt

12. Sima, C.A., Popescu, C.L., Popescu, M.O., Roscia, M., Seritan, G., Panait, C.: Techno-economic assessment of university energy communities with on/off microgrid. Renew. Energy **193**, 538–553 (2022)

13. Simoiu, M.S., Fagarasan, I., Ploix, S., Calofir, V.: Optimising the self-consumption and self-sufficiency: a novel approach for adequately sizing a photovoltaic plant with application to a metropolitan station. J. Clean. Prod. **327**, 129399 (2021). https://doi.org/10.1016/j.jclepro.2021.129399

14. Simoiu, M.S., Fagarasan, I., Ploix, S., Calofir, V.: Modeling the energy community members' willingness to change their behaviour with multi-agent systems: a stochastic approach. Renew. Energy **194**, 1233–1246 (2022)

15. Simoiu, M.S., Fagarasan, I., Ploix, S., Calofir, V., Iliescu, S.S.: General considerations about simulating energy communities. In: 2021 11th IEEE International Conference on Intelligent Data Acquisition and Advanced Computing Systems: Technology and Applications (IDAACS), pp. 1126–1131. IEEE, Cracow (2021). https://doi.org/10.1109/IDAACS53288.2021.9661053

16. Simoiu, M.S., Fagarasan, I., Ploix, S., Calofir, V., Iliescu, S.S.: Towards energy communities: a multi-agent case study. In: 2022 IEEE International Conference on Automation, Quality and Testing, Robotics (AQTR), pp. 1–6. IEEE, Cluj-Napoca (2022). https://doi.org/10.1109/AQTR55203.2022.9802060

17. Stephane Ploix: buildingenergy. https://gricad-gitlab.univ-grenoble-alpes.fr/ploixs/buildingenergy

18. Tostado-Véliz, M., Kamel, S., Hasanien, H.M., Turky, R.A., Jurado, F.: Optimal energy management of cooperative energy communities considering flexible demand, storage and vehicle-to-grid under uncertainties. Sustain. Cities Soc. **84**, 104019 (2022). https://doi.org/10.1016/j.scs.2022.104019

19. Twum-Duah, N.K., Amayri, M., Ploix, S., Wurtz, F.: A comparison of direct and indirect flexibilities on the self-consumption of an office building: the case of Predis-MHI, a smart office building. Front. Energy Res. **10**, 874041 (2022). https://doi.org/10.3389/fenrg.2022.874041

Toward a Sawmill Digital Shadow Based on Coupled Simulation and Supervised Learning Models

Sylvain Chabanet[✉], Hind Bril El Haouzi, and Philippe Thomas

Université de Lorraine, CRAN, CNRS UMR 7039, Campus Sciences, BP 70239,
54506 Vandœuvre-lès-Nancy Cedex, France
{sylvain.chabanet,hind.el-haouzi,philippe.thomas}@univ-lorraine.fr

Abstract. Digital Twins (DT) have been introduced as promising decision support tools in many different settings and serve a variety of purposes. Many challenges are raised by their development, including an efficient usage of their computational resources to balance performance on precision, computational cost and speed. This study is, in particular, concerned with Digital Shadows (DS), a concept derived from DT, applied to sawmills sawing production units. A method to combine a computationally intensive sawmill simulation model with a machine learning model is proposed to predict the set of lumbers sawed from logs. Numeric experiments are exposed, and the proposed method demonstrates improvements from 11% to 18% of the monitored couple regret from its baseline.

Keywords: Digital shadow · Active learning · Sawmill simulation

1 Introduction

The concept of digital shadows (DS) is derived from the concept of digital twin (DT) which is tracked back to NASA Apollo program, which built two identical vehicles one sent to space missions and the other remaining on earth [2]. The objective was to monitor the twin sent to space during mission and find solutions when faced with problems. Since then, the concepts and definitions proposed for DS and DT have evolved and still remains a field of active research. A broad, generally accepted definition encompassing both DS and DT concepts was proposed as "a set of digital models of physical entities usable for optimization and decision-making purposes based on real-time information" [19]. In particular, while both DS and DT are often associated with simulation, digital models based on many technologies have been proposed to support their respective capabilities. For example, authors of [7] propose a hybrid DT based on physics simulation augmented by a machine learning (ML) model to predict acceleration responses of a structure subject to vibrations. Similarly, [12] introduce a DT based on analytical models for the performance evaluation of a manufacturing system. While the difference between DS and DT is still subject to debates, an

© The Author(s), under exclusive license to Springer Nature Switzerland AG 2023

T. Borangiu et al. (Eds.): SOHOMA 2022, SCI 1083, pp. 59–70, 2023.
https://doi.org/10.1007/978-3-031-24291-5_5

interesting discussion is provided by [18]. DTs should be able to automatically influence their physical counterpart. As such, they require an extremely precise representation and detailed behaviour models. DS, on the other hand, are decision support tools designed to answer specific questions. It should be noticed, however, that this distinction is not always made in the scientific literature, and DS can hardly be studied independently from DT.

Due to the variety of technologies proposed to design them, DS might be better described based on their capabilities: to maintain a description of the physical counterpart (PC) based on real time or near real time data, to predict future states of the PC or compare various configurations, and to support decision making by returning instruction or optimized plans to the physical world. Such capabilities are naturally built on top of one another. Descriptive capabilities are required for predictive model to front-run production. Similarly, the output of predictive models can be fed as input to optimization models prescribing optimized plans (Fig. 1).

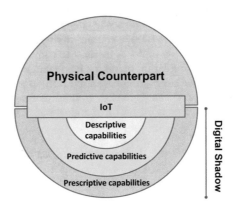

Fig. 1. The three main capabilities of a complete DS decision support tool

Digital shadows and digital twins have been studied in very different contexts, from hospitals [9] to sawmills [5]. Sawmills, in particular, might benefit greatly from DS for operational production monitoring and control. Sawing operations, the breaking of logs into lumber, is a process divergent and in co-production. From a single log, a sawmill will obtain simultaneously several pieces of lumber of different dimensions (Fig. 2). In the remaining of this paper, the set of lumber obtained from a single log is named a basket of products (BoP). Many factors introduce uncertainty on the yield of the sawing process. These factors include, for example, the heterogeneity of the shapes of logs, whose general characteristics may change from one harvesting site to another, or the emergence of automated sawlines which adjust sawing operations on a log per log basis.

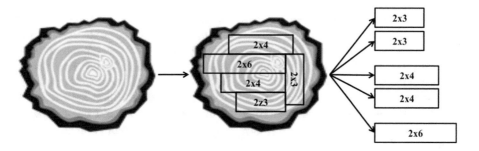

Fig. 2. Sawing a log produce several pieces of lumbers with various dimensions

All this makes it difficult to know in advance the mix of lumber that would be obtained from sawing a batch of logs. Solutions proposed in the literature to overcome this uncertainty include the use of specialized optimization methods, such as robust or stochastic optimization [23], or the usage of sawing simulators able to predict logs BoP based on information over their shape [13] or even internal structure. Information over log shape may, in particular, take the form of 3D scans, for example obtained from laser scanners which may be placed at the entrance of sawmill log yards or sawing units. General simulators able to flexibly model many sawmill configurations and predict BoP of logs based on their scans have been developed by researchers and industry. They include, for example, Sawsim[1] or Optitek [8]. Optitek, in particular, has been used in many academic studies [14,22]. Interestingly, both of these studies demonstrate the benefit of BoP predicted by either simulation or ML based metamodels of these simulators as input for planning decision support optimization models.

Such generic sawmill simulation tools may be important steps toward the development of flexible simulation-based sawmill DS. In particular, they would provide flexible yet precise models backing predictive capabilities. Several researchers, however, have noticed the important computational time associated with running these simulations for thousands of logs [14,16,22], which would impair their use for real or near real-time problems. The design of DS based on simulation and ML models used simultaneously, appears therefore, as an interesting solution to this problem. Such DT and DS mixing data-driven methods and other simulation techniques are sometimes referred to as "hybrid" in the literature, for example by [1]. The "hybrid" qualifier refers only to the variety of models involved in the predictive capabilities of the twin. An interest of these hybrid DS is that, in particular, predictions performed by a ML model are, in general, very fast. Such models may require computationally intensive training phases, but these can be performed offline when necessary. [1], for example, develops such a model to combine a simulation model based on finite element analysis with a ML model to optimize welding operations.

Several factors complicate, indeed, the use of DS based only on data-driven, ML-based predictive models. ML models, in particular, require large datasets of

---

[1] https://www.halcosoftware.com/software-1-sawsim, Last accessed on June, 2021

labelled examples to be trained on. While in some cases these training examples may be fully obtained from the physical twin, this is complicated for sawmills by the difficulty of tracking logs and lumbers through several diverging transformation processes. In a more general setting, simulation remains necessary when historical data are not accessible from the start to train ML models, for example after changing machine settings. A second reason for favouring a hybrid DT is that ML models may remain, on some prediction problems and depending on the simulation technology, more prone to error and less explainable than these simulations.

This, however, raises the question of how to couple a simulation and ML models performing the same prediction task with their respective advantages and drawback. In this paper we propose a method to automatically decide, on a log per log basis, whether its BoP should be predicted by a ML model or a simulator.

The remaining of this study is structured as follows. Section 2 first presents in more details the hybridization problem considered. Section 3 details our proposed method, for which an experimental proof-of-concept is given in Sect. 4. Section 5 concludes this paper and exposes future works.

2   Problem Description

The general problem considered in this study assumes the existence of a stream of data items. For every individual data item X, some quantity y has to be predicted by a DS. In this study, a data item is a set of information over a specific log, and y is its BoP, but it could be generalized to any other type of stream. The existence of a DS containing at least two digital models able to make this prediction is similarly assumed. The first model is a simulation model, very precise but computationally intensive. The second is a ML model, fast but overall not as precise as a simulation. This ML model may be based on any supervised learning algorithm, for example neural network or random forest. As such, it is data-driven and requires labelled examples to be trained on. The quality of a model prediction $\hat{y}$ is evaluated by a loss function $l(\hat{y}, y)$, so that a prediction $\hat{y}$ closest to the real value y will lead to a lower loss. No upper bound is assumed on the number of simulations that may be run in parallel. This may be an acceptable approximation if, for example, the simulation is run on a cloud service so that the amount of computational resources which can be punctually used is far greater than what is wished to be used on long term to maintain overall computational costs reasonable. However, an asymptotic simulation budget b is targeted, i.e., a limit of the simulation budget as time goes to infinity. It is defined here as the limit of the total accumulated simulation time divided by the time length of the stream:

$$b = \lim_{T \to +\infty} \frac{1}{T} \sum_{k=0}^{\infty} p_k \mathbf{1}_{\{a_k < T\}} . \tag{1}$$

Here, p_k is the time necessary to predict y for the k_{th} data item entering the stream, a_k its time of arrival and $\mathbf{1}_{\{...\}}$ the indicator function.

The objective is to propose a dispatching function $\mathbf{1}_{ToSimul(X)}$ taking value 1 if the prediction of the data item X has to be performed by the simulation model. Such a decision has to be taken immediately after arrival, without consideration for future data item. Additionally, the loss of a prediction from the ML model, $l(\hat{y}_{ML}, y)$ is, of course, unknown when the decision has to be taken and can't be used directly.

For the purpose of mathematical modelling, the following assumptions are also used. The prediction of the simulation model is supposed to be exact, i.e., $l(\hat{y}_{Simul}, y) = 0$ if $\mathbf{1}_{ToSimul(X)} = 1$. The prediction time of the ML model is neglected, i.e., $l(p_k, y) = 0$ if $\mathbf{1}_{ToSimul(X_k)} = 0$. The times of arrivals of data items are assumed to follow a Poisson process of rate μ. Simulation times are sampled from an exponential law of mean λ. These choices make the process recording the number of simulations being run simultaneously a continuous-time Markov Chain process, ensure the existence of a stationary limiting distribution and allow the usage of the ergodic theorem. However, the results described further of this section hold in a more general case, as long as stationarity and ergodicity can be assumed.

Following this modeling, consider $\alpha = \mathbf{P}(\mathbf{1}_{ToSimul(X)} = 1)$ the fraction of the stream whose prediction is obtained from the simulation model. If the decision $\mathbf{1}_{ToSimul(X_k)}$ is taken independently from past and future arrival, the process formed by the data items for which a prediction is requested from the simulation model forms a Poisson process of rate $\mu\alpha$. Noting $N(t)$ the number of simulations running in parallel at time t, the budget b can be rewritten as:

$$b = \lim_{T \to +\infty} \frac{1}{T} \int_0^T N(t) \, dt, \tag{2}$$

i.e., the target budget b is also the average number of simulation running in parallel at any time. Additionally, according to the ergodic theorem, it is equal to $\mathbf{E}(N)$ under the process stationary distribution. Little's law [11] yield, in turn:

$$\mathbf{E}(N) = \mathbf{E}(p)\mu\alpha = \lambda\mu\alpha. \tag{3}$$

Hence, to target the budget b, the simulation rate α must be $\alpha = \frac{b}{\lambda\mu}$.

3   Proposal

The problem considered in this study is similar to problems detailed in the active learning (AL) literature. AL is a field from the ML literature concerned with reducing the cost associated with labelling data to train an ML model. While unlabelled collection of data items X are, in many settings, easy to gather in huge quantities, their labels y may be costly or difficult to obtain. This is often the case when the Oracle providing the labels is, for example, a team of human experts. More precisely, AL is concerned with guiding the selection of a subset of unlabelled items that are considered the more useful to improve the performances of the ML model. It is often an iterative process where a ML model has already be initially trained on a small training set and is used to define a measure m of the utility of labelling a point. Data items maximizing this measure, or having it above some threshold are then selected from an unlabelled dataset or data stream and labelled. The ML model may be periodically retrained when a sufficient amount of new labelled data have been gathered, or when another condition is triggered such as the detection of changes in the data stream. For example, [21] propose an AL method based on a Random Forest (RF) model to detect surface defects in a stream of products.

Three main scenarios exist in the AL literature, depending on the way the active learner gains access to data:

- Pool-based active learning is concerned with cases where an unlabelled dataset may be accessed at once by the learner.
- Stream-based active learning considers cases where data items are not accessible at once but one by one from a data stream. The decision to label every item has then to be taken immediately for every data item, or for small batches of item, without considering the future of the stream.
- Membership Query Synthesis scenarios do not consider the existence of real unlabelled data. Instead, the learner may create synthetic data and request labels for these. This scenario is rarely investigated as synthetic data may be difficult to label, especially for human oracles.

All three scenarios may be of interest in the context of DS. In particular, simulation-based DS may be of great interest to label synthetic data and stream-based active learning may be used to manage real time data streams continuously collected by the twin. This study focuses on this second stream based case.

Many measures m of labelling utility have been introduced for both regression and classification tasks [10]. In particular several are based on heuristic or statistical measures of the uncertainty or confidence associated with a prediction. The method proposed in this study is to use such measures of uncertainty to design the dispatching function $\mathbf{1}_{ToSimul(X)}$, under the reasoning that such a measure may bring valuable insight over the unobservable loss $l(\hat{y}_{ML}, y)$, and that ML prediction whose confidence is considered too low should preferably be obtained from the simulation model.

In particular, consider q the quantile at level $1 - \alpha$ of $m(X)$. In general, this quantile may be estimated from observed past values from the stream. If the dispatching function is taken as $\mathbf{1}_{ToSimul(X)} = \mathbf{1}_{m(X)>q}$, then a simulation result is requested for the portion α of the stream for which the ML model is considered the least confident according to m.

The ML model used in this study is a RF model [3]. RF are machine learning models popular for their ease of use, overall good results and robustness, and have been used with success in many studies in industrial context [4,17]. They have also given good performances in past studies comparing ML models to predict BoP of logs [14,15]. RF are ensembles of decision tree models, where the prediction is given by $\hat{y} = \bar{t}(X) = \frac{1}{B}\sum_{i=0}^{B} t_i(X)$. Here, $t_i(X)$ are predictions of individual trees and B is the number of trees in the forest. Individual trees are trained on bootstrap samples selected from the whole training data set.

The measure m considered here is an estimator of the sampling variance of a random forest introduced by [20]. It evaluates how much the prediction may change if a similar model had been trained on a different dataset. This measure was used for AL for example in [6] which propose a method to predict alloys melting points. It is expressed as:

$$\sum_{i=1}^{k} Cov[J_i, t_.(x)]^2 + \sum_{i=1}^{k} (\bar{t}_{-i}(x) - \bar{\bar{t}}(x))^2 - \frac{ek}{B^2} \sum_{i=1}^{B} (t_i(x) - \bar{t}(x))^2 . \quad (4)$$

Here, k is the member of data items in the training set, e the Euler number, $\bar{t}_{-i}(x)$ the average of the prediction of trees which didn't use the ith data item for training, and J_i the number of times the ith item was present in the bootstrap sample used to train a tree $t_.$. Cov is the usual covariance estimate. Considering that [20] recommend to use forests with high number a trees, comparable with the number of data items used to train it, B is fixed to 500 in this paper experiments.

If the simulation model is precise enough, its predictions, collected along the stream may, additionally, be added to a training dataset used to periodically retrain and improve the performances of the ML model with more data.

A pseudo-code summary of the proposed solution is exposed in Algorithm 1.

Algorithm 1. General flow of the proposed strategy

Initialize *LabelledSet*
Initialize *MLmodel* on *LabelledSet*
for x in stream **do**
    Predict $\hat{y} \leftarrow$ ML prediction
    Evaluate $m(x)$, the labelling utility measure
    if $m(x) < q$ **then**
        $\hat{y} \leftarrow$ Simulation prediction
        add $\hat{y}$ to *LabelledSet*
    end if
    if Model update condition **then**
        Update ML model on *LabelledSet*
    end if
end for

---

4 Experiments and Results

This section first presents the dataset used during the experiments proposed in this study. Additional details over these experiments are also given, and the results are exposed.

4.1 Materials and Method

The dataset used in the experiments described in this section originates from the Canadian forest-product industry. It contains information about 2219 wood log.

In particular, both the 3D scans of every logs are given, as well as a collection of six describing features. An example of such a 3D scan is presented in Fig. 3. A scan point cloud, whose points are organized into rough ellipsoids spanning the log surface is represented.

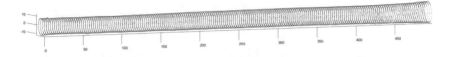

Fig. 3. 3D scan of a log

The six know-how features summarizing information over the shape of logs are the length, volume, diameter at both extremities, curvature and taper (a measure of the shrinking of a log from one end to another). While the scans were used to simulate the sawing of logs and obtain their BoP, these six know-how are required by the ML model to make their own predictions. 3D scans are, indeed, unordered collection of points and the number of points varies from scan to scan. They are, therefore, unstructured data and can't be used directly by commons ML models. To summarize scans by a few descriptive features implies a loss of information that cannot be used by the ML model to make predictions. Therefore, such a ML model cannot, in particular, be expected to perform as well as a simulation model.

The BoP of every log, obtained by simulating its sawing with the simulator Optitek have been also obtained. Simulation times, however, were not part of this dataset. They have been, therefore, generated randomly when required.

The sawmill simulated was able to produce up to 47 types of lumbers, characterized by their length, width and thickness. For this reason, BoP are represented as vectors of dimension 47, where the ith element encodes how many pieces of lumber of type i are present in the BoP. The machine learning prediction task is, therefore, modelled as a multi-output regression problem. Considering that the measure described in Eq. 4 is initially designed for single output regression problems, in practice a collection of 47 measures m_i have been used, one for every type of lumber, and their sum $m(X) = \sum_{i=1}^{47} m_i(X)$ too.

To simulate a stream for experimental purpose, the dataset is ordered randomly. The first 500 elements are used to initialize the RF model. Inter-arrival times in the stream were sampled from an exponential distribution with average $\lambda = 1$ and simulation times were sampled from an exponential distribution with rate $\mu = \frac{1}{10}$ so that the average simulation time is approximately 10 time greater than the average inter-arrival time. Several values of the budget were tested.

At the end of the stream, the total stream time and accumulated simulation time are collected to estimate the effective simulation budget. The fraction of the stream sent to the simulation model is also collected. Two evaluation scores are additionally collected. The first one is the ML regret:

$$\frac{1}{n}\sum_{l=1}^{n} l(\hat{y}_{ML,l}, y_l) \tag{5}$$

with n the number of data items in the stream, y_l the label of the lth arriving item, and $\hat{y}_{ML,l}$ its prediction performed by the ML model *whether it is sent to the simulator or not*. This score evaluates the performance of the ML model if it were used for every prediction.

Similarly, the coupling regret is defined as:

$$\frac{1}{n}\sum_{l=1}^{n} l(\hat{y}_{ML,l}, y_l)(1 - \mathbf{1}_{\{ToSimul(X_l)\}}) \tag{6}$$

and measures the performance of the hybridization. The performances of the proposed dispatching function will additionally be compared with the performances of a random dispatching requesting simulation depending on the result of a collection of i.i.d Bernoulli variables of parameter α. The ML model was retrained every time 100 new labelled items, or, in average, every 400 new arrivals.

To average out the impact of the exact ordering of a stream, and time sampling, this will be repeated 50 times with different random seeds.

4.2 Results

Results for three different values of the budget 1, 2.5 and 4 are presented in Table 1. In particular, this means that, according to Sect. 3, the use of the simulation model should be requested for approximately 10%, 25% and 40% of the logs from the stream to predict their BoP. This is, indeed what is observed in practice and is recorded in the column "simulated ratio". Additionally, the respective targeted asymptotic budgets are respected, as can be seen from the observed accumulated simulation times.

Table 1. Average and standard error over the 50 experiments of the simulation time over stream time, ratio of the stream sent to the simulator, and regrets

Budget	Method	Simulation time ratio	Simulated ratio	ML regret	Coupling regret
1	Random method	1.01 (0.10)	0.10 (0.01)	3.25 (0.13)	2.92 (0.12)
	AL-based method	1.12 (0.07)	0.11 (0.004)	3.17 (0.10)	2.59 (0.08)
2.5	Random method	2.49 (0.17)	0.25 (0.01)	3.16 (0.09)	2.38 (0.09)
	AL-based method	2.46 (0.15)	0.25 (0.004)	3.07 (0.07)	2.03 (0.06)
4	Random method	4.00 (0.21)	0.40 (0.01)	3.09 (0.08)	1.85 (0.07)
	AL-based method	3.96 (0.17)	0.39 (0.01)	3.01 (0.06)	1.52 (0.05)

It is interesting to notice that the ML regret which evaluates the performance of the ML model if it were used for all predictions is always slightly lower for the AL-based method than for the random method. It is, indeed, the initial objective of AL methods to improve the performances of ML models by guiding the selection of data items that should be labelled rather than sampling at random. Performing Welsh t-test to compare the averaged ML regrets over the various repetition of the experiment yields p-values of 0.001 for $b = 1$, 1.4×10^{-7} for $b = 2.5$ and 1.8×10^{-7} for $b = 4$. In every case, the null hypothesis of equality of the samples means can therefore be rejected. Despite being small, this difference is, therefore, considered statistically significant, w.r.t the repetition of the experiment. This includes randomness of the stream order, simulation time and inter-arrival time. It should be noticed that the randomness from the original data collection process itself is not considered by the test. These ML regrets remain, however, high with respect to the average number of products in a basket, at 4.6, and the quadratic error obtained by a predictor forecasting always the average BoP over the dataset, at 8.3. This might be partly caused by the small size of the datasets used to train these predictors.

Coupling regrets are, of course, always lower than ML regret, due to a fraction of the prediction being requested from the simulator and, therefore, considered to be exact. Once again, however, the coupling regret of the AL-based method is lower than the one of the random method, to a greater extent than what was already observed in the case of the ML regret. In particular, the regret is reduced by 11% for $b = 1$, 15% for $b = 2.5$ and 18% $b = 4$. Therefore, the sampling variance measure used in the AL-based method appears efficient at anticipating predictions from the ML model with higher loss, and requesting predictions from the simulation model instead.

5   Conclusion

Many challenges remain in the development of DS in general, and sawmill DS in particular. The development of methods to allow an efficient usage of computational resources balancing computational cost with analytic performances appears in particular important for both economic and sustainability reason. To this end, the development of DS based on hybrid simulation-ML models able to benefit from the advantages of both technologies while compensating their drawback.

This study introduces a method based on AL to couple a computationally intensive sawmill simulator with a less precise but fast ML model to predict BoP of wood logs. The main objective is to make the best use of the computational resources allocated to the simulation model by requested simulations only for these logs for which the ML model predictions appear the least certain according to some statistical measure. During experiments, this proposed method demonstrated reductions of the simulation ML couple regret ranging from 11% to 18% depending on the simulation budget. Additionally, when using the simulation results to improve the performances of the ML model itself demonstrates a slightly better effect with the proposed AL-based method than with a random dispatching of the prediction toward the simulator.

However, several limits exist in the results presented in this paper. Firstly, more experiments need to be run to assess the impact of various parameters, in particular the average simulation time and stream arrival rate. Secondly, the respect of the simulation budget is asymptotic and based on statistical assumptions, including stationary data streams. This mean, in particular, that this budget may be temporarily exceeded, and increase significantly if drifts were to happen in the stream. Therefore modifications of the proposed strategy enforcing the budget to always remain between acceptable bounds appear particularly important. Additionally, many more utility measures have been proposed in the AL literature. Further works comparing different such methods will therefore be necessary.

References

1. Asadi, M., Mohseni, M., Tanbakuei, M., Kashani, M.F., Smith, M.: Machine-learning-enabled digital twin of welded structures for rapid weld sequence design. In: 74th IIW on-line Assembly and International Conference (2021)
2. Boschert, S., Rosen, R.: Digital twin—the simulation aspect. In: Hehenberger, P., Bradley, D. (eds.) Mechatronic Futures, pp. 59–74. Springer, Cham (2016). https://doi.org/10.1007/978-3-319-32156-1_5
3. Breiman, L.: Random forests. Mach. Learn. **45**(1), 5–32 (2001). https://doi.org/10.1023/a:1010933404324
4. Carvajal Soto, J., Tavakolizadeh, F., Gyulai, D.: An online machine learning framework for early detection of product failures in an industry 4.0 context. Int. J. Comput. Integr. Manuf. **32**(4–5), 452–465 (2019)
5. Chabanet, S., Bril El-Haouzi, H., Morin, M., Gaudreault, J., Thomas, P.: Toward digital twins for sawmill production planning and control: benefits, opportunities, and challenges. Int. J. Prod. Res. (2022)
6. Farache, D.E., Verduzco, J.C., McClure, Z.D., Desai, S., Strachan, A.: Active learning and molecular dynamics simulations to find high melting temperature alloys. Comput. Mater. Sci. **209**, 111386 (2022)
7. Gardner, P., Dal Borgo, M., Ruffini, V., Hughes, A.J., Zhu, Y., Wagg, D.J.: Towards the development of an operational digital twin. Vibration **3**(3), 235–265 (2020)
8. Goulet, P.: Optitek: User's manual (2006)
9. Karakra, A., Fontanili, F., Lamine, E., Lamothe, J.: HospiT'Win: a predictive simulation- based digital twin for patients pathways in hospital. In: 2019 IEEE EMBS International Conference on Biomedical & Health Informatics. IEEE, New York (2019)
10. Kumar, P., Gupta, A.: Active learning query strategies for classification, regression, and clustering: a survey. J. Comput. Sci. Technol. **35**(4), 913–945 (2020)
11. Little, J.D., Graves, S.C.: Little's law. In: Chhajed, D., Lowe, T.J. (eds.) Building Intuition. International Series in Operations Research and Management Science, vol. 115, pp. 81–100. Springer, Boston (2008). https://doi.org/10.1007/978-0-387-73699-0_5
12. Magnanini, M.C., Tolio, T.A.: A model-based digital twin to support responsive manufacturing systems. CIRP Ann. **70**(1), 353–356 (2021)
13. Maness, T.C., Norton, S.E.: Multiple period combined optimization approach to forest production planning. Scand. J. For. Res. **17**(5), 460–471 (2002)
14. Morin, M., et al.: Machine learning-based models of sawmills for better wood allocation planning. Int. J. Prod. Econ. **222**, 107508 (2020)

15. Morin, M., Paradis, F., Rolland, A., Wery, J., Gaudreault, J., Laviolette, F.: Machine learning-based metamodels for sawing simulation. In: 2015 Winter Simulation Conference (WSC), pp. 2160-2171. IEEE, New York (2015)
16. Morneau-Pereira, M., Arabi, M., Gaudreault, J., Nourelfath, M., Ouhimmou, M.: An optimization and simulation framework for integrated tactical planning of wood harvesting operations, wood allocation and lumber production. In: MOSIM 2014, 10eme Conférence Francophone de Modélisation, Optimisation et Simulation (2014)
17. Ruiz, E., Ferreño, D., Cuartas, M., López, A., Arroyo, V., Gutiérrez-Solana, F.: Machine learning algorithms for the prediction of the strength of steel rods: an example of data-driven manufacturing in steelmaking. Int. J. Comput. Integr. Manuf. $33(9)$, 880–894 (2020)
18. Sapel, P., et al.: Towards digital shadows for production planning and control in injection molding. CIRP J. Manuf. Sci. Technol. 38, 243–251 (2022)
19. Savolainen, J., Knudsen, M.S.: Contrasting digital twin vision of manufacturing with the industrial reality. Int. J. Comput. Integr. Manuf. $35(2)$, 165–182 (2022)
20. Wager, S., Hastie, T., Efron, B.: Confidence intervals for random forests: the jackknife and the infinitesimal jackknife. J. Mach. Learn. Res. $15(1)$, 1625–1651 (2014)
21. Weigl, E., Heidl, W., Lughofer, E., Radauer, T., Eitzinger, C.: On improving performance of surface inspection systems by online active learning and flexible classifier updates. Mach. Vis. Appl. $27(1)$, 103–127 (2016). https://doi.org/10.1007/s00138-015-0731-9
22. Wery, J., Gaudreault, J., Thomas, A., Marier, P.: Simulation-optimisation based frame- work for sales and operations planning taking into account new products opportunities in a co-production context. Comput. Ind. 94, 41–51 (2018)
23. Zanjani, M.K., Nourelfath, M., Ait-Kadi, D.: Sawmill production planning under uncertainty: modelling and solution approaches. In: Stochastic Programming: Applications in Finance. Energy, Planning and Logistics, pp. 347–395. World Scientific, Singapore (2013)

A Digital Twin Generic Architecture for Data-Driven Cyber-Physical Production Systems

Miruna Iliuță[ID], Eugen Pop[(✉)][ID], Simona Iuliana Caramihai[ID], and Mihnea Alexandru Moisescu[ID]

Faculty of Automatic Control and Computers, University Politehnica of Bucharest, Bucharest, Romania

{miruna_elena.iliuta,eugen.pop,simona.caramihai, mihnea.moisescu}@upb.ro

Abstract. The integration of digital technology within organizations into their products, services, production, corresponding to various domains, which was started in the context of the Industry 4.0 initiative, has imposed the appearance of new emergent concepts and technologies. One of these is Digital Twin, which represents a virtual model of a physical object, which dynamically pairs the physical entity with its digital replica. The virtual system is connected to the real world through data transmission channels to acquire, analyse, process and simulate data within a virtual model. Thus, a Digital Twin improves the performance of the real entity; such systems are increasingly used today in various fields of activity. The progress realized in computation and communication enables digital representations of physical systems. This work proposes a generic Digital Twin architecture, together with main design guidelines and integrated technologies such as IoT, intended to be used in Cyber-Physical Production Systems for experimental applications. The proposed Digital Twin architecture offers the possibility to be accessed by a remote connection or locally. Several Digital Twin types, layers and applications, in the Data-Driven Cyber-Physical Production Systems context, are presented in the paper.

Keywords: Digital twin · Industry 4.0 · Cyber-physical production system

1 Introduction

Digital Twin represents a concept that integrates information and communication technologies, tools and applications and constitutes a digital replica of a physical object or of an intangible system. The more Digital Twin (DT) gathers information, the more complete is the realized model of the complex nature of the physical entity.

A Digital Twin is defined as a virtual model of a physical object with data transmission channels, that dynamically pairs the physical entity with its digital replica. It involves modern technologies like data analysis, artificial intelligence algorithms, smart sensors, used to improve system performance, detect, and prevent system failures, thus fostering

© The Author(s), under exclusive license to Springer Nature Switzerland AG 2023
T. Borangiu et al. (Eds.): SOHOMA 2022, SCI 1083, pp. 71–82, 2023.
https://doi.org/10.1007/978-3-031-24291-5_6

innovation. Digital Twin is more efficient when compared with usual simulation models [1]. The Digital Twin's main target is to obtain an optimized model of a physical object or service, which is able to be iteratively experimented in virtual environment until a desired performance level is achieved.

Digital models realize the simulation and prediction based on real time data, which is continuously received from the sensors. Afterwards, within a feedback loop, the values of the virtual model's parameters are sent back to the physical entities to optimize their behaviour [2].

Digital replicas have been used by industrial agents to model the information about an object or service functions during their lifecycle. Still, the model achieved doesn't have the consistency level which can be obtained through a DT [3].

The performances of Digital twin, which improve the information models, are mainly based on the digital technology progress and new achievements, architecture standardization, recent use cases and interactions [4]. Cloud computing, big data, 5G connectivity, virtual augmented reality, cyber security, IoT are examples of new technologies used by digital twins. The API for information access are deployed to ensure interoperability, security, data enrichment and synchronization within information models [5]. New use cases are also involved in the Industry 4.0 context, such as: information exchange within value chains, manufacturing analysis, predictive product design, real-time simulation.

The rest of this paper is organized as follows: Sect. 2 briefly presents the Digital Twin types, integration layers, complexity levels and architectures that are adequate for Industry 4.0. Section 3 describes Digital Twins applications which are representative for Industry 4.0, such as engineering, and automotive. Section 4 briefly highlights the Digital Twins' main design and development steps. In Sect. 5 the proposed Digital Twin experimental platform with the communication component is presented. Section 6 is reserved for conclusions.

2 Digital Twin Architecture

2.1 Types of Digital Twin

Several types of Digital Twins can be highlighted, as is presented in continuation.

Digital twin instance (DTI): A digital twin instance represents its corresponding physical entity along all its lifecycle [10]. The state of the physical twin is monitored continuously and the evolution or change of the physical twin behaviour is transmitted to its digital counterpart. In this assumption, a product, service or process is monitored from the beginning, throughout its lifecycle, and its evolution can be predicted. The performances of a product, object or process evolution are validated with the aid of a Digital Twin Instance.

Digital twin prototype (DTP): A digital twin prototype acquires and stores relevant data and parameter values regarding the physical twin.

The processed information can include data which may be useful for interconnecting the production chain agents within the manufacturing process in computer aided design, issuing bills of materials, etc. [6]. A DTP is able to accomplish quality control, testing, validation, simulate production scenarios before the real manufacturing or production process is started. Because possible risks or bottlenecks of physical twin can be identified before the production process is started, this approach decreases the manufacturing operational time and cost. Virtual experiments provide enough results, so the number of hardware prototypes, usually realized during a manufacturing process, is decreased. A virtual prototype is obtained, which can be considered an experimental Digital Twin. [7].

Performance digital twin (PDT): Within the real environment of the physical twin which is characterized by unpredictability, the PDT integrates and analyses product data and can monitor production processes. [8] The Performance Digital Twin processes the data gathered from its physical counterpart and, afterwards, provides useful data for maintenance and of design of the optimization strategy, defining conclusions concerning the product's performance level [9].

2.2 Digital Twin Integration Levels

Several integration levels for Digital Twin architectures, extending them from digital models to Digital Twins, are presented in the following.

- **Digital model**. The physical environment and the digital world are not interconnected automatically, so manual updating is necessary whenever changes arise. The digital model is not able to register the information traffic between the physical world and the digital world. In its main approach, a digital model does not involve an automatic functionality.
- **Digital shadow**. The digital shadow gathers the unidirectional, automatically generated data traffic from the physical world to the virtual world end [10]. This scenario is supported through a system based on sensors which acquire and interpret the data received from the physical model and transmit this information to the virtual model. The information flow can be obtained by applying interrupt methods or by polling.
- **Digital Twin**. Within this concept, the physical environment is totally bidirectional interconnected with the virtual world and the data flows automatically in both directions. This means that the Digital Twin is a fully integrated entity. In this scenario, the data gathered and processed by the virtual twin is used to support the evolution of its physical counterpart or to command new states or operations of actuators. Through reciprocity, the information from the physical twin automatically impacts the virtual twin. Following this behaviour, the physical twin current state and evolution are accurately represented by the virtual twin.

2.3 Digital Twin Levels of Complexity

The Digital Twin proposed in our paper can be configured according to three levels of complexity, as presented in the following. Digital Twins monitor, control, model and optimize the functionality of the physical objects. According to the business objectives, three scenarios of operations can be pointed out for a DT, which corresponds to three different levels of complexity [11].

Level 1 - Basic level: monitoring, targeting to achieve a Digital Twin

Data from the real environment object is acquired, representing real-time or stored information from sensors, maintenance information, etc. This information is afterwards introduced and processed by the Digital Twin software (e.g., AnyLogic software module) to create a digital model of the physical object. The data ca be fed automatically or manually within the software. These types of models are suitable for physical asset monitoring and for data acquisition. The Digital Level 1 operates mainly as intuitive interface for introducing real time or historical data, referring to the functionality and cost.

Level 2 - Middle level: "what-if" simulation

The middle level of Digital Twin offers the possibility to test the manufacturing processes or assets regarding the operational settings, and to identify optimal configurations. Basically, DT can be improved and can gain in efficiency through "What-if models" type. Various object or service parameters can be tested, in order to evaluate their impact on processes.

Level 3 - Advanced level: AI-enabled systems

Digital Twins can integrate machine learning algorithms. Data collected from sensors and historical data are used for ML algorithms' training. By detecting abnormal behaviours, these Digital Twins can propose or initiate corrective operations. DTs may be equipped with complete automatized functions.

2.4   Digital Twin Integration Levels

In our work we propose a DT architecture that can be used, continuously improved and extended to comply with some or all of the complexity levels presented above. In Fig. 1, the thin arrows represent the interconnections corresponding to Level 1, while the thick ones are associated with Level 2 and 3 of the Digital Twin architecture.

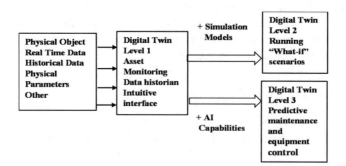

Fig. 1. Digital Twin architecture containing three levels

3   Digital Twin Applications

3.1   Engineering

Digital Twins are applied in various domain of activities like: industry control, predictive maintenance, automotive, lifecycle management [12, 13]. Among most recent applications of Digital Twins, the following can be mentioned: business, education, construction, communication and security, agriculture, healthcare [14, 15].

Emerged from Cyber Physical Systems, the Digital Twin concept is very important for applications and services regarding manufacturing, simulation, optimization, diagnosis, programming, modelling, management [16, 17]. Companies are enabled to realize predictions, informed planification and take rational decisions with increased accuracy [18–20].

Digital Twins may realize the prediction of physical system performances and behaviours, this representing an engineering facility of forecasting. Industrial, relevant big data is collected to support the decision-making process in self-adaptive systems. This represents an interrogative capability.

Although a DT does not involve in a necessary manner visual or spatial modelling, simulation based on augmented- or virtual-reality is an efficient approach for operating with hazardous environments and for system remote access [21].

From the documentation found in literature, 35% of DT engineering applications focus on production areas, 18% on design, 38% are applied in health management and prognostics (PHM) and 9% in other areas. Various engineering stages of a product lifecycle are covered by DT applications like design, manufacturing, logistics, distribution, usage, exploitation, end of life [22].

3.2   Automotive

In the automotive industry there are various DT configurations, each of which with impact on different phases of a vehicle lifecycle like: project design, engineering, testing, performance evaluation, exploitation and, finally, end of life.

Different type of content can be assigned to Digital Twins when an application is developed, as synthetically presented in Table 1.

Table 1. Content assigned to Digital Twins for applications development

	Product design	Product plan	Build
Information	Product lifecycle management (PLM)	Product lifecycle management	Product lifecycle management
Model	Functionality	Functionality	Functionality
Simulation	Design simulation	Functionality	HW loop - tests
Data model	Engineering data	Engineering data	Manufacturing data

DT appears to be an important enabler of the data driven automotive production. Among the relevant advantages provided by DTs for the manufacturing processes, the following can be mentioned: operational time savings, increased quality, decreased cost, increased productivity, reduced complexity, process optimization [6]. The global production system benefits of the data processed by DTs [23].

4 Digital Twin Development Steps

Each building block component of a Digital Twin must be designed, experimented, and tested before it can be considered that it functions correctly. The DT concept consists of the realization of the digital model for a real process, together with an analysis which is executed to prevent failures of the physical counterpart.

The digital twin facilitates digital simulations, providing the possibility of improving the monitored product or process, while preventing its failures and errors.

In the following, the digital twin creation is presented step by step, because this represents a complex task, used for in experimental digital twin designing. These steps are macro-processes of a Digital Twin design methodology, according to the stakeholder requirements for DT implementation.

4.1 Step 1 - Related to the Manufacturing Process Chain

This step defines the manufacturing process, by identifying its subsystems or components, investigating the domains and needs for improvement. Also, it realizes a functional diagram which represents the process chain flow.

A value chain represents the way an organization transforms the value of raw materials into value for customers. The process chain models the system inputs, tasks, and outputs decisions to be made. A well-performing process chain minimizes the manufacturing time.

4.2 Step 2 - Establishing the Digital Twin Functionality Needs

The functionality needs of a digital twin and the parameters that the system will focus on are: product quantity and quality; design and operation, process time, cost.

Design. AnyLogic software is a software tool that can be used for process simulation, and can be thus used to build the digital twin of a real process. The system's sensors can be specific devices, wearable or even humans. The parameters of an AnyLogic software-based simulation can be adjusted to analyse the effects of various process scenarios and time values. As process characteristics and parameters which should be taken into consideration from this point of view, the following are mentioned: utilization of certain subsystems, process capacity, waiting time, machine time, etc.

Operation. After designing the digital twin and its functions, it will be established if additional devices or components are necessary to be used or attached to the asset. Then, it is necessary to answer the following questions:

- Whether the main function of the digital twin is monitoring; if yes, the system will be chosen to be of Level 1;

- Whether the main functions of the digital twin do not realize the asset control, but the DT uses the data for planning and predictions if the conditions are able to be run; in this case this is a Digital Twin of Level 2;
- Whether the data gathered from the asset can be available for advanced, future analysis, to assist the predictive maintenance; if yes, this is a Digital Twin of Level 3.

4.3   Step 3 Choosing the Type and Level of Digital Twin

Before starting the DT design, it is necessary to choose its type. Afterwards, the steps to follow will be determined and what each step consists of. The four types of Digital Twin can be identified in literature as: *product, process, system,* and *process twinning.* We focus on system and process twinning, as our research targets the Industry 4.0 framework. By achieving the process optimization through digital twining, time optimization will be of interest. Three levels of digital twinning are currently available and were already presented above.

4.4   Step 4 - Augmentation

Digital twin augmentation opportunities appear following the interconnection of the real system with its digital counterpart. While the real processes is executed, data is transmitted in both directions, improving the operations of the Digital Twin.

5   Experimental Digital Twin Platform

The target for our development is to realize a platform that integrates sensor nodes with configuration and management functions. The physical twin sends the data collected from the surrounding environment sensors to its digital counterpart. The physical twin should be an assembly of sensors and actuators which executes simple tasks controlled by a Programmable Logic Controller (PLC).

In this architecture, to comply with the data exchange requirements in industrial systems, to control the physical twin and to gather the corresponding information, the OPC UA standard is used. This architecture is generic enough to be adapted to any case study, according to the development steps presented in the previous section. Compared with other architectures and solutions available that use communication protocols, OPC UA and IoT enable the Digital Twin to provide safe and secure interactions with the physical twin and a remote control [24]. The proposed DT generic architecture is able to integrate synchronously physical and virtual components to create a close loop. It implements a generic data data-driven Cyber-Physical Production System architecture specific to the engineering domain. The hardware and software tools can be integrated within a Sensor Configuration Platform (SCP). Data are acquired from the sensors and an OPC UA platform transforms the informational traffic in an XML structured data.

The SCP should satisfy the following main design requirements:

- *Communication*: supports Wi-Fi or other communication technologies, while adding new communication technologies and application layer protocols;

- *Functionalities*: Remote configuration of sensors over the Internet, support for I2C bus sensors, adding new sensors;
- *Configuration parameters*: sample rate, sensitivity, Wi-Fi settings (SSID, key, auth type), burst length and rate; data send rate;
- *User interface*: friendly, data visualization, possibility to download data;
- *Sensor nodes*: low energy consumption, led lights indicating the status, support for multiple communication technologies;
- *API*: sensors can be added, deleted, HTTP basic or token-based authentication.

The generic Digital Twin architecture, which is proposed in this paper integrates three main components: Physical Twin, IoT Gateway and the platform's Server. The information captured by the sensors from the surrounding environment refer to different parameter values which can be of interest for data-driven Cyber-Physical Production Systems, such as humidity, temperature, power consumption, vibrations, pressure, etc. IoT devices capture data according to their configuration, then the acquired information is synchronized relative to a common time reference within the IoT network. The IoT devices use ZigBee protocol for real time functionality.

The gateway, build in a single board computer, performs the following tasks:

- Gathers from the environment the information of interest for the Industry 4.0 context;
- Structures and assembles the data in an OPC UA XML file and transmits it to the platform Server, where the Digital Twin is installed and is running;
- The IoT gateways exchange data between the Server and Physical Twin through IoT devices and wired connection.

The Server platform (Internal Server) represents the hardware host where the Digital Twin is installed and executed. The Digital Twin is equipped with two access interfaces and panels providing: a) An accurate Physical Twin representation, facilitating a reliable actuation, based on precise values determination and process iterations; b) Decision-making capabilities, for management and business approach.

Two approaches were considered regarding the server. One vision presumes that the server is part of the workplace, so it benefits of the most important details to perform relevant operations of the processes. Parameters of AnyLogic based simulation can be adjusted and changed to analyse the effects of various processes scenarios. As process characteristics and parameters which should be taken into consideration from this point of view, the following can be mentioned: utilization of certain subsystems, process capacity, waiting time, machine time, etc. The other vision corresponds to a server installed remotely, outside the workspace. In this situation, the information and tasks relative to the processes are displayed and performed according to the user's demands. The experimental DT architecture is presented in Fig. 2.

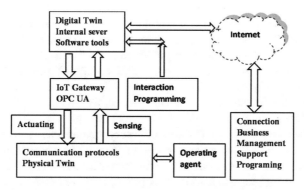

Fig. 2. Experimental Digital Twin architecture

In the architecture presented in Fig. 1, the Open Platform Communication Unified Architecture (OPCUA) is a M2M communication protocol which facilitates device connectivity and interoperability through different protocols. It complies with Industry 4.0 information modelling and Service Oriented Architecture (SOA) requirements [25].

OPC UA standard complies with the data exchange requirements in manufacturing systems; thus, it is included in the DT experimental architecture for data acquisition and Physical Twin control [26]. Various protocols are available to be used by distinct smart devices and systems to exchange information.

The interaction between the digital and physical components can be programmed. A specialized agent, engineer or stakeholder can operate the system. In the data-driven approach the basic information tables are as follows: Table 1 - Process variables in the DT functional block; Table 2 - Manipulated variables in the DT functional block; Table 3 - Related process disturbance faults in the DT functional block; Table 4 - Control schemes and performance indicators for the actuator failure.

The communication component of the experimental Digital Twin realizes the data transmission between the platform and other entities involved. The sensors' configuration is possible through Web interface, while the connection of the Digital Twin with other systems is realized through the platform's REST API. The access to the information transmitted by the sensor nodes is not allowed for the SCP. Data is stored in a specialized

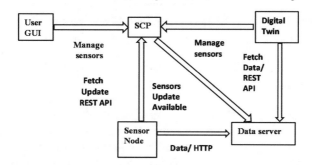

Fig. 3. Communication within the experimental Digital Twin platform

database or repository which can be an external software component of the Digital Twin. Users can access the information stored in the database. The communication facilities within the experimental DT platform is presented in the Fig. 3.

The configuration of sensor nodes is performed through Web user interface (WUI) or API and a Micro Python setting file is generated by the sensor node configuration. If the SDTP (Sensor Data Transmission Protocol), is used, the available updated information is sent to the data server using JWT authentication, through REST API.

JWTs (JSON web tokens JWTs) are sending cryptographically signed data between systems, using a standardized format. Any type of data cand be theoretically included, but mostly JWTs transmit information representing part of authentication, session handling, and access control mechanisms. Often this procedure is referred as ("claim"), which means sending of information about users.

Sensor nodes can obtain the available update information in two ways:

- Sensors require updates periodically, at configurable time intervals, through HTTP messages, from Sensors Configurator Platform;
- Sensors collect data for the server and, afterwards, updates are required from the given IP address through an HTTP request.

The first option corresponds to a system's main design principle: SCP should operate without interconnections with other entities as an independent system. This means that the first option does not involve any connection between the SCP and the data server. The second option is more efficient and that is why it was implemented. A connection with the server is opened by the sensor only when it has data available to be transmitted, following an available update received from the data server. A use case of the generic Digital Twin Generic Architecture is presented in Fig. 4.

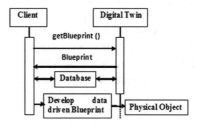

Fig. 4. A use case of the data driven Digital Twin Generic Architecture

6   Conclusions

A Digital Twin architecture for Data Driven Cyber-Physical Production Systems was proposed in this paper. The architecture complies with different levels of Digital Twins complexity: monitoring, "what - if" simulation, AI-enabled, but also with various types and integration levels. Several Digital Twins applications are presented, and the main steps of Digital Twins platform development are highlighted.

The main components of the Digital Twin architecture are represented by sensors, actuators, PLC, IoT gateway, and data server. The DT software processing tool can be represented by Any logic software. The DT content and results can be accessed remotely through a REST API interface. The DT architecture is designed to be used in a large variety of Industry 4.0 scenarios, business modelling, etc.

References

1. Marr, B.: What is Digital Twin Technology and Why Is It so Important? Forbes (2017). https://www.forbes.com/sites/bernardmarr/2017/03/06/what-is-digital-twin-technology-and-why-is-it-so-important/
2. Glaessgen, E.H., Stargel, D.S.: The digital twin paradigm for future NASA and U.S. Air force vehicles. Collect. Tech. Pap. - AIAA/ASME/ASCE/AHS/ASC Structures, Structural Dynamics and Materials Conference, pp. 1–14 (2012)
3. Fei, T., Weiran, L., Meng, Z., et al.: Five-dimension digital twin model and its ten applications. Comput. Integr. Manuf. Syst. 25(1), 1–18 (2019)
4. Negri, E., Fumagalli, L., Macchi, M.: A review of roles of digital twin in CPS-based production systems. Procedia Manufact. 11, 939–948 (2017). Digital_Twin_Architecure_2_ICCE_08662081_Artigo
5. Liu, Y., Peng, Y., Wang, B., Yao, S., Liu, Z.: Review on cyber-physical systems. IEEE/CAA J. Automatica Sinica 4(1), 27–40 (2017)
6. Singh, M., et al.: Digital twin: origin to future. Appl. Syst. Innov. 4, 36 (2021)
7. Dahmen, U., Rossmann, J.: Experimentable digital twins for a modeling and simulation-based engineering approach. In: Proceedings of the 2018 IEEE International Systems Engineering Symposium (ISSE), Rome, Italy, pp. 1–3 (2018)
8. Boschert, S., Rosen, R.: Digital twin - the simulation aspect. In: Hehenberger, P., Bradley, D. (eds.) Mechatronic Futures, pp. 59–74. Springer, Cham (2016). https://doi.org/10.1007/978-3-319-32156-1_5
9. Sharma, A., Kosasih, E., Zhang, J., Brintrup, A., Calinescu, A.: Digital Twins: State of the Art Theory and Practice, Challenges and Open Research Questions, arXiv:2011.02833 (2021)
10. Juarez, M., Botti, V., Giret, A.: Digital twins: review and challenges. J. Comput. Inf. Sci. Eng. 21, 030802 (2021)
11. Sidyuk, A.: Five Examples of Digital Twin Technology in Different Industries (Use Cases and Benefits) (2021). https://www.softeq.com/blog/5-digital-twin-examples-in-different-industries
12. Autiosalo, J., Vepsalainen, J., Viitala, R., Tammi, K.: A feature-based framework for structuring industrial digital twins. IEEE Access 8, 1193–1208 (2019)
13. Lim, K., Zheng, P., Chen, C.: A state-of-the-art survey of digital twin: techniques, engineering product lifecycle management and business innovation perspectives. J. Intell. Manuf. 31, 1313–1337 (2020)
14. Ran, Y., Lin, P., Zhou, X., Wen, Y.: A survey of predictive maintenance: systems, purposes and approaches. Comput. Sci. Eng. (2019). http://xxx.lanl.gov/abs/1912.07383. Accessed 9 July 2021
15. Sharma, M., George, J.P.: Digital Twin in the Automotive Industry: Driving Physical-Digital Convergence, White Paper; Tata Consultancy Services Ltd., Mumbai, India (2018)
16. Rathore, M., Shah, S., Shukla, D., Bentafat, E., Bakiras, S.: The role of AI, machine learning, and big data in digital twinning: a systematic literature review, challenges, and opportunities. IEEE Access 9, 32030–32052 (2021)

17. Quirk, D., Lanni, J., Chauhan, N.: Digital twins: answering the hard questions. ASHRAE J. **62**, 22–25 (2020)
18. Qi, Q., et al.: Enabling technologies and tools for digital twin. J. Manuf. Syst. **53**, 3–21 (2021)
19. Zhou, M., Yan, J., Feng, D.: Digital twin framework and its application to power grid online analysis. CSSE J. Power Energy Syst. **5**, 391–398 (2019)
20. Shengli, W.: Is Human Digital Twin possible? Computer Methods Programs Biomed. Update 2021, vol. 1, p. 100014 (2021)
21. Barricelli, B., Casiraghi, E., Gliozzo, J., Petrini, A., Valtolina, S.: Human digital twin for fitness management. IEEE Access **8**, 26637–26664 (2020). https://doi.org/10.1109/ACCESS.2020. 2971576
22. Laubenbacher, R., et al.: Using digital twins in viral infection. Science **371**, 1105–1106 (2021)
23. Laamarti, F., Badawi, H., Ding, Y., Arafsha, F., Hafidh, B., El Saddik, A.: An ISO/IEEE 11073 standardized digital twin framework for health and well-being in smart cities. IEEE Access **8**, 105950–105961 (2020)
24. Malakuti, S., Grüner, S.: Architectural aspects of digital twins in IIoT systems. In: Proceedings of 12th European Conference on Software Architecture Companion, ECSA 2018, pp. 1–2 (2018)
25. OPC Foundation: OPC UA Part 1 – Overview and Concepts 1.03 Specification (2015)
26. Haskamp, H., Orth, F., Wermann, J., Colombo, A., W.: Implementing an OPC UA interface for legacy PLC-based automation systems using the Azure cloud: an ICPS-architecture with a retrofitted RFID system, pp. 115–121. IEEE Industrial Cyber-Physical Systems (ICPS), St. Petersburg (2018)

Digital Twins in Industrial Systems

Modelling Manufacturing Systems for Digital Twin Through Communicating Finite State Machines

Lorenzo Ragazzini[✉], Elisa Negri, and Luca Fumagalli

Department of Management, Economics and Industrial Engineering, Politecnico di Milano, Milan, Italy
{lorenzo.ragazzini,elisa.negri,luca1.fumagalli}@polimi.it

Abstract. Manufacturing systems are impacted by frequent reconfigurations, increased uncertainty, and greater complexity. Such characteristics force the use of simulation for their design and analysis. New research trends in the field require further study to provide the capability to cope with the requirements of emerging disciplines. This work proposes a novel approach to simulation modelling to support the development of digital twins. It is based on a communicating hierarchical extension to finite state machines paradigm, extensively used both for representing discrete-event systems and to model agent behaviours. The approach has been validated through the development and implementation of two simulation models in Python.

Keywords: Digital twin · Discrete-event simulation · Agent-based model · Finite state machine

1 Introduction

The development of simulation models for the analysis and the design of manufacturing systems dates to the second half of the 20th century. With the advent of digital technologies, simulation is becoming increasingly integrated both with factory IT architecture and with data-driven models. Within manufacturing industry, simulation models may support decision-making at any company level, from the shop-floor to the top management [1].

The field of simulation modelling is also experiencing renewed interest by both scholars and practitioners. Previous research works have highlighted two main trends [2]:

- *Hybrid simulation.* It is based on the seamless integration of at least two disciplines of simulation models among which Discrete-Event Simulation (DES), Agent-Based Modelling (ABM), and System Dynamics (SD) [3]. Since this work is directed towards a support to field-level decisions, for the purpose of this paper the focus is on coupling DES and ABM models, ignoring SD.

© The Author(s), under exclusive license to Springer Nature Switzerland AG 2023
T. Borangiu et al. (Eds.): SOHOMA 2022, SCI 1083, pp. 85–95, 2023.
https://doi.org/10.1007/978-3-031-24291-5_7

– *Digital Twins (DT)*. The DT can be considered as new simulation paradigm supported by new digital technologies [4]. The DT in fact consists of a simulation model synchronized with the physical world [5].

Overall, this requires studying innovative approaches to the development of simulation models which can respond to the new needs brought by the trends discussed previously. New simulation models shall in fact facilitate a quick integration of field data to be synchronized as DT. This is particularly helpful in some DT applications, for instance production control, where the current state of the production system must be represented as precisely as possible [6]. Besides, to fully grasp the complexity of modern manufacturing systems, modelling tools are required to be modular to represent different levels of detail through co-simulation capabilities and to allow the creation of individual agents for including specific behaviour of production resources such as operators or AGVs.

The remainder of the document is organized as follows: Sect. 2 summarizes relevant literature, while Sect. 3 better specifies the objectives of this work. Section 4 describe the modelling methodology and Sect. 5 present sample applications. Finally, Sect. 6 draws the conclusion highlighting possibilities for future works.

2 Background

To better position this work within the scientific community, relevant gaps in three areas are analysed. We thus consider the simulation models on which DT are built, open-source simulation tools, and models for DES and ABM based on finite state machines.

2.1 Simulation Models for DT

Previous research has shown that DTs are often grounded on simulation models, especially those representing manufacturing systems [7, 8]. In fact, Melesse et al. suggested that the simulation capability of DT supports what-if and scenario analysis [9]. This should enable improving production system performances.

The review by dos Santos et al. focuses on the analysis of the new developments of DES and ABM and their applications as DT, emphasizing main issues in their development [10]. Juarez et al. studied the synergies between DT and multi-agent systems [11]. An uncommon example of DT based on hybrid modelling can be found in Mykoniatis et al. [12]. Despite using commercial simulation software, the authors worked towards a cheaper and more immediate DT creation. Moreover, the ease and the efficiency of creating simulation models for DT was stressed by Erye et al. [13]. Tiacci also highlighted the importance of developing new and general-purpose modelling tools and software to support DT [14].

2.2 Open-Source Code-Based Simulation Tools

This section of the work is devoted to the analysis of the main open-source code-based simulation tools. In fact, despite the evolution of simulation software, simulation-based decision support requires additional functionalities [15]. The authors noticed a growing interest in the integration of simulation models with other tools for analysis and optimization, which are often developed in Python [16]. Therefore, this review is based on the solutions developed in Python programming language. Table 1 summarizes the main applications reviewed.

Table 1. Simulation libraries for Python programming language

Type	Name	Description
DES	SimPy	It was the first DES library available in Python programming language. It is process-oriented and released under LGPL license, and can thus be exploited for proprietary projects [17, 18]
	Salabim	It is another library developed for Python for modelling and control of logistics and production processes, and also supports graphical animations [19]
	ManPy	It is a wrapper for SimPy devoted to the simulation of manufacturing systems, and was developed within the scope of DREAM European project [20]. Unfortunately, ManPy is not compatible to Python 3 as the last update released was in 2016, and only an unofficial version was ported to Python 3
ABM	MESA	It is the most relevant example of ABM in Python [21]
	Repast4Py	It was developed as a distributed ABM toolkit and is now provided with Python interfaces [22]

Within the reviewed open-source solutions for simulation, two significant gaps emerged:

- The DES libraries available are process-oriented as focus of these tools is on the description of the processes and their interactions only. Conversely, we still miss an event-oriented approach which better represents the activities performed by each modelled resource.
- Python simulation libraries are based on generators. For this reason, it is not possible to save nor restore the simulation state, hindering the possibility to synchronize the simulation with the current state of the system. This is particularly critical in the development of a DT as the initial state of the model should be set according to shop-floor data.

2.3 Communicating Finite States Machine-Based Models

Among the tools allowing to model manufacturing systems, various authors have adopted communicating finite state machines (CFSM) [23].

Regarding DES, some authors have developed CFSM for simulation and control of manufacturing systems [24]. In fact, CFSM are used to represent the logic of the system under analysis. Conversely, CFSM-based ABM are usually adopted to represent agents and to model their interactions within a multi-agent system. Few examples were surveyed by Spanoudakis [25]. Being used as basis for both DES and ABM, CFSM is a particularly promising solution for hybrid models coupling both simulation disciplines.

3   Research Objective

The previous section highlighted the need for a new specific approach to implement simulation models to develop DT. The literature in fact still lacks a modelling approach resulting in open, distributable model of manufacturing systems.

This work proposes a novel approach to simulation modelling based on Communicating Hierarchical Finite State Machines (CHFSM) and object-orientation. It is supported by a visual formalism and includes a software implementation in an open-source programming language. The ease to create models and to integrate data makes it suitable for DT development. It is based on DES but developed modelling resources as agents, and thus supports hybrid modelling. The properties of the CHFSM allow to model systems with different degrees of detail specifying multiple subsystems.

In particular, the research objective is to propose a simulation modelling tool suitable for designing DT which has the following fundamental capabilities:

- Facilitate the synchronization between the physical and virtual worlds. Its characteristics should allow the representation of the system in such a way to permit an agile real-time update of the model. By doing so, the DT of the shopfloor may achieve a high synchronization frequency.
- Hierarchically represent manufacturing systems, thus enabling co-simulation and modelling of subsystems. This also enables a bottom-up approach, particularly suitable for modelling systems characterized by complex interactions.
- Support the modelling of each resource as individual agents to cope with the growing interest in hybrid simulation, which allows to better represent complex systems.

To achieve the proposed research objective, we begin from considering a manufacturing system as a set of integrated resources that work together to fulfil the required production.

4   Modelling Methodology

4.1   Overview

The proposed modelling approach is designed to implement simulation using an object-oriented programming language. For this reason, each manufacturing resource represented in the simulation model consists of an object of a specific class, which represent a specific resource type.

The modelling method is based on the work by Harel that uses UML state charts [26]. Furthermore, it considers their generalization as CHFSM [27]. In particular, communication is performed through messages sent and received on specific interfaces acting as message ports.

The adopted paradigm can be also seen as an extension of the UML *state chart standard* thanks to the specification of message ports for each state chart. In fact, the main novelty proposed is the development and usage of communication through messages that allow defining the interactions between manufacturing resources.

4.2 Communicating Hierarchical Finite State Machines

The CHFSM previously mentioned are applied to represent manufacturing resources in the simulation model. For each type of manufacturing resource, a specific class must be defined inheriting from a general and empty CHFSM. For each CHFSM, two main elements are considered as crucial for the development of the simulation model:

- The set of possible *states* of each manufacturing resource. They allow to easily represent the condition of a resource.
- The *messages* through which the communication among resources is realized.

This can be considered one of the main strengths of the proposed approach towards the development of DT. In fact, a complete synchronization of the DT would only require considering the state of each resource and the messages in each queue. Conversely, this approach also has two major disadvantages. Being based on discrete events, the monitoring of continuously changing variables is difficult. In addition, more complexity raises when trying to model agents acting at global level such as system controllers.

States and Transitions

A state can be considered as a specific condition of a resource that lasts for a finite amount of time. At any moment, the resource is always in a state which fully characterizes its behaviour.

Transitions allow the resource to change from one state to another. For the purpose of this work, they involve three main elements:

- *Triggers* are the events which an agent is waiting for while it is in a specific state. Once the event happens, the transition may be started.
- *Guards* are the conditions that must be verified for the transition to take place.
- *Actions* are executed while the transition occurs. They allow to model the actions performed during a state change and terminate with the actual transition to the new state.

Messages and Communication

We can consider a message anything that can be sent, received, and stored. Messages are particularly helpful in representing passive and temporary objects, such as entities which flow through a system, although even agents themselves can be sent and received as messages. Messages are stored in queues with a positive, integer, but not necessarily finite capacity.

A set of different use cases may be defined according to the behaviours of the two agents involved in the communication, namely the sender and the receiver.

The sender may:

- Send a message to the receiver, without any feedback of reception.
- Send a message and wait until the message is received in the message queue of the receiver. Thus, the sender waits for a response from the receiver confirming that the message has been successfully received, which means space was available in the receiver queue. This allows, for instance, to model blocking phenomena of production lines.
- Send a message request to the receiver. In this case, once the message request is accepted and a response is returned, the sender decides whether to send the real message.

Similar behaviours apply also to agents receiving messages. The receiver may in fact:

- Wait for a message and get it as soon as it is available.
- Subscribe to a message queue. In this case, once a message is available, the agent may decide if reading and removing a message. If more than one message is available, it is possible for the receiver to select a specific message.

The virtual queue is a special case of the message queue with no capacity. Its peculiarity is that message reception is confirmed only when the message itself has been forwarded to another queue, in such a way that the virtual queue is always empty. This object is crucial to model logical spaces where messages can go through, but never stay for a finite amount of time, such as switches or other routing elements.

Finally, communication can be also realized directly by means of events. In fact, specific events can be associated with any agent and triggered by other agents to synchronize their behaviours. For instance, a failed machine waiting for the completion of the repair process will wait until the repairman triggers it.

4.3   Model Implementation

The implementation of a model applying the proposed methodology consist of three main steps, as shown in Fig. 1:

1. Definition of the classes for each manufacturing object type
2. Creation of an agent representing each manufacturing object as class instance
3. Definition of the connection among agents to enable communication

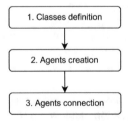

Fig. 1. Implementation steps

Agents' Definition

The definition of the agent classes allows to model each resource type specifying the set of its possible states. The resource is then instantiated as an agent and the communication with other agents is defined. For each manufacturing object type (i.e., machines, queues, AGVs) a class must be defined. This consists of characterizing the states the object may assume, and the messages it can process.

Agents' Creation

Once the resource types are defined, each resource (e.g., each machine included in the system under analysis) must be created as instance of a specific class. Being represented by a CHFSM, each manufacturing resource can act as a unique agent. Again, the agent is derived as instance of a specific class representing a certain manufacturing resource type. Furthermore, the agents are characterized by sets of variables and methods. Variables allow to better specify the properties of each agent and to store data. Methods allow the execution of all the actions required for the agent (e.g., computing the service time for a workstation or selecting the machine to repair for a repairman).

Agents' Connection

The definition of the relationships among the agents must be finally specified. For each agent, the connections with the other agents must be defined to allow communication between them. This allows for example moving entities between machines or workstation requesting operators to perform manual activities.

5 Examples of Model Implementation

In order to better explain the modelling principles introduced above, this section proposes a graphical representation of a simple simulation model. The syntaxis is derived from UML; for this reason, the triggering event is marked on the transition arrow, conditions (or guards) are included within square brackets while actions are preceded by the slash symbol ("/").

The UML formalism has been extended to support modelling messages in a way slightly different from previous works [28]. The content of the message is included in parentheses. Input and output message ports are addressed through the symbol ("@"). Similarly, an event triggered on another agent is put in parentheses and the target agent is preceded by the @ symbol. The graphical representation is functional to the encoding of the model. The proposed simulation framework has been then developed and implemented in Python[1]. For achieving this, the work is built as a wrapper of the well-known DES simulation library SimPy. The simulation environment and the event scheduler were in fact borrowed from the original library. The class defining the *State* is derived from SimPy *Event*, whereas the *CHFSM* class was built from scratch.

---

[1] Code is available at https://github.com/lorenzo-ragazzini/hsim.

5.1 Representation of a Single Server Queueing Model

The graphical representation of a simple queue-server model is depicted in Fig. 2. The queue forwards entities from its queue towards the server. The server switches from idle to working state as soon as it receives a message. Then, the entity is sent through the *next* port. If the message is received, the server returns idle, otherwise it enters a blocking state. This means that it stays idle and cannot accept other entities; it waits until the message has been received. For the sake of simplicity, the arrival process of the entities in the queue is omitted from the model.

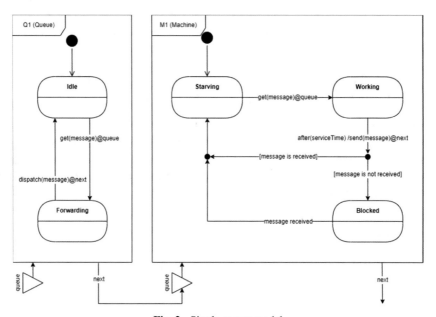

Fig. 2. Single server model

5.2 Simulation Model of a Manual Assembly Line

The model implemented is derived from an industrial case study of hybrid modelling developed by the authors. The system consists of a line of manual assembly stations where a set of operator works. Each operator is assigned to a range of station.

The advantage of defining the operator model allows to define policies according to which the operator selects which activity to perform according to certain priorities.

Figure 3 represents the models of operator and assembly station. The operator searches for any station available for working and selects it. Then, it triggers *operator entry* on the station and stays in working state until the operation is completed. Meanwhile, the assembly station transits from idle to working state. As soon as the process is completed, the assembly station triggers the *operation completed* event and the operator is free to move and work in another station.

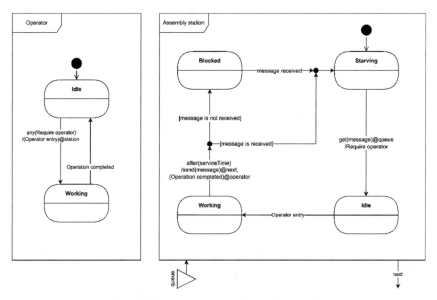

Fig. 3. Operator and assembly station models

After running the simulation, a detailed event log is provided, which describes the state change of each resource modelled specifying the state entry and exit time for each resource. Furthermore, the tool also provides the distribution of states for each resource, as represented in Fig. 4.

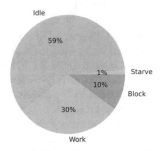

Fig. 4. States distribution of an assembly station

6 Conclusion

The proposed modelling approach allows to specify object-oriented models of manufacturing system based on CHFSM. Despite being grounded on event-oriented DES principles, it also supports hybrid modelling as each resource is defined as an agent.

The relevance of this work is twofold. From a scientific perspective, it proposes a new methodology to model and simulate manufacturing systems with a specific focus towards the development of DT and the support to hybrid models. Regarding industrial practitioners, the relevance lies in an easy and open-source simulation tool whose characteristics facilitate its use and integration within existing manufacturing systems.

Further works should validate the possibility to connect a simulation model designed with this tool to a real manufacturing system, and also to assess its capabilities to work as DT. In addition, a tool for automatically converting the graphical representation of the model as CHFSM into Python code ready for simulation will be developed.

References

1. Negahban, A., Smith, J.S.: Simulation for manufacturing system design and operation: literature review and analysis. J. Manuf. Syst. **33**, 241–261 (2014). https://doi.org/10.1016/j.jmsy.2013.12.007

2. Mourtzis, D.: Simulation in the design and operation of manufacturing systems: state of the art and new trends. Int. J. Prod. Res. **58**, 1927–1949 (2020). https://doi.org/10.1080/00207543.2019.1636321

3. Brailsford, S.C., Eldabi, T., Kunc, M., Mustafee, N., Osorio, A.F.: Hybrid simulation modelling in operational research: a state-of-the-art review. Eur. J. Oper. Res. **278**, 721–737 (2019). https://doi.org/10.1016/j.ejor.2018.10.025

4. Rosen, R., Von Wichert, G., Lo, G., Bettenhausen, K.D.: About the importance of autonomy and digital twins for the future of manufacturing. IFAC-PapersOnLine **28**, 567–572 (2015). https://doi.org/10.1016/j.ifacol.2015.06.141

5. Negri, E., Fumagalli, L., Macchi, M.: A review of the roles of digital twin in CPS-based production systems. Procedia Manufact. **11**, 939–948 (2017). https://doi.org/10.1016/j.promfg.2017.07.198

6. Ragazzini, L., Negri, E., Macchi, M.: A digital twin-based predictive strategy for workload control. IFAC-PapersOnLine **54**, 743–748 (2021). https://doi.org/10.1016/j.ifacol.2021.08.183

7. Abdoune, F., Cardin, O., Nouiri, M., Castagna, P.: About perfection of digital twin models. In: Borangiu, T., Trentesaux, D., Leitão, P., Cardin, O., Joblot, L. (eds.) Service Oriented, Holonic and Multi-agent Manufacturing Systems for Industry of the Future, pp. 91–101. Springer, Cham (2021). https://doi.org/10.1007/978-3-030-99108-1_7

8. Pystina, K., Sekhari, A., Gzara, L., Cheutet, V.: Digital twin for production systems: a literature perspective. In: Borangiu, T., Trentesaux, D., Leitão, P., Cardin, O., Joblot, L. (eds.) Service Oriented Holonic and Multi-agent Manufacturing Systems for Industry of the Future, pp. 103–117. Springer, Cham (2021). https://doi.org/10.1007/978-3-030-99108-1_8

9. Melesse, T.Y., Di Pasquale, V., Riemma, S.: Digital twin models in industrial operations: a systematic literature review. Procedia Manuf. **42**, 267–272 (2020). https://doi.org/10.1016/j.promfg.2020.02.084

10. dos Santos, C.H., Montevechi, J.A.B., de Queiroz, J.A., de Carvalho Miranda, R., Leal, F.: Decision support in productive processes through DES and ABS in the digital twin era: a systematic literature review. Int. J. Prod. Res. **60**, 2662–2681 (2022). https://doi.org/10.1080/00207543.2021.1898691

11. Juarez, M.G., Botti, V.J., Giret, A.S.: Digital twins: review and challenges. J. Comput. Inf. Sci. Eng. **21** (2021). https://doi.org/10.1115/1.4050244

12. Mykoniatis, K., Harris, G.A.: A digital twin emulator of a modular production system using a data-driven hybrid modeling and simulation approach. J. Intell. Manuf. **32**(7), 1899–1911 (2021). https://doi.org/10.1007/s10845-020-01724-5

13. Eyre, J.M., Lanyon-Hogg, M.R., Dodd, T.J., Lockwood, A.J., Freeman, C., Scott, R.W.: Demonstration of an industrial framework for an implementation of a process digital twin. In: ASME 2018 International Mechanical Engineering Congress and Exposition, vol. 2, Advanced Manufacturing, pp. 1–9 (2018). https://doi.org/10.1115/IMECE2018-87361

14. Tiacci, L.: Object-oriented event-graph modeling formalism to simulate manufacturing systems in the Industry 4.0 era. Simul. Model. Pract. Theory **99**, 102027 (2020). https://doi.org/10.1016/j.simpat.2019.102027

15. Steringer, R., Zörrer, H., Zambal, S., Eitzinger, C.: Using discrete event simulation in multiple system life cycles to support zero-defect composite manufacturing in aerospace industry. IFAC-PapersOnLine **52**, 1467–1472 (2019). https://doi.org/10.1016/j.ifacol.2019.11.406

16. Peyman, M., Copado, P., Panadero, J., Juan, A.A., Dehghanimohammadabadi, M.: A tutorial on how to connect python with different simulation software to develop rich simheuristic. In: Proceedings Winter Simulation Conference WSC, December 2021 (2021). https://doi.org/10.1109/WSC52266.2021.9715511

17. Castillo, V.: Parallel Simulations of Manufacturing Processing using Simpy, a Python-Based Discrete Event Simulation Tool, p. 2294 (2007). https://doi.org/10.1109/wsc.2006.323064

18. Ling, S., et al.: Computer vision-enabled HCPS assembly workstations swarm for enhancing responsiveness in mass customization. In: IEEE 16th International Conference on Automation Science and Engineering (CASE), pp. 214–219 (2020). https://doi.org/10.1109/CASE48305.2020.9216907

19. van der Ham, R.: salabim: discrete event simulation and animation in Python. J. Open Source Softw. **3**, 767 (2018). https://doi.org/10.21105/joss.00767

20. Olaitana, O., et al.: Implementing ManPy, a semantic-free open-source discrete event simulation package, in a job shop. Procedia CIRP **25**, 253–260 (2014). https://doi.org/10.1016/j.procir.2014.10.036

21. Kazil, J., Masad, D., Crooks, A.: Utilizing python for agent-based modeling: the mesa framework. In: Thomson, R., Bisgin, H., Dancy, C., Hyder, A., Hussain, M. (eds.) SBP-BRiMS 2020. LNCS, vol. 12268, pp. 308–317. Springer, Cham (2020). https://doi.org/10.1007/978-3-030-61255-9_30

22. Collier, N.T., Ozik, J., Tatara, E.R.: Experiences in developing a distributed agent-based modeling toolkit with python. In: Proceedings of PYHPC 2020 9th Workshop on Python High-Performance and Scientific Computing, pp. 1–12 (2020). https://doi.org/10.1109/PyHPC51966.2020.00006

23. Brand, D., Zafiropulo, P.: On communicating finite-state machines. J. ACM **30**, 323–342 (1983). https://doi.org/10.1145/322374.322380

24. Ramakrishnan, S., Thakur, M.: A SDS modeling approach for simulation-based control. In: Proceedings - Winter Simulation Conference (WSC), pp. 1473–1482 (2005). https://doi.org/10.1109/WSC.2005.1574414

25. Spanoudakis, N.I.: Engineering multi-agent systems with statecharts. SN Comput. Sci. **2**(4), 1–21 (2021). https://doi.org/10.1007/s42979-021-00706-5

26. Harel, D.: Statecharts: a visual formalism for complex systems. Sci. Comput. Program. **8**(3), 231–274 (1987). https://doi.org/10.1016/0167-6423(87)90035-9

27. Bordeleau, F., Corriveau, J.P., Selic, B.: A scenario-based approach to hierarchical state machine design. In: Proceedings of 3rd IEEE International Symposium on Object-Oriented Real-Time Distributed Computing, ISORC 2000, pp. 78–85 (2000). https://doi.org/10.1109/ISORC.2000.839514

28. Lange, J., Tuosto, E., Yoshida, N.: From communicating machines to graphical choreographies. In: Proceedings of 42nd Annual ACM SIGPLAN-SIGACT Symposium on Principles of Programming Languages, pp. 221–232 (2015). https://doi.org/10.1145/2676726.2676964

An Architecture for Integrating Virtual Reality and a Digital Twin System

G. S. da Silva, A. H. Basson[⊠], and Karel Kruger

Department of Mechanical and Mechatronic Engineering, Stellenbosch University, Stellenbosch, South Africa
{20935927,ahb,kkruger}@sun.ac.za

Abstract. Industry 4.0 is associated with the development of various technologies, including virtual reality and digital twins. The integration of these two technologies has the potential to aid with the visualising of complex digital twin data. A system architecture is presented here for such an integration. This architecture is implemented for the context of facilities management, which is representative of complex systems with a hierarchical physical nature and a large number of similar data sources. Preliminary evaluations considered latency and computational power requirements. The paper shows that the proposed system architecture supports the integration of virtual reality and digital twins, in addition to various other roles for digital twins.

Keywords: Digital twin · Virtual reality · Data visualization

1 Introduction

The growth in Industry 4.0, the Internet of Things (IoT) and cyber-physical systems (CPS) have led to digital twins (DTs) that capture large amounts of data. A DT is here taken to be a virtual representation of a physical system, from which the DT continually receives data. DTs can also include various other aspects, such as algorithms that are used to reflect their physical twin and make decisions for it [1]. However, a key feature of a DT is to provide historical and up-to-date data, in a consistent format, about its physical counterpart. For complex systems, this richness of data leads to the challenge of providing users the right data they need for decision making, without overwhelming the user, and in formats that aid interpreting the data.

Virtual reality (VR) can be used to aid the user's experience with various applications, such as the entertainment industry [2], and has seen increasing use in areas such as scientific data visualisations [3–5]. VR is a computer-generated experience that uses some of a person's senses to create a simulation where the user is able to interact with and explore an immersive 3D environment. This simulation senses the user's movements and responds accordingly by altering the simulation environment to match the user's movement, providing "a synthetic, interactive, and explorative information representation system that provides maximum intuitiveness to be exercised within an

© The Author(s), under exclusive license to Springer Nature Switzerland AG 2023
T. Borangiu et al. (Eds.): SOHOMA 2022, SCI 1083, pp. 96–106, 2023.
https://doi.org/10.1007/978-3-031-24291-5_8

artificial 3D environment" [6]. VR therefore has potential as visualisation medium for the data-driven decision making process.

The integration of VR and DTs offers the potential where VR's data visualisation potential is used to better exploit a DT's rich set of data representing a complex system and aid with the data driven decision making process [7]. However, little has been published on implementing such integration and, in particular, the architecture to be used to integrate VR and DTs.

This paper contributes to the integration of VR and DTs by presenting an architecture for such integration. The architecture is created using a design framework, developed by [8], for DT systems for complex problems. The VR and DT system is designed for the context of facilities management, which is representative of complex systems with a hierarchical physical nature and a large number of similar data sources. Preliminary evaluations are presented, focussing on latency and computational requirements.

Section 2 discusses related work in the area of VR and DTs, and the integration of the two. Section 3describes the VR and DT integration architecture and outlines the implementation used for the evaluation. Section 4 discusses the preliminary evaluations. Section 5 presents conclusions and suggests some future work.

2   Related Work

The DT concept has been applied in many contexts [9], but here only a few applications to complex systems are considered for the sake of brevity. A DT architecture was developed by [10] and [11] and a DT of the University of Cambridge was developed to evaluate their architecture. This architecture was for a building and city level DT. In unrelated work, the Six-Layer Architecture for Digital Twins with Aggregation (SLADTA) [12] was created for complex systems and was initially implemented for a manufacturing cell. Another implementation used a DT to optimise the energy usage of an assembly line [13].

The use of VR for data visualisation has also seen an increased interest. For example, [14] investigated how users reacted to employing VR to visualise data in comparison with using conventional PC displays for visualisation. The study found that users ranked VR higher than PC displays and concluded that the use of VR could aid with visualising large datasets. A VR application for visualising microclimatic data was developed by [3] where the data was displayed to a user in a 3D VR environment and the user was able to interact with a user interface (UI) to visualise desired information.

The concept of integrating VR and DTs, although in its infancy, has also seen some research. For instance, a VR training/design environment was developed by [15]. The research indicated that the case study robot's DT in the VR environment behaved realistically like the physical robot and engineers were able to use this during the design phase for the environment. A similar implementation was created by [16] where a human-robot collaborative (HRC) system was created. The purpose of this study was to use the DT of a robot and a VR environment to aid in the designing of a HRC system. [7] looked at the potential opportunities and challenges that are associated with the integration of VR and DTs and using VR as a data visualisation tool, and concludes that the opportunities of integrating VR and DTs outweigh the challenges presented by the integration of the two.

The architecture presented in this paper is for the facilities management context, a context that has not been considered by the related work. The implementations of the above-mentioned related work make use of DTs that are essentially only 3D models and simulations of a physical component. The 3D model and simulations are then used in the developed VR applications. The implementation in this paper uses DTs that receive constantly updating non-geometric real-world information that is then available in VR.

3   Virtual Reality and Digital Twin System Architecture

This section presents the VR and DT system architecture. The design process [8] starts with determining the user requirements and leads to a DT architecture and high-level implementation. The section ends with a case study implementation.

3.1   User Requirements

The Facilities Management (FM) Division at Stellenbosch University was used as target users because FM manages a large number and variety of physical resources, which constitute a complex system with complex data interpretation requirements. A university campus' DT should capture large amounts of data to support decision making. A project is already underway at FM to establish a campus DT, which offered the opportunity to explore the integration of VR and the DT technology.

Discussions with FM led to the decision to target use of VR to visualising utility usage (e.g., water or electricity) across the campus. The campus has hundreds of water and electricity meters, each generating a few data sets per hour, with complex usage patterns related to time of day, seasons, phases of the academic year, weather, etc. Utility usage is an important operational cost and input into carbon footprint monitoring. VR could, therefore, help FM to gain insight into these factors to detecting areas for reducing cost and the carbon footprint.

3.2   System Architecture

Employing the design framework developed by [8], the architecture design started with a needs and constraints analysis, and a physical system decomposition. The needs and constraints analyses were used to identify what services are required in the system, while DTs were chosen based on the physical system decomposition. The identified services were then allocated to the DTs or, if the data required by the service is not directly associated with a DT, to a services network.

The above-mentioned design framework starts with a reference architecture [17], which is then adapted. The resulting system architecture is shown in Fig. 1. The main differences from the reference architecture are the "VR application" node that replaces the more general user interface node and "External databases" that provides the source data. As intended by [17], the DTs in the "Aggregation hierarchy" maintain a reflection of the physical reality and each DT provides software services that rely on its part of the physical reality. The "Services network" hosts software that use data from multiple

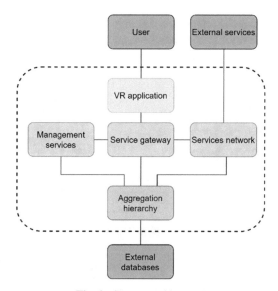

Fig. 1. System architecture

DTs and, potentially, external services. The separation of these roles enables modular refinement and extension of the digital twin system.

Figure 2a gives a typical aggregation hierarchy for FM, derived from the physical decomposition step of the design framework, and employing the SLADTA [12] aggregation approach. The internal architectures for the DTs are illustrated in Fig. 2b, following SLADTA [12].

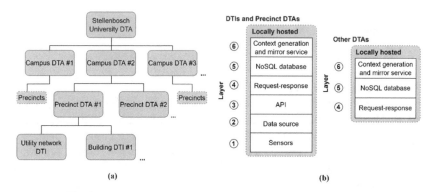

Fig. 2. a) DT aggregation hierarchy, and b) Internal DT structure

The DTs also contain various services in Layer 6 to address user requirements, while the other layers are aimed at reflecting physical reality. The DTs receive the utility usage data, via an API to external databases, from the various meters that are on the campus.

The services of the architecture, in addition to those offered by DTs are management services, a service gateway, a services network, and a UI (in this case, the VR application). The services and their interactions are illustrated in Fig. 3.

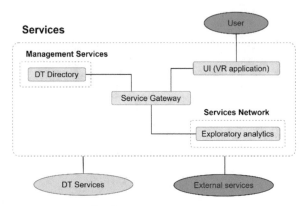

Fig. 3. Service components

It should be noted that the architecture is aimed at proving insight through VR that FM can use to make decisions. As shown here the architecture is able to receive information from the physical system through external databases. The architecture would need to be extended to allow information to flow between (from and to) the physical system and the VR (via the DTs). However, the use of SLADTA as internal architecture for the DTs makes such an extension straightforward and SLADTA has already been shown to accommodate such information flows [12].

3.3 Implementation

This section outlines the implementation of the system architecture in Figs. 1, 2 and 3.

Digital Twin and Services Implementation.
The DTs and services in the system were implemented using C# because Unity, the game engine used for the VR implementation, also uses this programming language. However, because the system components communicate using TCP/IP as described below, many other programming languages could be used to achieve the same results. The different DT types were developed as C# classes. These classes could then be instantiated with customised constructor parameters for the respective DTs required in the system. DTs communicate with one another using TCP/IP on a network. Each higher-level DT can request information from the DTs below it in the aggregation hierarchy. For example, a Precinct DTA is able to request data from a Building DTI. During operation, the DTs periodically update as new information from the utility usage meters becomes available. This is vital as up-to-date data is needed for the services a DT can offer.

Each DT also makes use of a local database (Layer 5 in Fig. 2b) to store the information necessary to provide that DT's services (Layer 6 in Fig. 2b). For the case study, these services are the "Mirror service" and the "Context generation service". Services in the Services network and other DTs can make use of these DT services to provide the users with the information they request.

Similarly to the DTs, the services components communicate using TCP/IP. For the case study, the Service gateway, a Directory service, and an Exploratory analytics service were developed. They are typically employed as follows: when the users select what information they would like to visualise in the VR application, this application sends a request to the Service gateway, which calls on other services to obtain the selected information from the DT aggregation hierarchy. The information is then sent back, via the service component and the Service gateway, to the VR application to update the display.

Virtual Reality Application Implementation.
The software architecture for this VR application, illustrated in Fig. 4, was implemented using the Unity game engine. A game engine like Unity has useful VR capabilities and functionalities [18] and allows a VR application to be created relatively easily. Another game engine, such as Unreal Engine, could have been used to achieve the same result, but Unity was selected because it has been used in research related to using VR as a data visualisation medium [15, 19, 20].

The VR application architecture in Fig. 4 shows that there are five main aspects to the VR application. The *"Visualisation"* aspect is the VR environment that the user will be able to see and interact with. The *"User movement"* aspect is responsible for interpreting the user's input, using the VR controller, and adjusting the visualisation accordingly so that the user is able to navigate the VR environment.

The *"UI functions"* aspect interprets the user's input, using the VR controllers, for the UI that is presented to the user in the VR environment. This UI is used to allow the users to select what information they would like to visualise. The UI is updated automatically according to which DTs are part of the system because the menu options presented on

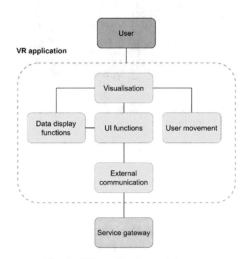

Fig. 4. VR application architecture

the UI are requested from the DT system. This is done when the user presses the menu button on the VR controller. The VR application requests, through the service gateway, a list of DTs at a specific hierarchy level. The service gateway obtains this list of DTs from the aggregation hierarchy and replies to the initial request from the VR application with the DT list. The information in the UI is, therefore, not fixed and can update dynamically as the DT system is changed.

The *"Data display functions"* are responsible for interpreting the information received from the service gateway and displaying the information to the users in the VR environment so that they are able to interpret the information. Finally, the *"External communication"* aspect is responsible for the external communication, using TCP/IP, with the service gateway, as the service gateway is used to communicate with the aggregation hierarchy and request the information desired by the user.

4   Preliminary Evaluation and Results

This section presents preliminary evaluations and their results. Two evaluation tests were conducted, i.e. the system's latency and computational resource utilisation.

Figure 5 shows a few images that a user sees when using the VR application. Figure 5a shows a user requesting information for the latest energy usage for the various aggregation levels for main campus at Stellenbosch University. In Fig. 5b, a green bar represents the latest energy usage for a building, the blue bar is for the latest energy usage for a precinct, and the red bar is the latest energy usage for the campus. The selected information is overlaid on a map of the campus to enrich the context of the information and reveal spatial relationships.

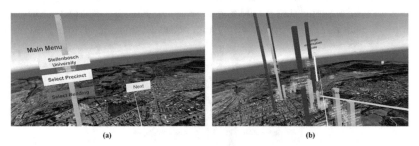

Fig. 5. a) Requesting campus-wide energy usage information. b) Campus-wide energy usage information being presented to the user.

4.1   Latency

Because VR is intended to aid a user in navigating and visualising complex data, the responsiveness of the system when a user requests a change of the data being viewed, is important. The 3-D manipulation of the views is provided by the game engine, but changes to the data that is viewed requires receiving that data from the DT system.

This evaluation was therefore aimed at evaluating the responsiveness of the architecture presented above.

Latency here refers to the amount of time elapsed from when the user initiates a request for information to when the information is displayed to the user. This includes the time taken for messages to be transferred to the Service gateway, for services in the service network to request data from the DT aggregation, and for the information to be sent back, interpreted, and displayed in the VR environment for the user.

Several test scenarios were evaluated, with various combinations of the number of DTs in the system and the amount of information requested from each DT. Each scenario was repeated three times for more accurate results to be obtained. Although several scenarios were tested, only one scenario's results are reported here, i.e. where the amount of requested information, one data point from each DT remains the same, but the number of DTs in the system is increased. In this test, all the system components are hosted on the same computer. The results of this test are illustrated in Fig. 6.

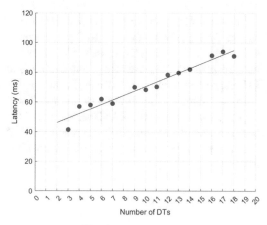

Fig. 6. Latency results

From the results in Fig. 6 it is evident that, as expected, the increase in DTs in the system results in an increase in the total time taken to retrieve and display the information. However, the magnitude of this latency is in the millisecond range, which is extremely fast for human perception. Therefore, the results indicate that the VR and DT architecture implementation has a low latency when a user requests information. This shows that the integration of VR and DTs produces a sufficiently responsive VR data visualisation experience.

4.2   Computational Power Usage

The purpose of this evaluation was to record the computational power usage of the VR and DT system, measured in terms of total system RAM used and the CPU percentage used. These measurements were taken during the information retrieval and display process for the latency evaluation. The measurements included the computational power usage for the VR application, the DT system, and the Services components.

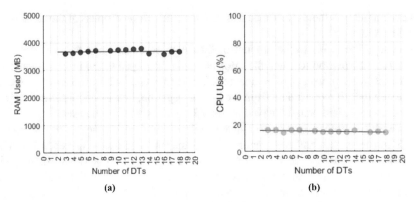

Fig. 7. a) CPU usage results. b) RAM usage results.

A computer with an Intel Core i7-10700 CPU and 64 GB of RAM running Windows 10 Pro was used for the tests. The results of the experiment (Fig. 7 show that, with an increase in DTs in the system, the amount of RAM used changes very little. The component that was identified to be using the most RAM, as expected, was the Unity VR application which is responsible for a very large portion of the overall RAM usage. The results for the CPU usage show a similar trend to the RAM usage, with the percentage of CPU used being fairly constant for the duration of the test. The CPU percentage of the system is also relatively low, and, like the RAM usage test, the Unity VR application required the majority of the overall CPU usage. These results show that the VR and DT system is not computationally expensive as the overall RAM and CPU usage is relatively low for a VR application.

5   Conclusion and Future Work

The paper presents an architecture where VR is integrated with DTs to add the benefits of VR to the functionality offered by a DT system. The paper shows that VR can be added as an additional UI service to a modular DT system that is built on a SLADTA aggregation architecture coupled to a services network. The VR application is able to make use of the services network of the DT architecture when requesting information. These services make use of the DT aggregation hierarchy of the architecture to obtain the data from the relevant DTs.

Preliminary evaluations show that the proposed integration architecture for VR and DTs has low latency and relatively low computational power requirements. In spite of the complexity of the digital twin system, neither the latency nor the computational load presented significant challenges. The DTs in the paper's implementation made use of local storage databases to reduce latencies when retrieving data. Further consideration should be given to latencies that will be incurred if the DTs' data is stored in cloud repositories.

The integration architecture presented above is highly modular. The modularity and the use of TCP/IP for communication between modules allows the distribution of the DTs and service components on different hardware. For a large-scale implementation, this distributed operation reduces the computational resource utilisation the components have on a single hardware device.

The modularity also simplifies extensions to the system's capabilities. For example, a new display that requires a new analysis of the existing DT data (e.g., energy cost) could be generated by a VR specialist adding software code related to the views in the VR application and by an FM domain specialist adding a new service in the services network. If appropriately coded, only configuration file changes would be required for the Management Services and Services Gateway, avoiding the need to alter the other components or individual DTs. Further research to quantify reconfiguration and further development effort is required.

The integration architecture presented here obtains the data of the physical system from an existing external data source. The use of SLADTA as internal architecture for the DTs makes it straightforward to implement a situation where information is exchanged (both ways) between the VR application and the physical system, via the DTs. However, this bi-directional flow still needs to be demonstrated.

Another advantage of the proposed architecture is that the DT and services system can perform other "regular" DT services, while also providing the information the VR application requires. On the other hand, the integration with VR can enrich the context of the DT information to aid interpretation, e.g., overlaying information on a map. Further research into VR displays that can aid in interpreting DT data is required.

Further research is required to compare the vendor-neutral SLADTA-with-services architecture to digital twin applications offered by major automation vendors. Also, further application contexts should be explored because, although the system was implemented in a facilities management context, the architecture can be applied more generally.

Acknowledgements. This work is based on the research supported in part by the National Research Foundation of South Africa (Grant Number 129360).

References

1. Kritzinger, W., Karner, M., Traar, G., Henjes, J., Sihn, W.: Digital Twin in manufacturing: A categorical literature review and classification. IFAC-PapersOnLine **51**, 1016–1022 (2018). https://doi.org/10.1016/j.ifacol.2018.08.474
2. Hu, M., Luo, X., Chen, J., Lee, Y.C., Zhou, Y., Wu, D.: Virtual reality: a survey of enabling technologies and its applications in IoT. J. Netw. Comput. Appl. **178** (2021). https://doi.org/10.1016/j.jnca.2020.102970
3. Kroupa, J., Tumova, E., Tuma, Z., Kovar, J., Singule, V.: Processing and visualization of microclimatic data by using virtual reality. MM Sci. J. 2621–2624 (2018). https://doi.org/10.17973/MMSJ.2018_12_2018104
4. Erra, U., Malandrino, D., Pepe, L.: Virtual Reality Interfaces for Interacting with Three-Dimensional Graphs. Int. J. Hum. Comput. Interact. **35**, 75–88 (2019). https://doi.org/10.1080/10447318.2018.1429061
5. Drouhard, M., Steed, C.A., Hahn, S., Proffen, T., Daniel, J., Matheson, M.: Immersive visualization for materials science data analysis using the Oculus Rift. In: Proceedings of the 2015 IEEE International Conference on Big Data, pp. 2453–2461 (2015). https://doi.org/10.1109/BigData.2015.7364040

6. Chandra Sekaran, S., Yap, H.J., Musa, S.N., Liew, K.E., Tan, C.H., Aman, A.: The implementation of virtual reality in digital factory—a comprehensive review. Int. J. Adv. Manuf. Technol. **115**(5–6), 1349–1366 (2021). https://doi.org/10.1007/s00170-021-07240-x

7. Da Silva, G.S., Kruger, K., Basson, A.H.: Opportunities for visualising complex data by integrating virtual reality and digital twins. In: International Conference on Competitive Manufacturing, COMA 2022, Stellenbosch, South Africa (2022)

8. Human, C.: A design framework for aggregation in a system of digital twins, Ph.D. Dissertation, Stellenbosch University, Stellenbosch, South Africa (2022)

9. Juarez, M.G., Botti, V.J., Giret, A.S.: Digital twins: review and challenges. J. Comput. Inf. Sci. Eng. **21**(3) (2021). https://doi.org/10.1115/1.4050244

10. Vivi, Q.L., Parlikad, A.K., Woodall, P., Ranasinghe, G.D., Heaton, J.: Developing a dynamic digital twin at a building level: using Cambridge campus as case study. In: International Conference on Smart Infrastructure Construction. ICSIC 2019 Driv. Data-Informed Decision, pp. 67–75 (2019). https://doi.org/10.1680/icsic.64669.067

11. Lu, Q., et al.: Developing a digital twin at building and city levels: case study of West Cambridge Campus. J. Manag. Eng. **36**, 05020004 (2020). https://doi.org/10.1061/(asce)me. 1943-5479.0000763

12. Redelinghuys, A.J.H., Kruger, K., Basson, A.: A Six-layer architecture for digital twins with aggregation. In: Borangiu, T., Trentesaux, D., Leitão, P., GiretBoggino, A., Botti, V. (eds.) SOHOMA 2019. SCI, vol. 853, pp. 171–182. Springer, Cham (2020). https://doi.org/10.1007/ 978-3-030-27477-1_13

13. Karanjkar, N., Joglekar, A., Mohanty, S., Prabhu, V., Raghunath, D., Sundaresan, R.: Digital twin for energy optimization in an SMT-PCB assembly line. In: Proceedings - 2018 IEEE International Conference on Internet Things Intelligent System. IOTAIS, pp. 85–89 (2019). https://doi.org/10.1109/IOTAIS.2018.8600830

14. Andersen, B.J.H., Davis, A.T.A., Weber, G., Wunsche, B.C.: Immersion or diversion: does virtual reality make data visualisation more effective? In: ICEIC 2019 International Conference on Electron. Information, Communication (2019). https://doi.org/10.23919/ELINFO COM.2019.8706403

15. Havard, V., Jeanne, B., Lacomblez, M., Baudry, D.: Digital twin and virtual reality: a cosimulation environment for design and assessment of industrial workstations. Prod. Manuf. Res. **7**, 472–489 (2019). https://doi.org/10.1080/21693277.2019.1660283

16. Malik, A.A., Masood, T., Bilberg, A.: Virtual reality in manufacturing: immersive and collaborative artificial-reality in design of human-robot workspace. Int. J. Comput. Integr. Manuf. **33**, 22–37 (2020). https://doi.org/10.1080/0951192X.2019.1690685

17. Kruger, K., Human, C., Basson, A.: Towards the integration of digital twins and service-oriented architectures. In: Borangiu, T., Trentesaux, D., Leitão, P., Cardin, O., Joblot, L. (eds.) SOHOMA 2021. Studies in Computational Intelligence, vol 1034, pp. 131–143. Springer, Cham (2022). https://doi.org/10.1007/978-3-030-99108-1_10

18. Unity: Unity Manual VR overview. https://docs.unity3d.com/540/Documentation/Manual/ VROverview.html. Accessed 06 July 2022

19. Donalek, C., et al.: Immersive and collaborative data visualization using virtual reality platforms. Proceedings - 2014 IEEE International Conference on Big Data, pp. 609–614 (2015). https://doi.org/10.1109/BigData.2014.7004282

20. Liagkou, V., Salmas, D., Stylios, C.: Realizing virtual reality learning environment for Industry 4.0. Procedia CIRP **79**, 712–717 (2019). https://doi.org/10.1016/j.procir.2019.02.025

Integrating Lean Data and Digital Sobriety in Digital Twins Through Dynamic Accuracy Management

Nathalie Julien and Mohammed Adel Hamzaoui[✉]

Lab-STICC, University of South Brittany, 17 Bd Flandres Dunkerque, 56100 Lorient, France
{nathalie.julien,mohammed.hamzaoui}@univ-ubs.fr

Abstract. In this paper we focus on the design, development, and responsible use of the flagship technology of Industry 4.0, namely the digital twin (DT). After a brief review of the literature on industrial applications of the digital twin in a smart manufacturing environment, we look at usual misconceptions about this emerging concept, which often hinder the democratization of the digital twin or pushes to put too many resources at stake compared to the needs. Therefore, we propose to integrate lean data and digital sobriety at the design phase of the DT through a generic architecture including dynamic accuracy management. We present here the principle of our approach, an application example, as well as some adjustment capabilities.

Keywords: Digital Twin · Digital sobriety · Dynamic accuracy management · Smart manufacturing systems

1 Introduction

The advent of the fourth industrial revolution has led to the emergence of new technologies, each one as disruptive and innovative as the next, like the digital twin (DT) often considered a pillar of the smart factory. The notion of DT was originally introduced in 2003 at Michael Grieves' executive training on product lifecycle management (PLM) [1]. Even though the literature has more than 50 distinct definitions (Julien & Martin, 2021), the one provided by the new ISO standard 23247 [2] appears to be the most thorough and exact for manufacturing applications: "*a fit for purpose digital representation of an observable manufacturing element with a means to enable convergence between the element and its digital representation at an appropriate rate of synchronization*".

The digital twin is considered in a sense to be the spearhead of Industry 4.0, as it is transversal in terms of the tools, technologies and skills implemented (AI, Internet of Things, Data, as well as electronic engineers, automation engineers, data scientists, psycho-ergonomists, etc.) but also in terms of the issues dealt with (maintenance, control, product quality management, supply chain management, prognostic and fault detection, etc.).

In this paper, we focus on this technology and the methodologies for its development in line with the objectives of sustainable growth and human-centred technological

© The Author(s), under exclusive license to Springer Nature Switzerland AG 2023
T. Borangiu et al. (Eds.): SOHOMA 2022, SCI 1083, pp. 107–117, 2023.
https://doi.org/10.1007/978-3-031-24291-5_9

development of society. Through this work, we look at the contribution that dynamic accuracy management of digital twins can bring to the sober development and use of this technology, as well as the means and methods that can be used to achieve it.

2   The Digital Twin, a Polymorphic Technology

The digital twin is clearly distinguished by its capacity to address a broad range of issues and in a variety of application domains. Indeed, in the various fields that aim to upgrade technologically by keeping pace with the global technological revolution, the digital twin receives an unparalleled interest. In what follows we present the different applications of the digital twin in these areas to address various concerns.

1. *Healthcare*: Healthcare 4.0, which refers to the capability of customizing healthcare in real time for patients, professionals, and caregivers, is perfectly suited to the digital twin. DT health applications may be found in a variety of settings, including (i) digital patients, (ii) the pharmaceutical industry, (iii) hospitals, and (iv) wearable devices [3].
2. *Smart cities*: Due to the extensive use of information and communications technology (ICT) to monitor municipal operations, cities have become smarter during the past several years. These operations (including traffic and transportation, energy production, utility provisioning, water supply, and trash management) generate a large amount of data that can be used to build data-based models and use data-driven decision support tools such as AI solutions to regulate and optimize city activities [4]. A number of governmental and academic research initiatives are working to create digital twins of cities with various levels of complexity (Dublin's Docklands area [4], the "Virtual Singapore" [5], the Zurich city digital twin [6]).
3. *Transportation and logistics*: A number of logistics-related issues, including those involving internal logistics (like warehousing and material handling) [7] and external logistics (like air traffic management, railway management systems, elevators and vertical transportation systems, and even pipeline DTs) [8–10]) are being studied for their potential application of the digital twin.
4. *Energy*: When it comes to energy production, transit, storage, and even consumption, digital twins can offer a disruptive edge at several levels (control, optimization, and conception). A number of useful applications and use cases can also be found here, including energy-efficient manufacturing systems [11], energy consumption optimization [12], implementation of digital twins in energy cyber-physical systems [13], and DT based development of smart electrical installations [14].
5. *Smart manufacturing*: Even if the digital twin has been exported to different domains, the fact remains that its adopted home is in industry, which is why the most notable advances for this technology can be found there. In the following, we briefly present the essence of digital twin applications in smart manufacturing.

The applications of the digital twin in industry are very varied, however they can be classified into four DT lifespan phases: design, manufacturing, service, and retirement as proposed by Liu et al. [15].

– Regarding the first phase, they identified iterative optimization; data integrity; and virtual evaluation and verification as the main applications: automated flow-shop production system customization [16], optimal-selection in product design [17], virtual validation in electronics [18].

– With regard to the second phase, they have instead identified seven main usages: real-time monitoring, production control, workpiece performance prediction, human-robot collaboration and interaction, process evaluation and optimization, asset management, and production planning. These can be declined according to different applications: autonomous manufacturing [19], DT implementation in Manufacturing Execution Systems (MES) [20], start-up phases acceleration through DT-based product modifications management [21].

– Concerning the third phase, five applications for Digital Twins were identified: predictive maintenance, fault detection and diagnosis, state monitoring, performance prediction, and virtual test. The work in the literature supporting this deals with concerns as time-based maintenance (TBM) policies optimization [22], Remaining Useful Life (RUL) assessment [23].

– Finally, Liu et al. considered that the fourth and final phase was generally not considered. Knowledge of a system's or product's behaviour is frequently lost when it is decommissioned. When a system or product is developed upon, it usually suffers from the same flaws that might have been avoided with the use of past knowledge. In the retire phase, the DT may be inexpensively kept in virtual space since it possesses the physical twin's whole lifespan information [15]. We can however mention the work of [24] who identified disassembly, reusage, disposal, retrospect, and upgrade as five major applications in the retirement phase.

This ability of a digital twin to fit into very diverse application domains and to deal with a variety of problems requires it to have a very important capacity for adaptation. Hence the need to set up models and standards or even generic development and deployment methodologies, which, without taking away the adaptive characteristic of the twin, allow it to be correctly sized and dimensioned to the needs of the different situations it may face. It is for this reason that we are interested in this paper in the appropriate dimensioning of the digital twin, either at the level of the technologies brought into play in the DT during the design process or at the level of the operation modes. Here, we will discuss the integration of lean data and digital sobriety in digital twins through dynamic accuracy management.

3   Usual Misconceptions About the Digital Twin

Not only is the digital twin an emerging concept, but, because of the lack of standards and formalization, it is often accompanied by implicit assumptions that fluctuate depending on the speaker, making its definition and design very application-specific and technology-dependent.

Among these assumptions, we identify several usual misconceptions that lead the designer to over-size the architecture of the digital twin, making its deployment unnecessarily complex, consuming, and expensive.

- *Big Data and Cloud Computing*

Generating, processing, manipulating, storing, and transferring large amounts of data requires significant material and financial resources, which can be very distressing for a company and a brake on its development. Turning to data warehousing solutions may seem like a logical next step for any company that wants to upgrade technologically and keep up with the digital revolution, but it can be largely counterproductive considering the technical constraints and financial burdens whether it is done via cloud computing technologies or even via local servers. It is often necessary to opt for digital sobriety and lean data instead of big data by knowing how to properly size the means to be put in place in relation to established needs and available capacity.

- *Real time*

The digital twin is often associated with real time computing which implies a very high level of synchronization. However, this is usually not necessary and sometimes not even feasible. In some cases, feedbacks at longer intervals may clearly be sufficient, as well as asynchronous updates (when the value or state of some variable changes, or even when some thresholds are exceeded).

- *High fidelity and 3D Model*

Another misconception about the digital twin is that it must be represented graphically (mainly via a 3D model). This is clearly not necessary, and may even be inadvisable, as it requires a considerable amount of time and computational resources, as well as the use of often very expensive proprietary software, without bringing any real added value apart from the aesthetic aspect. With the exception of cases where the digital twin must be visualized graphically by a human operator (clearly expressed need, and clearly identified added value) the integration of a 3D model is certainly not recommended; it is quite sufficient to implement models adapted to the needs (Tao et al. [25] have proposed different types of models to be implemented in a digital twin).

- *IoT and over instrumentation*

It is also necessary to properly size one's instrumentation needs so as not to over-instrument the installations (except for sensitive installations governed by regulations which require certain redundancy in equipment and measurements). A good needs sizing is necessary in order to correctly identify the data to be collected, the collection points, the sensitivity of the measurements as well as their precision and reliability (it is not necessary to use a microscope when a magnifying glass can do the job).

Moreover, the excessive use of IoT can only disrupt the objectives of digital sobriety and the costs incurred by the digitization process of any installation. For if there is no need to implement a distributed architecture of computing resources or a strong horizontal and transversal interconnectivity (or even DT networks [26]), there is clearly no need to excessively use IoTs when simple sensors may suffice.

- *Artificial Intelligence*

One of the disruptive contributions of the digital twin is to be able to offer decision support tools to human agents in order to make the best use of their assets (control, optimization, maintenance, safety and security, etc.). This implies the presence of an "intelligent" element in this technology. However, this does not necessarily imply the presence of complex artificial intelligence algorithms that require a lot of data, resources, and computing time. In some cases, simple statistical models can be sufficiently accurate and reliable to achieve the fixed objectives; again it is all about precise dimensioning.

- *Autonomous Digital Twin*

An implicit inference from the assumed characteristic of the digital twin cited above (high intelligence) is that the digital twin must have a high degree of autonomy. This is not only a question of evaluating the resources to be brought into play in relation to the needs, but also of the allocation of decision-making and cognitive operations between the human agent and the technical agent (the digital twin). The human must necessarily be in the decision loop (closed, open, or mixed [27]). Kamoise et al. [28] presented a work that proposes the use of CWA (Cognitive Work Analysis) for the allocation of decision making tasks between the human agent and the digital twin in maintenance-related applications.

The first step to digital sobriety is to be aware that all these misconceptions may lead to DT oversizing (in design and use). This leads designers to provide the DT with useless functions, wasting more energy and producing more - irrelevant - data, more processes and calculations, supplemental tools, and even external expertise that is not really necessary for the actual requirements and needs. Such significantly, that one could state that the soberest Digital Twin is the one that does not exist! However, we assume that in some cases and provided that the cost of development and operation of the Digital Twin is controlled, it can globally reduce the overall consumption of a system. We therefore propose to size the digital twin as accurately as possible according to its uses, especially by integrating into its architecture a device that can adjust its precision through DAM (Dynamic Accuracy Management).

4 Dynamic Accuracy Management

Our research approach is to develop a DT design methodology based on genetic architectures. These architectures are defined through a typology characterizing the used needs via various parameters [29].

The DAM (Dynamic Accuracy Management) is an approach to integrate in the generic DT architecture from the design stage associated with policies suited to the application and its usages as represented on Fig. 1. It is inserted between the Physical Element (PE) and the Virtual Element (VE) that is the set of models.

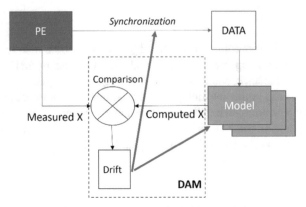

Fig. 1. DAM in the generic architecture

4.1 Principle

We decide to act on two parameters of our generic architecture: *synchronization* (data acquisition rate) and *models*, both having a direct impact on the volume of data to be processed by the digital twin.

It is therefore necessary for the application under consideration to list the different models by order of precision and data volume as well as the different possible synchronizations (Table 1).

Table 1. List of models and synchronizations

List of Synchronizations			List of Models		
Synchronization	Frequency	Accuracy	Models	Parameters	Accuracy
Asynchronous	$F = 0$ (event-based)	Low	Model 1	Few	Low
Synchronous 1	F_1	Medium	Model 2	Several	Medium
Synchronous 2	$F_2 > F_1$	High	Model 3	Many	High
Synchronous 3	$F_3 > F_2$	Best			

To define the accuracy policy, we first determine the various states corresponding to the operational modes characterized by a pair (model, synchronization) as in Table 2.

We then need to precise the transitions between these states i.e., the condition to pass from one operational mode to another. To do this, we must define which physical variable X to monitor, and to which are compared the variables computed by the digital twin. The transitions are defined as thresholds qualifying the difference between the physical value and the virtual one, these drifts representing the correctness of the digital twin. This threshold can be expressed in absolute or relative value, as the example in Table 3, considering that ΔX is the difference between the measured and computed values.

Table 2. Operational modes

States	Model	Synchronization	Data volume
State 1	Model 1	Asynchronous	Least consuming
State 2	Model 1	Synchronous 1	
State 3	Model 1	Synchronous 2	
State 4	Model 2	Synchronous 2	
State 5	Model 2	Synchronous 3	
State 6	Model 3	Synchronous 3	Most consuming

Table 3. Transition table example

State	State 1	State 2	State 3	State 4	State 5	State 6
ΔX	$\Delta X < V_1$	$V_1 < \Delta X < V_2$	$V_2 < \Delta X < V_3$	$V_3 < \Delta X < V_4$	$V_4 < \Delta X < V_5$	$V_5 < \Delta X < V_6$
ΔX	$\Delta X < 1\%$	$\Delta X < 2\%$	$\Delta X < 3\%$	$\Delta X < 4\%$	$\Delta X < 5\%$	$\Delta X < 10\%$

Therefore, all the elements characterizing the accuracy policy are defined. The application to a simple case of battery state of charge monitoring will illustrate our approach.

4.2 Battery Digital Twin

The application consists in monitoring a lithium battery by collecting its temperature, voltage and current together with its State of Charge (SoC) which is the physical value we want to predict with the Digital Twin.

We can rely on two different models: Model 1 is a Thevenin model composed of 4 electrical elements; such a simplicity enables Edge Computing as indicated in Table 4. The second model is more detailed with 12 electrochemical parameters and requires Cloud computing.

Table 4. Models for battery Digital Twin

Models		Parameters	Accuracy	Computing
Model 1	Thevenin	4 parameters	Low	Edge
Model 2	Electrochemical	12 parameters	High	Cloud

We choose to apply three levels of synchronization (Table 5) and three operational modes (Table 6).

Table 5. Synchronizations for battery Digital Twin

Synchronization	Frequency	Accuracy
Asynchronous	F = 0 (event-based)	Low
Synchronous 1	$F_1 = 1/10s$	Medium
Synchronous 2	$F_2 = 1/1s$	High

Table 6. Operational modes for battery Digital Twin

States	Synchronization	Synchronization	Data volume
State 1	Model 1	Asynchronous	Least consuming
State 2	Model 1	$F_1 = 1/10s$	
State 3	Model 2	$F_2 = 1/1s$	Most consuming

By default, we are in State 1 with the Model 1 in an asynchronous mode, so we are waiting to see a drift between the measured and computed SoC to act, as represented in Fig. 2. This state typically corresponds to the monitoring when we are controlling that measured and predicted values are coherent i.e. less than 1%.

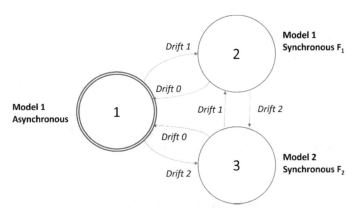

Fig. 2. State diagram for battery Digital Twin

In case of slow variation with a difference of less than 5% (Drift 1), we switch to the State 2 by keeping the same Model 1 but commute the acquisition into a synchronous mode and collecting data at the rate of 10s to have more information and adjust the computed SoC; if the drift persists, we can modify the model parameters either manually or through Machine Learning.

In case of strong variation, if ΔX, the relative difference between measured and computed SoC, is more than 5%, we switch on the Model 2 and raises the synchronization rate to 1s to ensure the most detailed information and work onto the model parameters until the drift decreases under 5%, see Table 7. Such an approach also allows tuning the model along with the aging of the battery and environmental context. Of course, we can implement more elaborated policies with various models and frequencies, but it is important to guarantee that the architecture stay efficient with no useless complexity.

Table 7. Transition table for battery Digital Twin

State	State 1	State 2	State 3
Drift	Drift 0	Drift 1	Drift 2
ΔX	$\Delta X < 1\%$	$1\% < \Delta X < 5\%$	$5\% < \Delta X < 10\%$

4.3 Dynamic Accuracy Adjustment

With dynamic accuracy management we can act on the data volume both on a temporal way by tuning the synchronization and on a spatial way by choosing the model along with its number of parameters and complexity. All this approach can be done either off-line where the DT user can set the application-specific accuracy required, or online to follow hazards and malfunctions, alert the user, and modify the model along the physical element aging.

When doing health monitoring for maintenance of an equipment, one can first use a simple model with few parameters and low accuracy; then, when the equipment is aging, the data number can be increased to better anticipate the warning signs of failure. For example, if health monitoring for an engine conditional maintenance is performed, one can first monitor only the vibrations (Model 1) as long as the engine does not show signs of aging, and then the temperature can be added (Model 2) to be more efficient in preventing the failure.

We also can define a policy depending on the data dynamics by raising the frequency rate to be more responsive to important variations while the model switching is more convenient to prevent unpredicted events.

Another way to use this approach, as shown in Fig. 3, is to first adjust the DT accuracy as close as possible to the physical element with the higher synchronization rate (State 4) and then to reduce it progressively while keeping the drift in the predefined threshold (State 3). If possible, we can also switch on a simpler model (State 2 or 1). Each time the drift is increasing, we can use spatial or temporal adjustment while modifying manually or automatically the model parameters.

Finally, the dynamic accuracy adjustment can be determined by the DT usage. For the same equipment, a simple model (Model 1) can be used for monitoring, a more precise model and synchronization for production optimization and the highest accuracy level for preventive maintenance.

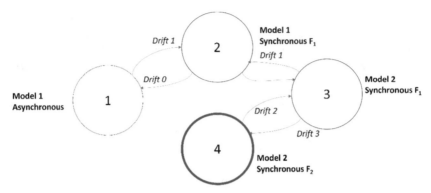

Fig. 3. Dynamic accuracy adjustment

5   Conclusion

Through this work, we have proposed some good practices to apply when developing a digital twin. These practices are based on dynamic accuracy management to ensure digital sobriety, while highlighting all the preconceived notions around the digital twin that encourage a certain digital gluttony when it comes to setting up a digital twin. As a complementary future work, we aspire to implement these operating mechanisms in real applications in order to evaluate their relevance and their actual impact under real conditions.

References

1. Grieves, M.: Digital Twin: Manufacturing Excellence through Virtual Factory Replication (2014).
2. International Organization for Standardization. Automation systems and integration - Digital twin framework for manufacturing (ISO No. 23247) [Internet] (2021). https://www.iso.org/obp/ui/#iso:std:iso:23247:-1:ed-1:v1:en. Accessed 5 Jul 2022
3. Erol, T., Mendi, A.F., Dogan, D.: The digital twin revolution in healthcare. In: 4th International Symposium on Multidisciplinary Studies and Innovative Technologies (ISMSIT'20), Istanbul, Turkey, pp. 1–7 (2020). https://ieeexplore.ieee.org/document/9255249/
4. White, G., Zink, A., Codecá, L., Clarke, S.: A digital twin smart city for citizen feedback. Cities **110**, 103064 (2021). https://doi.org/10.1016/j.cities.2020.103064
5. Farsi, M., Daneshkhah, A., Hosseinian-Far, A., Jahankhani, H. (eds.): Digital Twin Technologies and Smart Cities. IT, Springer, Cham (2020). https://doi.org/10.1007/978-3-030-18732-3
6. Shahat, E., Hyun, C.T., Yeom, C.: City digital twin potentials: a review and research agenda. Sustainability **13**(6), 3386 (2021). https://doi.org/10.3390/su13063386
7. Martínez-Gutiérrez, A., Díez-González, J., Ferrero-Guillén, R., Verde, P., Álvarez, R., Perez, H.: Digital Twin for Automatic Transportation in Industry 4.0, Sensors **21**(10), 3344 (2021)
8. Bhatti, G., Mohan, H., Raja Singh, R.: Towards the future of smart electric vehicles: digital twin technology. Renew. Sustain. Energy Rev. **141**, 110801 (2021)
9. Wang, S., Zhang, F., Qin, T.: Research on the construction of highway traffic digital twin system based on 3D GIS technology. J. Phys. Conf. Ser. **1802**(4), 042045 (2021). https://doi.org/10.1088/1742-6596/1802/4/042045

10. Sleiti, A.K., Al-Ammari, W.A., Vesely, L., Kapat, J.S.: Carbon dioxide transport pipeline systems: overview of technical characteristics, safety, integrity and cost, and potential application of digital twin. J. Energy Resources Technol. **144**(9), 092106 (2022)
11. Li, H., Yang, D., Cao, H., Ge, W., Chen, E., Wen, X.: Data-driven hybrid petri-net based energy consumption behaviour modelling for digital twin of energy-efficient manufacturing system. Energy **239**, 122178 (2022)
12. Vatankhah Barenji, A., Liu, X., Guo, H., Li, Z.: A digital twin-driven approach towards smart manufacturing: reduced energy consumption for a robotic cell. Int. J. Comput. Integr. Manuf. **34**(7–8), 844–859 (2021)
13. Saad, A., Faddel, S., Mohammed, O.: IoT-based digital twin for energy cyber-physical systems: design and implementation. Energie **13**(18), 4762 (2020)
14. Fathy, Y., Jaber, M., Nadeem, Z.: Digital twin-driven decision making and planning for energy consumption. JSAN **10**(2), 37 (2021)
15. Liu, M., Fang, S., Dong, H., Xu, C.: Review of digital twin about concepts, technologies, and industrial applications. J. Manuf. Syst. **58**, 346–361 (2021)
16. Liu, Q., Zhang, H., Leng, J., Chen, X.: Digital twin-driven rapid individualised designing of automated flow-shop manufacturing system. Int. J. Prod. Res. **57**(12), 3903–3919 (2019)
17. Xiang, F., Zhang, Z., Zuo, Y., Tao, F.: Digital twin driven green material optimal-selection towards sustainable manufacturing. Procedia CIRP **81**, 1290–1294 (2019)
18. Howard, D.: The digital twin: virtual validation in electronics development and design. In: 2019 Pan Pacific Microelectronics Symposium Kauai, HI, USA, pp. 1–9. IEEE (2019). https://ieeexplore.ieee.org/document/8696712/
19. Zhang, C., Zhou, G., He, J., Li, Z., Cheng, W.: A data- and knowledge-driven framework for digital twin manufacturing cell. Procedia CIRP **83**, 345–350 (2019)
20. Cimino, C., Negri, E., Fumagalli, L.: Review of digital twin applications in manufacturing. Comput. Ind. **113**, 103130 (2019)
21. Zhu, Z., Xi, X., Xu, X., Cai, Y.: Digital Twin-driven machining process for thin-walled part manufacturing. J. Manuf. Syst. **59**, 453–566 (2021)
22. Savolainen, J., Urbani, M.: Maintenance optimization for a multi-unit system with digital twin simulation: example from the mining industry. J. Intell. Manuf. **32**(7), 1953–1973 (2021)
23. Werner, A., Zimmermann, N., Lentes, J.: Approach for a holistic predictive maintenance strategy by incorporating a Digital Twin. Procedia Manuf. **39**, 1743–1751 (2019)
24. Falekas, G., Karlis, A.: Digital Twin in electrical machine control and predictive maintenance: state-of-the-art and future prospects. Energies **14**(18), 5933 (2021)
25. Tao, F., Zhang, M., Nee, A.Y.C.: Digital Twin Driven Smart Manufacturing. Academic Press, London (2019). ISBN: 9780128176313
26. Hamzaoui, M.A., Julien, N.: Social cyber-physical systems and Digital Twins networks: a perspective about the future digital twin ecosystems. IFAC-PapersOnLine **55**(8), 31–36 (2022). https://doi.org/10.1016/j.ifacol.2022.08.006
27. Julien, N., Martin, E.: Typology of manufacturing Digital Twins: a first step towards a deployment methodology. In: Borangiu, T., Trentesaux, D., Leitão, P., Cardin, O., Joblot, L. (eds.) Service Oriented, Holonic and Multi-agent Manufacturing Systems for Industry of the Future. SOHOMA 2021, Studies in Computational Intelligence, vol. 1034. Springer, Cham (2021). https://doi.org/10.1007/978-3-030-99108-1_12
28. Kamoise, N., Guerin, C., Hamzaoui, M., Julien, N.: Using Cognitive Work Analysis to deploy collaborative digital twin. In: European Safety and Reliability Conference, Aug 2022, Dublin, Ireland (2022). https://hal.archives-ouvertes.fr/hal-03774847
29. Blanchet, A., Julien, N., Hamzaoui, M.A.: Typology as a deployment tool for digital twins: application to maintenance in industry. In: European Safety and Reliability Conference, Aug 2022, Dublin, Ireland

Handling Uncertainties with and Within Digital Twins

Farah Abdoune[1], Leah Rifi[2], Franck Fontanili[2], and Olivier Cardin[1(✉)]

[1] Nantes Université, École Centrale Nantes, CNRS, LS2N, UMR 6004, 44000 Nantes, France
olivier.cardin@univ-nantes.fr

[2] Department of Industrial Engineering, Toulouse University, IMT Mines Albi, Route de Teillet, 81013 Albi Cedex 9, France

Abstract. The Digital Twin (DT) is often used in environments characterized by uncertainty and complexity, where operating conditions are prone to variability based on external and internal factors. Thus, the literature about DT emphasizes the importance, limitations, and absence of uncertainty quantification. However, there is no explicit review discussing uncertainty in complex systems and within the digital twin model. Such an explicit review could improve the conception, construction, and utilization of DT in environments that are both dynamic and stochastic. Thus, this article aims to (1) describe how a DT can help manage uncertainties in a dynamic system, and (2) explain how DT should deal with uncertainties inside the model.

Keywords: Digital Twin · Dynamic system · Aleatoric uncertainty · Epistemic uncertainty

1 Introduction

Multiple fields paid close attention to digital twins (DTs) during the last decade. The Digital Twin has been extensively considered in the field of manufacturing [1] such as autonomous manufacturing [2] and additive manufacturing [3], and has even been extended to cyber-physical systems [4]. It has also been explored for asset management activities like damage detection and condition monitoring [5] as well as uncertainty quantification [6], and risk assessment [7], and also in the healthcare field [8] for patient pathway tracking in hospitals [9] and even inside operating room [10].

Such systems are dynamic, highly complex, and exhibit stochastic characteristics [11]. They evolve with time, are hard to model, and can deviate from their initial configuration because of random events. Thus, their DT is intended to capture the real system dynamics and adapt to changes that might occur in an uncertain environment. Regardless of how accurate a DT is or how much data is gathered from the monitored system, a DT will always be an imperfect representation. Indeed, a DT systematically suffers from modelling errors and parameter uncertainty. Even when these parameters are known and correct, there is always a mismatch between the real behaviour of the system and the simulated behaviour of the DT because it is only an approximation of the real system.

© The Author(s), under exclusive license to Springer Nature Switzerland AG 2023
T. Borangiu et al. (Eds.): SOHOMA 2022, SCI 1083, pp. 118–129, 2023.
https://doi.org/10.1007/978-3-031-24291-5_10

For all these reasons, the literature abounds with work related to the notion of DT, their associated enabling technologies, and their problems.

The DT is often used in environments characterized by uncertainty and complexity, where working circumstances might change based on external and internal variables [9]. As a result, Z. Liu et al. [13] suggest the definition of 'a living model that continually adapts to change in the environment or operation using real-time sensory data and can forecast the future of the corresponding physical assets', whereas Zhuang et al. [14] present the concept as 'a dynamic model in the virtual world that is fully consistent with its corresponding physical entity in the real world and can simulate its physical counterpart's characteristics, behaviour, life, and performance in a timely fashion'.

The literature about DT emphasizes the importance, limitations, and absence of uncertainty quantification [15–17]. However, to the best of the authors' knowledge, there is no clear review regarding uncertainties in the digital and the physical worlds. Such an explicit review could improve the conception, construction, and utilisation of DT in environments that are both dynamic and stochastic. Thus, this article aims to (1) describe how a DT can help manage uncertainties in a dynamic system, and (2) explain how one should deal with uncertainties inside the DT model.

According to this motivation, we investigate the literature to answer the following research questions:

- **RQ1:** What are the types of uncertainties identified in the literature?
- **RQ2:** What is the difference between uncertainties in the physical world and the DT model?
- **RQ3:** What are the used methods to deal with the uncertainties with and within the DT?

The remainder of this paper is structured as follows. In Sect. 2, we present several uncertainty categories. Then, in Sect. 3, we investigate where these uncertainties appear in both the physical system and DT model. In Sect. 4, we describe the methods used to manage these uncertainties with and within the DT. Finally, Sect. 5 is dedicated to the Discussion and Sect. 6 proposes concluding remarks.

2 Types of Uncertainty

The scientific community provides field-specific typologies for uncertainties [18]. Milliken [15] characterizes *uncertainty* in organization theory with state uncertainty as an inability to "understand how components of the environment might be changing", *effect uncertainty* as "an inability to predict what the nature of the impact of a future state of the environment or environmental change will be on the organization", and *response uncertainty* as "an inability to predict the likely consequences of a response choice". In [19], process design and operations uncertainties are divided into model-inherent uncertainty, process-inherent uncertainty, external uncertainty, and discrete uncertainty. The two first are epistemic uncertainties, and the two lasts are aleatory uncertainties. The article [18] defines uncertainty as "any departure from the unachievable ideal of complete determinist". They distinguish three dimensions of uncertainty in decision support systems: the uncertainty's location (the place where the uncertainty occurs), the uncertainty's level (the amount of knowledge available), and the uncertainty's nature (whether the uncertainty is epistemic or aleatory). Finally in the operating room, [20] makes a difference between natural variability and artificial variability (which we would both call aleatory uncertainty in this article). Natural variability is the result of clinical variability, patient flow variability, and professional variability. It cannot be avoided as it is inherent to the healthcare environment. Artificial variability is human-induced (e.g., patient and staff preferences, planning, and scheduling). It is non-random and yet non-predictable.

Other definitions are more global. The paper [21] states that uncertainty is "an individual's perceived inability to predict something accurately" and that it "refers to the psychological state of doubt about what current events mean or what future events are likely to occur". [22] describes it as "a cognitive state of the individual, resulting from the evaluation of the number of alternatives available to predict future behaviour or alternatives available to explain a past behaviour". In the field of risk management, the ISO 31000:2009 standard on risk management vocabulary [23] describes uncertainty as "the state, even partial, of deficiency of information related to, understanding or knowledge of, an event, its consequence, or likelihood." These references agree with the idea that uncertainty is a perceptive phenomenon in which the access to information is crucial [24]. In various scientific fields (e.g., computational engineering [25], decision support [18], and electromechanics [26]), we can distinguish two main types of uncertainty: aleatory uncertainty and epistemic uncertainty.

Aleatory uncertainty is inherent and specific to any physical phenomenon which displays a random behaviour. It is the impact of the input natural variation on the outputs. It is irreducible without changing the studied system itself and is commonly treated with probability theory. Different classifications can be found in the literature depending on the area of application; in production planning, aleatory uncertainty is divided into demand forecast uncertainty, external supply process uncertainty, and internal supply process uncertainty [27]. In the supply chain domain, it includes supply uncertainty, process uncertainty, and demand uncertainty [28]. In the operating room environment [29] identifies four main uncertainties: patient arrival, surgery duration, care requirement, and resource uncertainty.

Epistemic uncertainty stems from imperfect knowledge of the system studied. It can thus be reduced by retrieving more information about the system. Different epistemic uncertainty categorization exists; for instance modelling errors and parameters uncertainties [25], or model uncertainty and data uncertainty [26]. More specifically, uncertainties can be designated as structured if they just affects the values of the model's parameters and not its structure, and as non-structured if it impacts the model's structure [30]. Kennedy and O'Hagan [31] state that epistemic uncertainties stem from two sources: parameters and models.

- Uncertainty due to the parameters refer to:

  - *Parameter uncertainties*: models invariably contain parameters that are quantifiable but are not completely understood or accessible in most circumstances.
  - *Parametric variability*: because inputs cannot be entirely controlled or described, the circumstance in which the model is used may change. A model, on the other

hand, may necessitate the definition of a single deterministic value, which should be adjusted depending on process information.

- Uncertainty due to the model are:

  - *Model discrepancy*: it is understood that even when the parameters are deterministic and 'really' known, there will still be mismatches between the model output and the 'true' physical process.
  - *Residual variability*: given the same set of inputs, the process may yield various results due to a chaotic or stochastic character. This is frequently caused by insufficiently specified inputs, the inherent random nature of the process, measurement noise, or caused by a lack of understanding or knowledge of the observed system.

[32] suggests another epistemic uncertainty typology and states that it can be divided into model uncertainty and data uncertainty. Model uncertainty includes parameter uncertainty, solution approximation errors, and model form uncertainty. Data uncertainty includes measurement uncertainty and sparse or imprecise. Sometimes aleatory uncertainty is referred to as "variability" and epistemic uncertainty as "uncertainty".

To illustrate the difference between epistemic and aleatory uncertainty, we suggest a simple example. When throwing a fair coin, we know that the results can be head (50% probability) or tail (50% probability). However, we cannot predict for sure what will be the output of a specific throw. We call aleatory uncertainty the variability output due to the inherent randomness of the studied system. In case the coin is not fair, there will also be epistemic uncertainty as we will not have perfect knowledge of the head/tail frequency. Thus, we will have to model the probability distribution that predicts the output. As a side note, the authors of [33] consider aleatory uncertainty as not uncertainty but as pure variability, and epistemic uncertainty as general uncertainty. This emphasizes the idea that defining the terms used is extremely important for a good comprehension of the problem.

Differentiating these two types of uncertainty is not always necessary. On one hand, in the context of probability theory and mathematical statistics of uncertainty quantification, the same tools are used for: (1) the stochastic modeling of uncertainties, (2) the analysis of uncertainty propagation in a computational model, and (3) the solving of inference or estimation. Thus, it is not necessary to deal with each type of uncertainty separately [26]. On the other hand, aleatory uncertainty is unavoidable without changing the real system, while epistemic uncertainty can be reduced by acquiring more knowledge about the system. When wanting to reduce the uncertainties in a model, one should focus on epistemic uncertainty only. Consequently,"epistemic uncertainty directly supports decisions about data collection and model improvement" [21].

3 Digital Twins and Uncertainty

DT's features allow them to be a powerful tool for modelling and monitoring dynamic systems [34]. To begin with, a DT can collect data from the physical world in real-time, which could support the development of dynamic models [35]. Indeed, a DT could provide at any moment information about the state of the physical system and assess the operational conditions of an uncertain situation. Moreover, the incorporation of data processing technologies such as artificial intelligence techniques into the DT may enable more effective predictive analysis and decision-making processes for uncertain events [36]. For this reason, in the next two sub-sections, we present an explicit overview of the aleatoric and epistemic uncertainties. First, we exploit the role of the DT in a dynamic system prone to uncertain events through literature. Then, we examine uncertainties within the DT model.

3.1   Uncertainty with Digital Twin in Dynamic Systems (Aleatoric)

There are numerous articles on asset management under uncertainties. A recent study in structural damage detection [37] describes DT implementation procedures with an approach combining physics-based models and machine learning. To quantify the uncertainties and develop a more robust method, a probabilistic approach in which a stochastic model is built and calibrated during the offline phase. In the same field, a technique for intelligent mission planning based on the DT is presented [38] while quantifying the uncertainties in the three components of the approach namely damage diagnosis, damage prognosis, and mission optimization considering aleatory and epistemic uncertainty sources using probabilistic methods. The structural health condition of an aircraft is known for its variability from one aircraft to another due to material qualities, mission history, pilot variability, and so on. The concept of dynamic Bayesian networks (DBN) is used in [39] to develop a health monitoring model for the diagnosis and prognosis of each aircraft. A probabilistic DT aircraft model is designed to achieve the integration of diverse uncertainty sources during the whole life of the aircraft wing, and decrease uncertainty in model parameters.

In the domain of safety and risk assessment, the authors of [34] emphasize how the inherent capability of DT in acquiring data from the physical world combined with the potentiality of reliable data processing could support a more effective diffusion of dynamic risk assessment models for improving safety. In this context, the authors of [40] propose the use of DTs to better identify high-risk scenarios and thus improve the risk assessment process. Another study [41] presents a framework for a DT based on machine learning and prognostics algorithms model to assess and predict the risk probability rate of an oil pipeline system by focusing on detecting a failure precursor and calculating risk conditions to determine remaining usable life (RUL).

In the manufacturing sector, several active areas can be identified that suffer from uncertainties, one of which is dynamic scheduling. Because of unknown occurrences in the manufacturing process, such as new task insertions, order cancellations, worker absences, and machine malfunctions, dynamic scheduling methods are needed and crucial to manufacturing systems. DT can aid in the detection of disruptions by continually comparing physical and virtual space and activating a rescheduling policy directly afterward a disturbance. In this context, [42] proposes reinforcement learning; (RL)-driven DT allows for effective collaborative scheduling between production and maintenance departments and assists manufacturers in improving real-time decision-making under uncertainties. Similarly, to address the aforementioned issues, the study in [43] presents a DT-enhanced dynamic scheduling strategy. Real data from physical machines and simulated data from their virtual models are fused to assist machine availability prediction. Furthermore, to identify disruptions effectively, the DT's virtual entity, which is constantly updated alongside its physical counterpart, is used as a dynamic reference to quantify the deviation of the real production from the expected plan. Another study [44] provides an architecture for robust scheduling that is applied to a flow shop scheduling problem and is connected with the field through the DT. To deal with uncertain events, a module is dedicated to the real-time estimation of the equipment's failure probability, using data-driven statistical models on collected data.

In material resource planning (MRP), DT is also sought to deal with uncertain events. The study [45] describes how to use DT to improve MRP. During planning, numerous MRP factors are unknown; machine learning is used to predict these parameters. Nonetheless, no prediction is flawless, and the unpredictability of some parameters may have a significant influence on the final result. Thus, the optimization methods and probability distributions are used to consider these uncertainties.

The authors of [46] use hidden Markov models to approximate uncertainties in the design of manufacturing systems DT. The suggested DT is made up of two parts: a model component and a simulation component. Using discrete states and associated transition probabilities, the model component creates a Markov chain that embodies the dynamics underlying the phenomena. The simulation component uses a Monte Carlo simulation approach to reproduce the events. [47] presents a DT-based methodology for simulating all the effects in an artifact-based machine tool calibration process, starting with the machine and its expected error ranges and progressing to the artifact geometry and uncertainty, artifact positions in the workspace, probe uncertainty, compensation model, and so on.

To conclude, most studies rely on probabilistic approaches to model uncertainties due to the inherent variability of dynamic systems and the occurrence of external disturbances (aleatory uncertainty).

3.2 Uncertainty Within the Digital Twin Model (Epistemic)

A system's model is an expression that approximates how the real system behaves. Regardless of how accurate a model is or how much data is provided, it will always be an imperfect and incomplete representation of the real system. Indeed, according to [30], there is always some discrepancy between the modeled and the real behaviours when modelling a dynamic process to monitor its behaviour. This gap is called residual resulting in epistemic uncertainty. Disturbances and noise compromise the accuracy and reliability of the DT to monitor the physical system correctly.

The sources of uncertainty may include: linearizing nonlinearities to simplify the model, calibrating the model poorly and causing errors in parameters and model form, and considering some parameter's value constant even though it might change because of external effects such (e.g., environmental effects), or parameters having different values in different units (tolerance).

To overcome the aforementioned uncertainties, the authors of [48] address the design of the DT model for machinery process deterioration. They present an updated technique based on parameter sensitivity analysis. In [45], machine learning is used to fine-tune the DT's parameters to reduce the model discrepancy. [50] suggest a reinforcement learning approach to adjust the model and data errors. In addition to reinforcement learning, [51] uses unsupervised learning to detect anomalies coming from measurement. Other studies use fuzzy logic to improve the performance of the DT by updating it based on measurements [52].

Overall, regardless of the emphasis put on the importance of uncertainty quantification in the DT, and the efforts put to fill the gap between the DT model and the physical system, the literature in this context remains poor and lacks more research and elaboration to tackle the different source of uncertainties and enrich DT functionalities.

4 Methodology

Model uncertainty, noise measurement, external disturbance, and stochastic outcomes of real systems all contribute to the use of uncertainty in modelling and inference. Thus, this section illustrates the approaches used to overcome the two types of uncertainty described previously, whether it is in the DT model or the real system. Figure 1 summarized the following information. First, because of the real process and its inherent disruptions, a physical system fed with identical inputs may display output variability (aleatory uncertainty). Second, uncertainty within the DT may be classified into two types: those depending on model parameters, known as model-parameter uncertainties, and those generated by modelling errors (epistemic uncertainty). Strategies for dealing with stochastic behaviours have been developed based on the type of uncertainty targeted.

– **Stochastic approaches**: Also called probabilistic approaches, all share a common mathematical foundation. These approaches aim to anticipate the occurrence of uncertainties by modelling them with probability densities. To model stochastic transitions in system dynamics with a DT one can use Hidden Markov Models [46] or Artificial Intelligence methods based on Probabilistic Graphical Models, e.g., Bayesian Networks (BNs) or Dynamic Bayesian Networks (DBNs) [39].
– **Non-probabilistic approaches**: Other non-probabilistic methods have been used to quantify and handle uncertainties. The goal of these approaches is to minimize the effects of uncertainty; they usually deal with epistemic uncertainties using fuzzy logic [52], rule-based techniques, and expert systems[53]. In the field of artificial intelligence, reinforcement learning is also being used [46].
– **Robust approaches**: Given the noise and uncertainty inherent in the system's monitoring, robust approaches aim at decoupling nonzero residuals caused by modelling errors/noise and disturbances of the physical system. Instead of establishing assumptions about stochastic distributions, it is feasible to use inference techniques to reduce the noise [54] or machine learning techniques to learn the normal pattern considering modelling errors and noise and still detect anomalies coming from the physical system [55]. The main objective of this approach is to obtain a DT which can listen to aleatory uncertainty from the physical system without being affected by epistemic uncertainties (model and parameters uncertainty).

Figure 1 summarizes the different uncertainty types and the approaches used to quantify them. In green, we represented a physical system monitored with a digital twin.

The observed input is the same for both systems. We measure the difference between the observed output and the expected output; the difference is called the residual. In red are represented the aleatoric uncertainties of the physical system due to its variability or external uncertain events occurrence connected with stochastic approaches to deal with them. In yellow are shown the epistemic uncertainties of the DT due to parameters, modelling errors or measurement noise connected to non-probabilistic approaches to deal with them. And then robust approaches are drawn in purple, connected to both epistemic and aleatoric uncertainties.

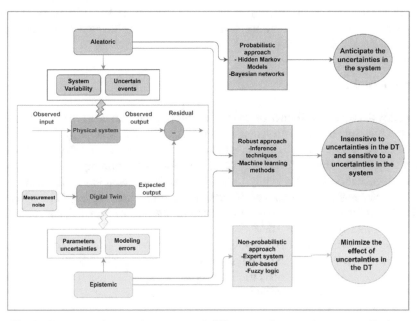

Fig. 1. Summary of uncertainty types and different used approaches to quantify them

5   Discussion

This article's contribution is twofold: (1) highlighting the capacity of a DT to quantify and manage uncertainties in dynamic systems, and (2) emphasizing the necessity of properly handling the uncertainty sources within the DT model itself. Overall, this article's outcome could help improve the use of DT in dynamic systems.

According to [56], uncertainty quantification is implicit in implementing a DT. That is, to allow risk-informed decisions, the best available knowledge about the DT features should be represented in the form of probability distributions or other uncertainty quantification metrics. However, as we saw in this article, aleatory uncertainty is irreducible whereas epistemic uncertainty can be reduced: the more knowledge we have about the real system, the less we need to model epistemic uncertainty. Indeed, if we have enough data from the physical system (e.g., by increasing the amount of data gathered and by improving the data collection methods), there is no need to use probability distributions to model uncertain events. Thus, the DT would be fed in real-time by the occurring events and be synchronized with them. On the opposite, if we are not certain about the behaviour of the physical system, the model will approximate its behaviour through a probability distribution.

This is the main difference between the simulation and the emulation capabilities of a DT. A DT based on simulation approximates the behaviour of its environment with uncertainty quantification; it is not entirely in sync with the real world (static feature). On the contrary, a DT based on emulation only changes its state when it receives appropriate information from the real world; it is entirely in sync with the real world (dynamic feature) [12]. Thus, to reduce epistemic uncertainty, more efforts should be done to develop the emulation capabilities of a DT.

6 Conclusion

The ability of the Digital Twin to be updated in real-time has earned a lot of attention from both the scientific and industrial communities. As complex systems generally display stochastic behaviours, uncertainty should be a crucial part of DT modelling. However, many elaborations and works on the DT do not consider uncertainty. Thus, until today, the implementation of DT's ability to adapt to random events in the physical system is a core challenge. Two sources of uncertainties must be investigated: (1) the inherent variability of the studied system (aleatory), and (2) the lack of knowledge regarding the aforementioned system (epistemic).

In this sense, this paper first proposed a review of the concept of uncertainty. Then it clarified their presence in both the physical system and the Digital Twin model. Finally, it discussed the different methods used to estimate them. The paper's contribution is an overview of what has been written on the topic and the creation of the first basis for future research works.

References

1. Wagg, D., Worden, K., Barthorpe, R., Gardner, P.: Digital twins: state-of-the-art future directions for modelling and simulation in engineering dynamics applications. ASCE-ASME J. Risk Uncertainty Eng. Syst. Part B Mech. Eng. **6** (2020). https://doi.org/10.1115/1.4046739
2. Rosen, R., von Wichert, G., Lo, G., Bettenhausen, K.D.: About the importance of autonomy and digital twins for the future of manufacturing. IFAC-PapersOnLine **48**(3), 567–572 (2015)
3. Knapp, G.L., et al.: Building blocks for a digital twin of additive manufacturing. Acta Mater. **135**, 390–399 (2017). https://doi.org/10.1016/j.actamat.2017.06.039
4. Cerrone, A., Hochhalter, J., Heber, G., Ingraffea, A.: On the effects of modeling as-manufactured geometry: toward digital twin. Int. J. Aerosp. Eng. vol. 2014 (2014). https://doi.org/10.1155/2014/439278
5. Seshadri, B.R., Krishnamurthy, T.: Structural health management of damaged aircraft structures using the digital twin concept (2017). https://doi.org/10.2514/6.2017-1675
6. Li, C., Mahadevan, S., Ling, Y., Choze, S., Wang, L.: Dynamic bayesian network for aircraft wing health monitoring digital twin. AIAA J. **55**, 1–12 (2017). https://doi.org/10.2514/1.J05 5201

7. Islavath, S.R., Deb, D., Kumar, H.: Life cycle analysis and damage prediction of a longwall powered support using 3D numerical modelling techniques. Arab. J. Geosci. **12**(14), 1–15 (2019). https://doi.org/10.1007/s12517-019-4574-y

8. Erol, T., Mendi, A.F., Doğan, D.: The digital twin revolution in healthcare. In: 2020 4th International Symposium on Multidisciplinary Studies and Innovative Technologies (ISMSIT), pp. 1–7 (2020). https://doi.org/10.1109/ISMSIT50672.2020.9255249

9. Karakra, A., Fontanili, F., Lamine, E., Lamothe, J.: HospiT'Win: a predictive simulation-based digital twin for patients pathways in hospital. In: 2019 IEEE EMBS International Conference on Biomedical and Health Informatics (BHI), pp. 1–4 (2019). https://doi.org/10.1109/BHI.2019.8834534

10. Patrone, C., Galli, G., Revetria, R.: A state of the art of digital twin and simulation supported by data mining in the healthcare sector. In: Advancing Technology Industrialization Through Intelligent Software Methodologies, Tools and Techniques, pp. 605–615 (2019). https://doi.org/10.3233/FAIA190084

11. Ullah, A.S.: Modeling and simulation of complex manufacturing phenomena using sensor signals from the perspective of Industry 4.0. Adv. Eng. Inform. **39**, 1–13 (2019)

12. Semeraro, C., Lezoche, M., Panetto, H., Dassisti, M.: Digital twin paradigm: a systematic literature review. Comput. Ind. **130**, 103469 (2021). https://doi.org/10.1016/j.compind.2021.103469

13. Liu, Z., Meyendorf, N., Mrad, N.: The role of data fusion in predictive maintenance using digital twin. AIP Conf. Proc. **1949**(1), 020023 (2018). https://doi.org/10.1063/1.5031520

14. Zhuang, C., Liu, J., Xiong, H.: Digital twin-based smart production management and control framework for the complex product assembly shop-floor. Int. J. Adv. Manuf. Technol. **96**(1–4), 1149–1163 (2018). https://doi.org/10.1007/s00170-018-1617-6

15. Zhang, M., Selic, B., Ali, S., Yue, T., Okariz, O., Norgren, R.: Understanding uncertainty in cyber-physical systems: a conceptual model. In: Wąsowski, A., Lönn, H. (eds.) ECMFA 2016. LNCS, vol. 9764, pp. 247–264. Springer, Cham (2016). https://doi.org/10.1007/978-3-319-42061-5_16

16. Schleich, B., Anwer, N., Mathieu, L., Wartz, S.: Shaping the digital twin for design and production engineering, CIRP Ann. **66**, 141–144 (2017). https://doi.org/10.1016/j.cirp.2017.04.040

17. Morse, E., et al.: Tolerancing: managing uncertainty from conceptual design to final product. CIRP Ann. **67**(2), 695–717 (2018). https://doi.org/10.1016/j.cirp.2018.05.009

18. Walker, W.E., et al.: Defining uncertainty: a conceptual basis for uncertainty management in model-based decision support. Integr. Assess. **4**(1), 5–17 (2003). https://doi.org/10.1076/iaij.4.1.5.16466

19. Pistikopoulos, E.N.: Uncertainty in process design and operations. Comput. Chem. Eng. **19**, 553–563 (1995). https://doi.org/10.1016/0098-1354(95)87094-6

20. Zonderland, M.E., Boucherie, R.J., Litvak, N., Vleggeert-Lankamp, C.L.A.M.: Planning and scheduling of semi-urgent surgeries. Health Care Manag Sci. **13**(3), 256–267 (2010). https://doi.org/10.1007/s10729-010-9127-6

21. Milliken, F.J.: Three types of perceived uncertainty about the environment: state effect, and response uncertainty. Acad. Manage. Rev. **12**(1), 133–143 (1987). https://doi.org/10.5465/amr.1987.4306502

22. Bradac, J.J.: Theory comparison: uncertainty reduction, problematic integration, uncertainty management, and other curious constructs. J. Commun. **51**(3), 456–476 (2001). https://doi.org/10.1111/j.1460-2466.2001.tb02891.x

23. Purdy, G.: ISO 31000:2009 - setting a new standard for risk management. Risk Anal. **30**(6), 881–886 (2010). https://doi.org/10.1111/j.1539-6924.2010.01442.x

24. Silva, A.A., Ferreira, F.C.M.: Uncertainty, flexibility and operational performance of companies: modelling from the perspective of managers. Acad. Manag. Rev. **18**, 11–38 (2017). https://doi.org/10.1590/1678-69712017/administracao.v18n4p11-38
25. Lin, L., Bao, H., Dinh, N.: Uncertainty quantification and software risk analysis for digital twins in the nearly autonomous management and control systems: a review. Ann. Nucl. Energy **160**, 108362 (2021). https://doi.org/10.1016/j.anucene.2021.108362
26. Mullins, J., Ling, Y., Mahadevan, S., Sun, L., Strachan, A.: Separation of aleatory and epistemic uncertainty in probabilistic model validation. Reliab. Eng. Syst. Saf. **147**, 49–59 (2016). https://doi.org/10.1016/j.ress.2015.10.003
27. Graves, S.C.: Uncertainty and production planning. In: Kempf, K.G., Keskinocak, P., Uzsoy, R. (eds.) Planning Production and Inventories in the Extended Enterprise. ISORMS, vol. 151, pp. 83–101. Springer, New York (2011). https://doi.org/10.1007/978-1-4419-6485-4_5
28. Angkiriwang, R., Pujawan, I.N., Santosa, B.: Managing uncertainty through supply chain flexibility: reactive vs. proactive approaches. Prod. Manuf. Res. **2**(1), 50–70 (2014). https://doi.org/10.1080/21693277.2014.882804
29. Zhu, S., Fan, W., Yang, S., Pei, J., Pardalos, P.M.: Operating room planning and surgical case scheduling: a review of literature. J. Comb. Optim. **37**(3), 757–805 (2018). https://doi.org/10.1007/s10878-018-0322-6
30. Escobet, T., Bregon, A., Pulido, B., Puig, V. (eds.): Fault Diagnosis of Dynamic Systems. Springer, Cham (2019). https://doi.org/10.1007/978-3-030-17728-7
31. Kennedy, M., O'Hagan, A.: Bayesian calibration of computer models. J. Roy. Stat. Soc. B **63**, 425–464 (2001). https://doi.org/10.1111/1467-9868.00294
32. Der Kiureghian, A., Ditlevsen, O.: Aleatory or epistemic? Does it matter? Struct. Saf. **31**(2), 105–112 (2009)
33. Begg, S.H., Welsh, M.B., Bratvold, R.B.: Uncertainty vs. variability: what's the difference and why is it important? (2014). https://doi.org/10.2118/169850-MS
34. Agnusdei, G.P., Elia, V., Gnoni, M.G.: A classification proposal of digital twin applications in the safety domain. Comput. Ind. Eng. **154** (2021). https://doi.org/10.1016/j.cie.2021.107137
35. Bouloiz, H., Garbolino, E., Tkiouat, M., Guarnieri, F.: A system dynamics model for behavioral analysis of safety conditions in a chemical storage unit. Saf. Sci. **58**, 32–40 (2013). https://doi.org/10.1016/j.ssci.2013.02.013
36. Varshney, K.R.: Engineering safety in machine learning. In: 2016 Information Theory and Applications Workshop (ITA), pp. 1–5 (2016). https://doi.org/10.1109/ITA.2016.7888195
37. Ritto, T.G., Rochinha, F.A.: Digital twin, physics-based model, and machine learning applied to damage detection in structures. Mech. Syst. Signal Process. **155**, 107614 (2021). https://doi.org/10.1016/j.ymssp.2021.107614
38. Karve, P.M., Guo, Y., Kapusuzoglu, B., Mahadevan, S., Haile, M.A.: Digital twin approach for damage-tolerant mission planning under uncertainty. Eng. Fract. Mech. **225**, 106766 (2020). https://doi.org/10.1016/j.engfracmech.2019.106766
39. Li, C., Mahadevan, S., Ling, Y., Wang, L., Choze, S.: A dynamic Bayesian network approach for digital twin. In: 19th AIAA Non-Deterministic Approaches Conference, American Institute of Aeronautics and Astronautics (2017). https://doi.org/10.2514/6.2017-1566
40. Dröder, K., Bobka, P., Germann, T., Gabriel, F., Dietrich, F.: A machine learning-enhanced digital twin approach for human-robot-collaboration. Procedia CIRP **76**, 187–192 (2018). https://doi.org/10.1016/j.procir.2018.02.010
41. Priyanka, E.B., Thangavel, S., Gao, X.-Z., Sivakumar, N.S.: Digital twin for oil pipeline risk estimation using prognostic and machine learning techniques. J. Ind. Inf. Integr. 100272 (2021). https://doi.org/10.1016/j.jii.2021.100272
42. Yan, Q., Wang, H., Wu, F.: Digital twin-enabled dynamic scheduling with preventive maintenance using a double-layer Q-learning algorithm. Comput. Oper. Res. **144**, 105823 (2022). https://doi.org/10.1016/j.cor.2022.105823

43. Zhang, M., Tao, F., Nee, A.Y.C.: Digital Twin Enhanced Dynamic Job-Shop Scheduling. J. Manuf. Syst. **58**, 146–156 (2021). https://doi.org/10.1016/j.jmsy.2020.04.008

44. Negri, E., Pandhare, V., Cattaneo, L., Singh, J., Macchi, M., Lee, J.: Field-synchronized Digital Twin framework for production scheduling with uncertainty. J. Intell. Manuf. **32**(4), 1207–1228 (2020). https://doi.org/10.1007/s10845-020-01685-9

45. Luo, D., Thevenin, S., Dolgui, A.: A digital twin-driven methodology for material resource planning under uncertainties. In: Dolgui, A., Bernard, A., Lemoine, D., vonCieminski, G., Romero, D. (eds.) APMS 2021. IAICT, vol. 630, pp. 321–329. Springer, Cham (2021). https://doi.org/10.1007/978-3-030-85874-2_34

46. Ghosh, A.K., Ullah, A.S., Kubo, A.: Hidden Markov model-based digital twin construction for futuristic manufacturing systems. AIEDAM **33**(3), 317–331 (2019). https://doi.org/10.1017/S089006041900012X

47. Iñigo, B., Colinas-Armijo, N., LópezdeLacalle, L.N., Aguirre, G.: Digital twin-based analysis of volumetric error mapping procedures. Precis. Eng. **72**, 823–836 (2021). https://doi.org/10.1016/j.precisioneng.2021.07.017

48. Wang, J., Ye, L., Gao, R.X., Li, C., Zhang, L.: Digital Twin for rotating machinery fault diagnosis in smart manufacturing. Int. J. Prod. Res. **57**(12), 3920–3934 (2019). https://doi.org/10.1080/00207543.2018.1552032

49. Sapronov, A., et al.: Tuning hybrid distributed storage system digital twins by reinforcement learning. Adv. Syst. Sci. Appl. 18(4), Art. nº 4 (2018). https://doi.org/10.25728/assa.2018.18.4.660

50. Cronrath, C., Aderiani, A.R., Lennartson, B.: Enhancing digital twins through reinforcement learning. In: 2019 IEEE 15th International Conference on Automation Science and Engineering (CASE), pp. 293–298 (2019). https://doi.org/10.1109/COASE.2019.8842888

51. Müller, M.S., Jazdi, N., Weyrich, M.: Self-improving models for the intelligent digital twin: towards closing the reality-to-simulation gap. IFAC-PapersOnLine **55**(2), 126–131 (2022). https://doi.org/10.1016/j.ifacol.2022.04.181

52. Alves de Araujo Junior, C.A., et al.: Digital twins of the water cooling system in a power plant based on fuzzy logic. Sensors **21**(20) (2021). https://doi.org/10.3390/s21206737

53. Luo, W., Hu, T., Zhu, W., Tao, F.: Digital twin modeling method for CNC machine tool, p. 4 (2018). https://doi.org/10.1109/ICNSC.2018.8361285

54. Sleiti, A.K., Kapat, J.S., Vesely, L.: Digital twin in energy industry: proposed robust digital twin for power plant and other complex capital-intensive large engineering systems. Energy Rep. **8**, 3704–3726 (2022). https://doi.org/10.1016/j.egyr.2022.02.305

55. Balta, E.C., Tilbury, D.M., Barton, K.: A Digital twin framework for performance monitoring and anomaly detection in fused deposition modeling. In: 2019 IEEE 15th International Conference on Automation Science and Engineering (CASE), Vancouver, BC, Canada, pp. 823–829 (2019). https://doi.org/10.1109/COASE.2019.8843166

56. Millwater, H., Ocampo, J., Crosby, N.: Probabilistic methods for risk assessment of airframe DT structure. Eng. Fract. Mech. vol. 221 (2019). https://doi.org/10.1016/j.engfracmech.2019.106674

Requirements for a Digital Twin
for an Emergency Department

Guillaume Bouleux[1], Hind Bril El Haouzi[2], Vincent Cheutet[1(✉)],
Guillaume Demesure[2], William Derigent[2], Thierry Moyaux[1],
and Lorraine Trilling[1]

[1] University Lyon, INSA Lyon, Univ. Jean Monnet Saint-Etienne,
Univ. Claude Bernard Lyon 1, Univ. Lyon 2, DISP UR4570, Villeurbanne, France
`{guillaume.bouleux,vincent.cheutet,thierry.moyaux,`
`lorraine.trilling}@insa-lyon.fr`
[2] University Lorraine, CNRS, CRAN UMR 7039, Nancy, France
`{hind.haouzi,guillaume.demesure,william.derigent}@univ-lorraine.fr`

Abstract. The Emergency Department (ED) represents the first stage
of the path for some patients in a hospital. The JUNEAU project aims
to propose a Digital Twin (DT) for the ED to both visualise the service
behaviour in quasi real-time, forecast and anticipate its behaviour to control the "Emergency throughput time" indicator. This DT will be centred
on the "patient pathway" view, which needs to first precise necessary data
and information and decision support. The project will tackle this issue
with a particular attention on exploring the DT architecture (with the
help or holonic architecture), on coupling centralised and decentralised
approaches to obtain a model as close as possible to the system dynamics
and on managing DT evolution in a complex environment.

Keywords: Digital Twin · Emergency Department · Holonic system ·
Agent-based modelling

1 Introduction

The Emergency Department (ED) is a strategic sector in the hospital chain. It
welcomes some patients as their first stage of care in the hospital. Nevertheless,
EDs have many limits, especially with the management of the flow of arrivals
which strongly determines patient pathway. Optimising the patients' pathway
is a topic of growing interest addressed through data-driven approaches [9] or
by integrating multi-agent systems into workflows [1]. Unfortunately, most published models are not implemented and used in EDs [5]. This lack of integration
into operational systems shows the limits of such models, which have a good
representation of flows and computing capacity, but are not connected to the
physical system in order to monitor and control it.

These flows and priority management issues have been widely studied in the
industrial field. In the current "Industry 4.0" dynamic, the approach based on

© The Author(s), under exclusive license to Springer Nature Switzerland AG 2023
T. Borangiu et al. (Eds.): SOHOMA 2022, SCI 1083, pp. 130–141, 2023.
https://doi.org/10.1007/978-3-031-24291-5_11

Digital Twins (DT) is known to meet a need for control as close as possible to the system and also to better anticipate behaviour thanks to the integration of simulation and artificial intelligence. A DT may be seen as an extension of the digital model of a product in the use phase. The notion of DT is currently being developed in the literature for production systems [23,28]. Nevertheless, Melese *et al.* [22] present challenges that are still open for the design and deployment of DT on production systems, including: difficulties in building, understanding and controlling systems with precise multi-scale models and reliable, the lack of knowledge and methodology on these models, the difficulty of predicting the behaviour of complex systems, the difficulty of accessing the relevant data, the lack of synchronisation between the physical and digital worlds and the difficulty of maintaining the models and check and test their validity.

The application of a DT to a healthcare system context is currently arousing great interest. The benefits of using DTs are nevertheless clearly affirmed and demonstrated in these works, such as the reduction in the waiting time of patients or their length of stay, more efficient and precise planning of resources and so on. However, the challenges cited in [22] remain open in this context, with specifics related to the nature of EDs.

Therefore, the JUNEAU project[1] aims to propose a DT for the ED allowing both to trace the data necessary to visualise the behaviour of the service in near real time and also to predict and anticipate its behaviour, by adding a simulation engine coupled to optimisation to respond to the hazards inherent in this type of service, in order to "enslave" and control the indicator "passage time for patients' emergencies". This indicator has been particularly observed by numerous research works on the modelling and simulation of emergency departments [27]. The objective of this paper is to discuss the requirements for such a DT and to present its architecture as proposed in the JUNEAU project.

The outline is as follows. Section 2 proposes an overview of the literature applied to ED organisation management on one hand, simulations and DT applied to ED on the other hand. Section 3 aims at synthesising the main requirements we identify for the DT architecture applied to ED. Section 4 gives details on specific scientific issues that will be tackle on this project and Sect. 5 will conclude.

2   State of the Art

2.1   ED Organisation and Management

A particularly interesting indicator for the ED is the "average waiting time in hospital emergency services". Our experience in the hospital world has shown that the waiting time in the emergency room is one of the symptoms of the establishment's maturity in the management of its processes, activity management and data management. This macro-indicator is multidimensional. Indeed,

---

[1] In French, JUNEAU stands for "JUmeau Numérique pour un sErvice d'Accueil des Urgences", *i.e.*, DT for an ED.

it is the reflection of the internal organisation of not only ED but also technical platforms (imaging, biology, operating room), related functions (stretcher among others) and intra-department organisation like procedure for taking care of patients in the ED, taking into account degrees of urgency.

It is also a mean of assessing the organisation of local primary care and the methods of recourse of the population to this care and to emergencies. The patient pathway in the ED can therefore be described (without being exhaustive) by Fig. 1.

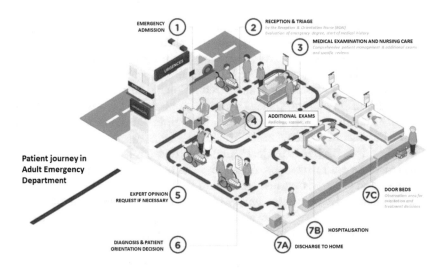

Fig. 1. Illustration of the patient's pathway in an ED (adapted from chl.lu)

The organisational complexity of the ED as well as its dependence on the availability of hospital beds could not be understood or explained by a study based only on the evaluation of static indicators. If we clearly perceive that the time of stay in the emergency room is one, even the parameter to be monitored for an effective management of the ED, it seems obligatory to consider this system in a dynamic form thus making it possible to model the decisions taken by the doctors, interns, external or other nurses, as well as their interactions.

For example, the current operation of the ED of Saint-Étienne University Hospital (France) is based on decentralised decision-making, generally based on priority management. Each caregiver selects his next patient based on the ED dashboard. As a first approximation, this choice is based on the level of severity of the patient: patients with severity levels 1 or 2 correspond to most urgent ones, where a doctor must be found as soon as possible, whereas patients with severity levels 3 to 5 will wait in the waiting room until a caregiver takes them by making a compromise (specific to each caregiver) between level gravity and order of arrival. More precisely, the different categories of caregivers all consult the same dashboard in which they select his next patient according to the care needed, the condition of the patient, etc.

In this context, the ED actors (doctors, nurses, patients and manipulators) are constantly brought to interact with each other and to learn from past experiences, thus influencing decision-making and the environment. In fact, the management of the patient journey in the ED is complex because, for example, the sequence of care is not known when the patient arrives, but is defined as the patient receives care and take exams. On the other hand, as described above, decision-making is highly decentralised as each caregiver selects his next patient based on the ED dashboard. For our project, it is therefore necessary to first clearly define the decision support that can be provided to each agent/carer.

2.2 Review of Simulations and Digital Twins for Healthcare Systems

Simulation has been widely reported in the literature as a means to improve patient pathways, especially in EDs. This literature primarily includes Discrete Event Simulation (DES) [4] and Agent-Based Simulation (ABS) [20], with possible hybridisation DES+ABS [16]. Yet all these works on simulation mostly propose "what-if" scenarios for improvement, whereas these results are offline and the input of data into simulation models is known to be a very difficult and tedious task. As stated in the introduction, the DT approach seems promising to further develop the use of simulations directly connected to physical systems.

Kritzinger *et al.* [18] define a DT as a digital system with an automatic data flow both from and to a Physical Twin (PT). In fact, we think that a DT is the convergence of an Information System (IS) and a simulation or, in other words, the concept of DT increases the dynamism of an IS by not only recording the past and present states of the PT, but also forecasting its future by simulation.

The application of DT in the context of healthcare systems is currently garnering keen interest. While some authors focus on developing a DT for specific patients to improve their health in interaction with healthcare systems [2], we focus here on developing a DT for the ED system alone. Few works are in this field, for instance [7,8,15]; hence, the novelty of our work. The benefits of using DTs are nevertheless clearly affirmed in the aforementioned works, such as the reduction of patients' waiting time or length of stay, the more efficient and precise planning of resources, and so on.

From [25], we understand that all previous work focuses on the fact that **simulation is the heart of a DT**, with data input in real-time from the PT through the internet of things. **Yet, no details on the architecture of DTs and the interactions with decision makers are provided.** Moreover, with the definitions proposed by Kritzinger *et al.* [18], one can say that simulation models described at the beginning at this section are **Digital Models** whereas the research works claiming to be Digital Twin are closer to **Digital Shadow**.

3   Requirements for a DT

Figures 2a, 2b and 2c recall how Kritzinger *et al.* [18] respectively define a digital model, a digital shadow and a DT, depending on how the digital system is connected to the physical system.

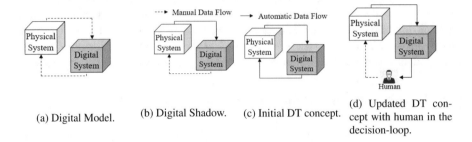

(a) Digital Model.   (b) Digital Shadow.   (c) Initial DT concept.   (d) Updated DT concept with human in the decision-loop.

Fig. 2. Typology adapted from Kritzinger *et al.* [18].

In these definitions, physical system and PT are synonymous, but a digital system may not be a DT. Moreover, we think that Fig. 2c deals with automation because the time delay is too short for humans to provide feedback. Since we want a DT for operational decisions – with a time range between minutes and weeks – we propose Fig. 2d as an evolution of Fig. 2c in order to include a human decision maker in the decision loop, because:

- *All models are false*: The digital system relies on a model which, by definition, is not a reproduction but a simplified representation of the reality. Some events occurring in PT (*e.g.*, failures, arrivals of patients, etc.) may be wrong or absent from the digital system, in which case the digital system has to resynchronize on the PT. On the other hand, if the digital system simulates events not present in the PT, then this system is wrong which, again, has to resynchronize on PT, not the other way around. This explains why the solutions generated by decision support systems (for planning, scheduling, etc.) are rarely applied to physical systems directly, and are often manually adapted beforehand.
- *Two systems cannot control each other*: If two systems control each other, then they need to do exactly the same thing, and may hence block each other and nothing happens in both of them. As a consequence of this point and the previous bullet points, PT has to be the master and the digital system a slave. This slave may suggest actions to PT, but cannot control it. This master/slave relationship complies with the next bullet point.
- *The digital system must be as non-invasive as possible*: This may not be the case for all kinds of production systems, but we assume that a healthcare system requires tools that suggest actions – rather than give orders – and these suggestions need to be updated whenever the digital system detects that they

are not followed in PT. In fact, care givers – especially highly qualified staff such as doctors – already make decisions and may thus sometimes disagree with the suggestions of the digital system.

When including a human decision maker in the decision loop, we aim at not only providing a framework making explicit all information needed by the decision maker to make his/her decision (such a situation corresponds to a Digital Shadow), but also at capturing the complete decision inside the system PT+DT.

Attached to this definition of DT, other requirements are necessary. A DT is based on a strong knowledge of the processes in order to reproduce the behaviour of the PT as accurately as possible, which can be provided in different formalisms like UML or BPMN (Business Process Model and Notation) [28]. A DT is also based on an exchange of real data which assumes (1) the existence of these data in the IS, (2) their consistency and (3) their usability. In addition to the Process, Resources and Organisation views recommended by [3], we plan to enrich our model with a data view.

Because of the nature of an ED, its DT must be seen as a decision support tool and it is up to the various decision-makers of the ED to take advantage of this environment. Thus we aim to integrate the Human into the loop. This ambition is reflected here in the desire to offer hybrid control, which will combine centralised and distributed control.

At the same time, the DT should be as non-invasive as possible. This may be the case for many other types of production systems. To this end, we assume that a care production system requires tools that suggest actions—rather than give orders—and which update these suggestions each time the DT detects that its suggestions are not valid. not enforced in the physical twin. Indeed, caregivers are already making decisions and therefore may sometimes disagree with the suggestions of the DT.

We do not wish in this project to limit ourselves to the design phase of the DT but to integrate a complete vision of the life cycle of such a system. A strong challenge for a system such as the ED is to be able to trust the DT in a highly dynamic, uncertain and evolving context. One of the ambitions of the project is therefore to propose an environment for controlling differences in twinning and management associated with both organisational and technical developments.

4   Main Scientific Questions

As previously stated, the JUNEAU project aims to propose a DT for an ED allowing to both trace the data necessary to visualise the behaviour of the department in near real time, and also predict and anticipate its behaviour by adding a simulation engine coupled to optimisation. As summarised in Fig. 3, such an ED should respond to the hazards inherent in this type of departments in order to control the indicator "time spent by patients in ED".

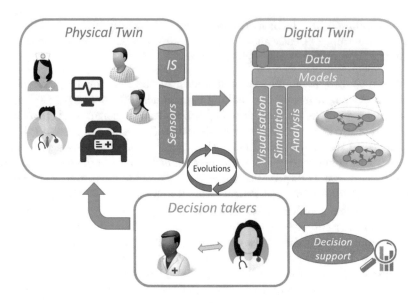

Fig. 3. Overview of Project JUNEAU.

4.1 DT Architecture

To meet the needs expressed above, we propose the digital system architecture of Fig. 4. It is based on the distinction between two types of events triggering the transitions of the digital system:

- External events go up from the physical system to cause a re-synchronisation of the digital system on the physical;
- Internal (i.e. inside the DT) events allow the operation of the digital system by causing it to simulate the possible future of the system.

Enabling transitions triggered by either or both of these event types causes our digital system to operate in one of three modes depicted in Fig. 4:

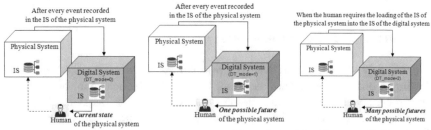

(a) Digital Shadow: The digital system only uses ds_* transitions.

(b) Synchronised DT: The digital system uses both ds_* and sim_* transitions.

(c) Exploratory DT: The digital system only uses sim_* transitions.

Fig. 4. The three operation modes of the digital system in Fig. 2d.

- **Digital Shadow** (Fig. 4a): In this mode, only transitions triggered by external events (i.e. `ds_*` ones) are activated in the digital system. Therefore, it only displays to the human decision maker the current state of the physical system or, in other words, this mode makes it work like a dashboard.
- **Synchronised DT** (Fig. 4b): Transitions triggered by both transition types are enabled in this mode of the digital system. Consequently, the internal events cause it to simulate in accelerated mode a possible future of the physical system, until an external event re-synchronises the digital system on the present state of the physical system, then the internal events start again to simulate the future in accelerated mode starting from the new present state of the physical system until the arrival of the next external event and so on. In summary, a large number of internal events simulate the future between two external events and these external events re-synchronise the digital system with the physical one.
- **Exploratory DT** (Fig. 4c): This mode only enables transitions triggered by internal events (i.e. `sim_*` ones). It therefore requires initialising the digital system on the current state of the physical one, then the internal transitions simulate the future of the physical system in accelerated mode without re-synchronisation at each external event going from the physical to the digital. This allows the decision maker to explore the variability of possible futures without being bothered by the regular re-synchronisations of the Synchronised DT mode.

Each of these operating modes requires the support of a detailed architecture, capable (1) of representing the behaviour of all the elements of the system, (2) of supporting the interaction with the various IS and physical elements. of the hospital and (3) to interact with human decision-makers, in a way that is generic enough to support scaling. For this purpose, Project JUNEAU relies on the holonic paradigm initially proposed by Koestler [17]. Holonic approaches are well suited for the representation of production or even hospital systems, as shown in [13] on the adaptation of holonic systems to the drug circuit of a hospital. In the current state of the project, we imagine adapting a classic holonic architecture to digitise the ED. Compared to the literature, this architecture will propose innovations in the sense that it (1) will integrate the Human as a decision-maker and actor and (2) will have to allow the simulation of the studied system (through a specific holon for example, such as in the PROSIS approach [26]) for the realisation of what-if scenarios, used in modes (b) and (c) of our DT.

4.2 Hybrid and Non-intrusive Control

The architecture of the digital system depends on the organisation of the PT. Given the specificities of our PT, our second scientific objective is the proposition of a hybrid control by coupling centralised approaches (often modelled as NP-hard problems) and decentralised approaches (*e.g.*, based on multi-agent or holonic approaches, or some variant of game theory) in order to address the issue of the patient pathway. This will allow us to aim at the combination of the best of both approaches: the efficiency of centralisation when there are no disturbances and the reactivity and resilience of the decentralisation to react after each disturbance [6]. Our hypothesis is that the application of this coupling to the management of patient pathway in an ED will make it possible to effectively manage this flow while coping with unpredicted events.

We plan to study the coupling of a centralised and a decentralised organisation to create a hybrid one able to switch between its centralised and decentralised modes. Switching from centralised to decentralised decision-making is easy because the agents start from the latest decisions received from the CA (Central Authority) as the starting point of their interactions. These inter-agent interactions correspond to the operation of the decentralised mode and represent the discussions of the agents agreeing on who does what and when. On the contrary, the switch in the other direction is more complicated because moving from decentralised decisions to centralised decisions requires that the CA takes into account what the agents are currently doing. For example, if the agents have agreed in a decentralised way to do tasks 1 to n for the next ten minutes, then (i) the CA has ten minutes to propose a new plan to the agents, (ii) this plan must take account of the fact that tasks 1 to n will have been carried out or will be in progress during the switch back to centralised decision-making and (iii) the agents will not all make this switch at the same time, but after everyone has completed its task (in the set of tasks 1 to n). In other words, the dynamic nature of the system to be controlled – *i.e.*, PT – is one of the main difficulties to be managed in the hybridisation. This is due to the fact that the dynamics of PT causes the model used by CA (in the digital system) to be late in comparison to the actual state of the caregivers in PT. Indeed, the decision support provided to CA at a given time is based on a previous state of the caregivers. During the duration of time between the evaluation of the state of ED and the decision support offered by CA, the (decentralised) decision-making carried out by the caregivers biases the perception that the CA has of the state of the ED and, consequently, may render the CA proposals inapplicable.

Then, the feedback from local systems (caregivers in the context of the ED) is another lever allowing hybridisation. This feedback raises questions about the quantity or volume of information to be reported, its delay, or even the absence of feedback and the consequences on the decision support offered by the CA.

Finally, the human dimension must necessarily be considered in a system in which the human is at the centre of the decision. For example, the acceptability of the system by humans (not taking into account and/or not questioning the decisions proposed, not reporting information, etc.), the human's understanding of the decisions proposed by the human or taking into account the autonomy of caregivers remain scientific obstacles that are difficult to resolve, and not only within the ED but also in other types of production systems.

4.3   DT Evolution Management

A complex system such as a DT evolves during its life, as much by the specific dynamics of the physical twin as by the evolution of the modelling hypotheses of the DT. The question of mastering such a system is strong, especially if we take into account the application system concerned.

Thus, through the study of the interactions between all the agents of the system and the understanding of the associated dynamic system, it is possible to obtain a modelling and organisational control. Organisational morphology should be understood as the representation of the global state of the multi-agent system in the form of a representation in a geometric space [14]. The evolution of this representation makes it possible to quantify, through local interactions between agents, a global behaviour known as organisational. An application of this approach to a crisis situation has been proposed by Lachtar [19]. This idea makes it possible to approach modelling in an unusual form and linked to physics. Indeed, a certain number of physical approaches - notably those related to the re-normalisation group - make it possible to infer organisational behaviour. From a less dynamic system but more algebraic point of view, many works have focused on a characterisation most often via a spectral study of the Laplacian matrix of the graph representing the interactions between agents. We wish here to give geometric and/or topological invariants as they were introduced for example by [21] to characterise the cooperative system between agents in their decision-making.

On the other hand, to take advantage of this massification of data and ensure the evolution of DT over time, it is crucial to update the simulation models. In this context, the DDDS (Dynamic Data Driven Simulation) paradigm [10] interests us. The principle is that the simulation is continuously compared to the data of the real system in order to ensure its accuracy with respect to the real system and to obtain a better quality of prediction and analysis. DDDS approaches address two issues related to this principle: (1) when to perform the update and (2) how to perform it? Updating the simulation model requires evaluating the difference between the real and simulated systems at an instant t in time. This evaluation can for example be done via an adaptation function [11] or a control chart [24]. When this difference becomes significant, the update of the model can be done according to several modes. This can be (1) the reconstruction of a new simulation model if it was obtained by "black box" approaches of the neural network type or (2) the change of the parameters of the initial model if this one is made up of known primitives [12], via the assimilation of data reported from the field. However, this second type of approach is less able to automatically evolve the topology of the model. We therefore wish to propose a methodological approach ensuring the accuracy over time of the simulation models proposed by the JUNEAU Project, according to a procedure coupling adaptation of the simulation model by "white box" methods as long as this remains possible, then by "black box" in order to offer the user new topologies when the initial one is no longer representative for the real system.

5   Conclusion

The Emergency Department is a complex organisation that needs improved ways to be managed and controlled in order to enhance the patient pathway. Digital Twin, defined as a digital, accurate replication of a physical system with decision support capabilities for a set of stakeholders seems to be a promising paradigm to tackle this issue. The JUNEAU project's aim is to contribute to the digital transformation efforts of the hospital by providing key constituting elements of such DT, in collaboration with a University Hospital. The expected results are various and with different dimensions: methodological, technical and economic.

Acknowledgement. The JUNEAU Project is going to be supported by the Agence Nationale de la Recherche of the French government between October 2022 and October 2026.

References

1. Ajmi, F., Zgaya, H., Othman, S.B., Hammadi, S.: Agent-based dynamic optimization for managing the workflow of the patient's pathway. Simul. Model. Pract. Theory **96**, 101935 (2019)
2. Angulo, C., Ortega, J.A., Gonzalez-Abril, L.: Towards a healthcare digital twin. In: Artificial Intelligence Research and Development, pp. 312–315. IOS Press (2019)
3. Augusto, V., Xie, X.: A modeling and simulation framework for health care systems. IEEE Trans. Syst. Man, Cybern. Syst. **44**(1), 30–46 (2013)
4. Ben-Tovim, D., Filar, J., Hakendorf, P., Qin, S., Thompson, C., Ward, D.: Hospital event simulation model: arrivals to discharge-design, development and application. Simul. Model. Pract. Theory **68**, 80–94 (2016)
5. Boyle, L.M., Marshall, A.H., Mackay, M.: A framework for developing generalisable discrete event simulation models of hospital emergency departments. Eur. J. Oper. Res. **302**(1), 337–347 (2022)
6. Cardin, O., Trentesaux, D., Thomas, A., Castagna, P., Berger, T., Bril El-Haouzi, H.: Coupling predictive scheduling and reactive control in manufacturing hybrid control architectures: state of the art and future challenges. J. Intell. Manuf. **28**(7), 1503–1517 (2017)
7. Chase, J.G., et al.: Digital twins in critical care: what, when, how, where, why? IFAC-PapersOnLine **54**(15), 310–315 (2021)
8. Croatti, A., Gabellini, M., Montagna, S., Ricci, A.: On the integration of agents and digital twins in healthcare. J. Med. Syst. **44**(9), 1–8 (2020)
9. Curtis, C., Liu, C., Bollerman, T.J., Pianykh, O.S.: Machine learning for predicting patient wait times and appointment delays. J. Am. Coll. Radiol. **15**(9), 1310–1316 (2018)
10. Darema, F.: Dynamic data driven applications systems: a new paradigm for application simulations and measurements. In: Bubak, M., van Albada, G.D., Sloot, P.M.A., Dongarra, J. (eds.) ICCS 2004. LNCS, vol. 3038, pp. 662–669. Springer, Heidelberg (2004). https://doi.org/10.1007/978-3-540-24688-6_86
11. Frazzon, E.M., Kück, M., Freitag, M.: Data-driven production control for complex and dynamic manufacturing systems. CIRP Ann. **67**(1), 515–518 (2018)
12. Goodall, P., Sharpe, R., West, A.: A data-driven simulation to support remanufacturing operations. Comput. Ind. **105**, 48–60 (2019)

13. Huet, J.C.: Proposition d'une méthodologie de réingénierie pour le contrôle par le produit de systèmes manufacturiers: application au circuit du médicament d'un hôpital. Ph.D. thesis, Université Blaise Pascal-Clermont-Ferrand II (2011)
14. Itmi, M., Cardon, A.: Model for self-adaptive SoS and the control problem. In: 2010 5th International Conference on System of Systems Engineering, pp. 1–6. IEEE (2010)
15. Karakra, A., Fontanili, F., Lamine, E., Lamothe, J., Taweel, A.: Pervasive computing integrated discrete event simulation for a hospital digital twin. In: 2018 IEEE/ACS 15th International Conference on Computer Systems and Applications (AICCSA) (2018). https://doi.org/10.1109/AICCSA.2018.8612796
16. Kaushal, A., et al.: Evaluation of fast track strategies using agent-based simulation modeling to reduce waiting time in a hospital emergency department. Socioecon. Plann. Sci. **50**, 18–31 (2015)
17. Koestler, A.: The Ghost in the Machine. Hutchinson, London (1967)
18. Kritzinger, W., Karner, M., Traar, G., Henjes, J., Sihn, W.: Digital twin in manufacturing: a categorical literature review and classification. IFAC-PapersOnLine **51**(11), 1016–1022 (2018)
19. Lachtar, D.: Contribution des systèmes multi-agent à l'analyse de la performance organisationnelle d'une cellule de crise communale. Ph.D. thesis, Ecole Nationale Supérieure des Mines de Paris (2012)
20. Liu, Z., Rexachs, D., Epelde, F., Luque, E.: An agent-based model for quantitatively analyzing and predicting the complex behavior of emergency departments. J. Comput. Sci. **21**, 11–23 (2017)
21. Markdahl, J.: A geometric obstruction to almost global synchronization on riemannian manifolds. arXiv preprint arXiv:1808.00862 (2018)
22. Melesse, T.Y., Di Pasquale, V., Riemma, S.: Digital twin models in industrial operations: a systematic literature review. Procedia Manuf. **42**, 267–272 (2020)
23. Negri, E., Fumagalli, L., Macchi, M.: A review of the roles of digital twin in CPS-based production systems. Procedia Manuf. **11**, 939–948 (2017)
24. Noyel, M., Thomas, P., Thomas, A., Charpentier, P.: Reconfiguration process for neuronal classification models: application to a quality monitoring problem. Comput. Ind. **83**, 78–91 (2016)
25. Patrone, C., Galli, G., Revetria, R.: A state of the art of digital twin and simulation supported by data mining in the healthcare sector. In: Advancing Technology Industrialization Through Intelligent Software Methodologies, Tools and Techniques, pp. 605–615. IOS Press (2019)
26. Pujo, P., Broissin, N., Ounnar, F.: PROSIS: An isoarchic structure for HMS control. Eng. Appl. Artif. Intell. **22**(7), 1034–1045 (2009)
27. Salmon, A., Rachuba, S., Briscoe, S., Pitt, M.: A structured literature review of simulation modelling applied to emergency departments: current patterns and emerging trends. Oper. Res. Health Care **19**, 1–13 (2018)
28. Semeraro, C., Lezoche, M., Panetto, H., Dassisti, M.: Digital twin paradigm: a systematic literature review. Comput. Ind. **130**, 103469 (2021)

A Digital Twin System to Integrate Data Silos in Railway Infrastructure

G. Christiaan Doubell[1], Anton H. Basson[1], Karel Kruger[1(✉)],
and Pieter D. F. Conradie[2]

[1] Department of Mechanical and Mechatronic Engineering, Stellenbosch University,
Stellenbosch, South Africa
kkruger@sun.ac.za

[2] Department of Industrial Engineering, Stellenbosch University, Stellenbosch, South Africa

Abstract. Many contexts can be found where data reside in silos, which presents challenges for supporting decision making that relies on data from multiple silos. Digital twins (DTs) have the potential to integrate heterogeneous data sources and overcome these challenges, but little has been published in this regard. This paper proposes a DT system architecture aimed at DTs that integrate data from different data silos. The architecture combines a hierarchy of DTs and a services network. A case study in railway infrastructure serves as an implementation example for preliminary evaluation.

Keywords: Digital twin · Aggregation · Data silos · Software architecture

1 Introduction

Complex systems generate large amounts of data [1], but often this is associated with segregated groups of data, likely stored in multiple enterprise applications [2]. These segregated groups of data can be referred to as "data silos" and impair an organisation's ability to utilise this data for effective management [2–4]. A 2019 study showed that data silos (or the lack of findable, accessible, interoperable and reusable data) invokes costs to the European economy of at least €10.2 billion per year [5].

The concept of digital twin (DT) in the context of the built environment (and thus also of railway infrastructure), can be defined as *a digital representation of a physical entity that integrates static and dynamic data from various sources*, such as that acquired from building information modelling (BIM) and condition monitoring systems [6]. Even though the integration of *data silos* through DTs are often implied in implementations in literature, it is mostly portrayed as a means to *prevent the creation of more data silos in the future* rather than a means to *integrate existing data silos* [1, 2].

Even though data integration is one of the most mentioned requirements for DTs in literature [7], papers on the explicit integration of data silos are hard to find – as opposed to the integration of data sources that are just heterogeneous in nature (such as temperature, acceleration and GPS data). Though similar, the issue of data silos revolves around the accessibility of the data for various use cases, instead of the nature (or degree

© The Author(s), under exclusive license to Springer Nature Switzerland AG 2023
T. Borangiu et al. (Eds.): SOHOMA 2022, SCI 1083, pp. 142–153, 2023.
https://doi.org/10.1007/978-3-031-24291-5_12

of heterogeneity) of the data itself. Further, a recent literature review on DTs suggests that one of the barriers to the advancement of the DT applications is the lack of publicly available DT implementation details [8].

This paper contributes to addressing these two research gaps by presenting a DT system architecture that can be implemented not only to integrate data from data silos, but to also provide a digital representation of a physical system that remains effective over an extended period. The architecture can adapt to different and dynamic stakeholder needs as use cases are added in a typical industrial digitisation context [7].

The railway infrastructure domain is used as context for this paper because this is a typical domain where such data silos are encountered – with a large and complex system to be managed, with concerns such as maintenance scheduling and implementation, infrastructural inspections, asset operation and supply chain, all generating data that pertain in some way to the infrastructure at hand.

The paper presents a brief literature review in Sect. 2, followed by the main contribution in the form of a DT system architecture (in Sect. 3). The architecture is then verified through implementation (presented in Sect. 4) and a preliminary evaluation thereof is presented in Sect. 5.

2 Literature Review

Grieves and Vickers, early authors on the concept of the DT [9] present organisational data silos as "probably the biggest obstacle to the Digital Twin", since a DT "requires a homogeneous perspective of this information that persists across functional boundaries" [9]. This homogenisation (or integration) of data silos is a general issue, with literature offering solutions such semantic profiling [10], Big Data Lakes [2], and mobile data management platforms [11]. However, very little has been published on the integration of data silos *with DTs*. Here the challenge is to not just create another data silo, or a so-called "solution silo" [3, 7].

Researchers from the ABB AG Corporate Research Centre in Germany presented a series of articles proposing a cloud-based architecture with a common information meta-model for defining the DTs [3, 7, 13]. The presented architecture provides a variety of APIs with which DT data can be requested from the DTs, over the entire lifecycle of the physical entity. The architecture focusses on the integration of data from the different silos, so as to have sensible, scalable and customisable relations between the data. However, the solution only supports the HTTP/REST protocol for communications. Data sources that require other interfaces, such as an SQL database or a local Microsoft Excel file, will need to be converted to a format that the given system can ingest data using the HTTP/REST protocol.

The architecture proposed in this paper expands on the reference architecture of [1], and the "Six Layer Architecture for Digital Twins with Aggregation" (SLADTA) [13]. The reference architecture presented by [1] was developed for a system of DTs with a services network (SN). The SN presented by [1] is a refinement of that of [14]. The reference architecture does not explicitly consider data silos, but the design framework that it forms a part of provides guidance for interfacing with proprietary technologies, retrofitting, different levels of technological maturity and integrating with existing information systems.

In SLADTA, the internal architecture of a DT comprises six layers: Layer 1 refers to the physical measurement devices, Layer 2 to the local controllers/data sources and Layer 3 to local (or short-term) data repositories. Layers 4, 5 and 6 refer to the IoT Gateway, cloud-based (or long term) data repositories and services, respectively.

3   Proposed Digital Twin System Architecture

This section presents an overview of the architecture and a description of the components of the architecture. The architecture's data integration roles are then discussed.

3.1   Overview

The architecture, illustrated in Fig. 1, is derived from the reference architecture provided by [1] with the internal architecture of the DTs according to SLADTA [13]. The architecture comprises four component groups: *management services* (MSs), the *services network* (SN), the *DT hierarchy* and the *wrapper layer* (WL). The first three are inherited from the similarly-named component groups proposed by [1]. The WL was added for interfacing with data silos, as will be discussed in Sect. 3.3.

The major differences to the reference architecture presented by [1] are the addition of the WL and the removal of some distinctions between DT instances (DTIs) and DT aggregates (DTAs). The WL contains wrappers for each external data source (the silos) and the wrappers provide a consistent interface for DTs, irrespective of the details of the data source. In the presented architecture, a DT's behaviour is consistent regardless of where it is situated in the hierarchy (irrespective of whether it can be considered to be a DTI or a DTA) – any DT gets data from sources (either wrappers or other DTs) and provides data to system components (DTs or services in the SN) that request it, through services on its Layer 6. A DT's placement in the hierarchy is primarily a reflection of how it fits into a physical decomposition of the system.

3.2   Architectural Components

Data Sources

The *data sources*, existing in different data silos, exist outside the scope of the architecture. This is because these data sources already exist in the form of either proprietary systems or established organisational systems, which are managed and maintained for the purpose they were created for. It is often impossible or undesirable to alter these data sources. This architecture does not require any modifications to the data sources, but only a known interface to them. If, for example, a data source provides data through a web API, the IP address, authentication information and the message ontology for that web API would be required.

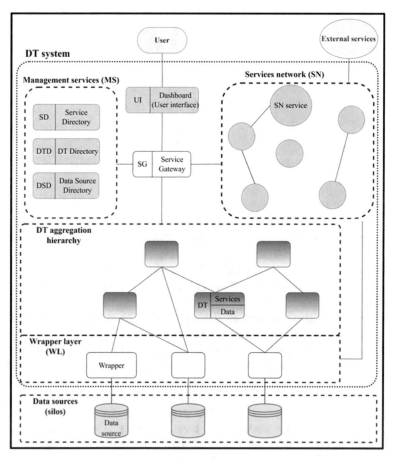

Fig. 1. General DT architecture for integrating data silos

Wrappers

The *wrapper* components in the WL fulfil the role of providing a consistent interface to the data sources, through a custom-developed wrapper for each data source. The wrappers' internal architecture is illustrated in Fig. 2. These components translate the unique API (or interface) of the data source to a common API for the DT system, so that SN services and DTs could request data from a wrapper.

The wrapper filters the data requested from the bulk of data that exist in the data source before transmitting the data to the component that requested it. In addition, the wrapper also monitors the data source for new data and, when it is detected, the wrapper notifies its connected components that rely on it for data provision. Queries to the data source often involve large data transfers, incurring significant latencies. The wrapper therefore contains a temporary, local storage component, to reduce the number of repeated queries that need to be made to the data source.

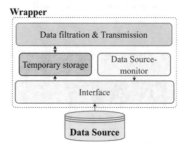

Fig. 2. Internal architecture of the wrapper component

SN Services and External Services

SN services (distinct from DT services, which will be discussed later) are implemented as microservices in a service-oriented architecture (SOA). These services perform functions that use data provided by DT services. SN services can request DT services, as well as other SN services, and are not limited to the number of requests they can make, i.e., a SN service can request multiple DT and SN services in the process of performing its own function/service. In general, SN services do not store data for significant periods, but rather use the data or services available in the DT hierarchy or, if the data is not representing a physical system element, an external service.

External services are services that are offered by other systems (external to the DT system), such as weather forecast services, financial systems or inventory systems. Such data is not provided by the DT hierarchy because the data is unrelated to the physical entities represented by the DTs. These external data sources can either be interfaced directly from the SN service, or through a wrapper that can be created for that data source (which will located in the WL). The latter option makes the data source available to both the DT hierarchy and services in the SN. However, the former option might often be faster if a limited interface is required.

Digital Twin Hierarchy

The DT components in this architecture are responsible for providing the digital representation of the physical entities. Each DT stores the relevant data that it obtains from the wrappers to ensure the availability of the data regardless of the data sources' characteristics. Therefore, DTs contain the data related to the physical entity they represent. The DTs also provide services to make that data available to other software components that require this data. *Data* here also includes static data relevant to the DT, such as its attributes (e.g., the date of entity installation, the entity's OEM, etc.). This information is not necessarily obtained from data sources through the wrapper, but allocated to the DT upon component start-up from a configuration file.

The internal architectural design of the DTs is based on SLADT. However, SLADT was developed for a context where the DTs are built "from the ground up" – from the sensors (Layer 1) to services (Layer 6). Data silos typically encapsulate the functions of the first few layers of SLADT already and, in the proposed DT system architecture, DTs obtain data from sources that have already stored and/or aggregated that data, and not from the sensors in Layer 1. The storage can be considered to be Layer 3, if the long-term availability of the data is not certain, or Layer 5 if the data will be available long term and the access latency is acceptable. However, as will be explained in Sect. 3.3, the integration of data silos would require some form of (at least temporary) storage to be used for data management.

It is important to note that, as in the original design of SLADT, data from the data sources is gathered into the DT through the IoT Gateway (Layer 4 in SLADT). In contrast to the original SLADTA, the rest of the DT system can only gain the information from a DT through its services on Layer 6. No provision is made in this architecture for aggregation data flows through Layer 4. Both Layers 4 and 6 would most probably require manipulations to the data – Layer 4 for calculating new parameters to store, performing data cleaning, etc. before storage and Layer 6 to perform requested services.

The services performed in Layer 6 of the DT are subsequently referred to as *DT services* (as opposed to SN services). These services transform the data stored in, or acquired by, the DT to information (e.g., by adding context) and provide a port through which this information can be retrieved.

A DT can gather data from more than one wrapper. From Layer 4, the data is stored in either or both short-term and long-term storage (Layer 3 and 5). From here the data can be accessed for use in service provision to requesting components. This Layer 3 and/or Layer 5 is what is designated as *Data* in the DT representation in Fig. 1. *Services* communicate the Layer 6 functionality.

Since DTs are the components responsible for the provision of information relating to the physical entities represented in this system, they would rarely be removed from the system. If, for example, a physical entity is removed from the actual system (e.g., as result of routine replacement), that DT component should stay within the system to ensure that the data pertaining to that asset will remain available. This DT will not receive new data, but will still be able to provide the data it obtained during its lifetime.

Management Services

The role of the *management services* in the general architecture (Fig. 1) is to manage the interactions between components in the DT system. The management services in this architecture are all *directory services* that keep record of the existing components in the system and connection information (e.g., IP address and port number).

The Digital Twin Directory (DTD), Data Source Directory (DSD) or SD (Services Directory), respectively, keep track of the DTs, wrappers (data sources) and SN services that exist within the DT system for use in component interactions. To indicate their availability to other system components, the WL and SN components in the DT hierarchy are required at start-up to register themselves (ID and information needed for interaction, such as socket address and authentication data) in their respective directories.

Service Gateway and Dashboard
The final two components - the *Service Gateway* (SG) and Dashboard (*UI*) are implementation-specific components. The former is responsible for directing all messages between components in different groups, while the latter is responsible for providing the user interfaces for the system.

Data Integration
The challenges associated with automatic data integration between silos are:

1. Using the data from the data silos without compromising the performance of existing applications that are associated with that data silo.
2. Creating and maintaining the relationship(s) between data records found in different data sources.
3. Handling inconsistent features when data was originally intended for different purposes, e.g., asset locations (containing static, latitude and longitude values), versus asset operational parameters (containing dynamic technical parameters).
4. Handling inconsistent storage patterns, e.g., when data pertaining to the same parameter is stored as different data types in different sources, or using slightly different fieldnames (e.g., stored as an integer in data source A, but as a string in data source B; or stored as "Track Gauge" in data source A, but as "TRACK_G" in data source B).
5. Managing different interfaces to the data (e.g., REST API, versus SQL).
6. Handling uncontrolled changes to the data source(s) (e.g., the proprietary interface changes, or the internal structure/field names changes, etc.).

The wrappers play a key role in addressing these challenges because they are the only parts of the architecture that interface directly with the data silos. The silos can differ substantially and, therefore, each data source's wrapper is likely to require custom software development. To limit the custom development, the wrappers are kept simple: they receive a request from the *requestor* (a DT or service in the SN) for a specific parameter for a specific time period, withdraw or filter the requested data from their source, homogenize the data to the format expected by the requestor and pass the data to the requestor. These data sets will typically each be a table of timestamps with one other parameter corresponding to the time stamp. The wrappers are responsible for filtering the requested data from the source, but not for integrating different parameters. The wrappers therefore prioritise challenges 1, 4, 5 and 6 mentioned above.

The DTs are intended for handling challenges 2 and 3, because the DTs encapsulate the required domain knowledge. DTs therefore will be able to integrate different parameters from one data source or from different sources, and handle of inconsistent features and data types, before storing the data in the DT's long-term repository.

As an example, consider a DT that represents railway track starting 10 km from the main station and extending for 0.4 km. The DT is configured to store the data that it receives from wrappers in multiple data tables in a relational database. The DT requires two different types of data, i.e., the condition of the track and the coordinates of the track. One wrapper is responsible for providing the DT with both the condition and the coordinate data. Integration will initiate when the wrapper notifies the DT that new data has been received. The DT will first check whether it is configured to receive coordinate data from this wrapper. If it is, the DT will ask the wrapper for *only the data fields related to its own coordinates* (it will ask for the latitude and longitude values where the track is between 10–10.4 km from the main station). The wrapper will filter for these records and provide it to the DT. When the DT receives the data, it knows that the data it just received is coordinate data, and since the communication from the wrapper has a known structure, the DT can correctly store the data in its appropriate data table according to a predefined set of operations. The predefined set of operations will ensure, inter alia, that the records are not duplicates of data that is already in the data table. The whole process is then repeated for condition data and the DT is responsible for integrating the condition data with the relevant locations of the track.

The above process can be characterised as integration upon data ingestion. However, integration can also occur upon service provision. The second scenario is where data is not obtained from a wrapper, but from another source such as an external service, a DT service or a SN service (e.g., when a SN service requests data from multiple DTs for the purpose of performing some function on that data). In these cases:

1. The *requestor* (typically a service in the SN) will request the data that it needs from the *provider* (DT with a service that can provide the data or another service in the SN). The requestor will have to select a service offered by the provider that will deliver all the data the requestor requires, even if the service provides other data that is not required by the requestor.
2. The provider receives this request, performs the service and provides the resulting data/information to the requestor.
3. After the requestor receives the data, the requestor has to filter the data it requires from the data received and relate the data records to one another through a set of operations. The code to execute these types of integrations would only need to be implemented within the script(s) that execute the specific service.

Both types of integrations require regular inspection to ensure that the data offered by the providers are still consistent with what was anticipated at the time of implementation.

4   Case Study: Rail Infrastructure Maintenance Management

As a case study, the architecture presented in Sect. 3.1 was implemented for the electrical and permanent way infrastructure between Cape Town main station and Claremont station in the Western Cape, which is managed by the Passenger Rail Agency of South Africa (PRASA). Due to confidentiality considerations and page restrictions, details of the implementation cannot be given here, but the use of the architecture is briefly outlined.

DTs for the individual rails, for two sections of catenary wire (between Cape Town and Salt River, and Salt River and Claremont), and for the complete track and ballast infrastructures, were included. The Track DT gets data related to the geometry of the track (such as track gauge, *superelevation*, etc.) from one wrapper, while also being able to request other data (such as crown wear) from the individual rails that form part of the DT from another wrapper. Figure 3 illustrates a part of the implementation. The remainder of the architecture is omitted because it closely resembles Fig. 1. The only implementation feature not shown in Fig. 3 is that the only service implemented in the SN was a service that counts the number of exceedances of a quality metric in a selected part of the rail network.

The entire DT system, except for the UI component, was implemented as a Python program, packaged with all of its dependencies into a Windows executable file. All the components, except for the SN service (which is used as is), are instances of their class – each DT an instance of the DT class, each wrapper an instance of the wrapper class, each directory service an instance of the directory service class, and the SG an instance of its own class. Note that these were all design choices and the system should have worked the same if a different class/script was written for each wrapper.

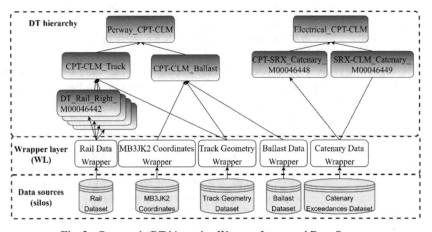

Fig. 3. Case study DT hierarchy, Wrapper Layer and Data Sources

A Python *entry point* program reads a JSON configuration file, which specifies all of the components of the DT system, except for UI, with their constructor arguments. Each component is then run as a separate thread on the computer that the entry point program is executed. The UI is implemented as a reactive web application, created with the Svelte web framework. With each component running in its own thread, communication between the components was done using TCP/IP sockets, implemented using the event-driven Socket.io library (*socket.io* in the Python components, and *socket.io-client* in the UI).

The DTs were all implemented with the long-term storage using PostgreSQL. This database was selected since there is no cloud backup for the data in the data sources yet. This presents a danger of data loss in the case of theft, data corruption, damage, etc. and thus the design decision was made to duplicate the data from the sources within the DT system.

The data sources in the case study were Microsoft Excel files that were stored on a hard drive. To simulate the nature of the problem addressed with this architecture, some of the MS Excel files were hosted as data on web servers from which their data can be obtained using REST APIs, while others were migrated to a PostgreSQL database, through which the data should be accessed with authentication, using SQL queries.

5   Preliminary Evaluation

Considering the architecture as presented in Sect. 3 and the case study presented in Sect. 4, this section presents a preliminary evaluation of the proposed DT system architecture.

To fulfil its intended role, the DT system should maintain an up-to-date digital representation of physical reality (to the extent possible with the available data) that can provide the user with information relating to that reality upon request. Because complex systems normally evolve over time, the architecture should be adaptable to changing contexts, such as changes in the configuration of the physical infrastructure, new data sources, or changing user information needs. Complex systems also have various stakeholders, each with their own data requirements. The DT system should thus be able to service different needs and changes to one stakeholder's needs should this not affect the services provided to another stakeholder. A DT system with such capabilities will be able to integrate between silos without itself becoming a solution silo.

The combination of a DT hierarchy and SN hold particular advantages for the above. The DT hierarchy provides an intuitive subdivision of concerns and data where they relate to physical system elements, because physical systems can usually be decomposed (disassembled) hierarchically. The range of DT system components that are affected by any change in the physical system can intuitively be determined, thereby simplifying the reconfiguration of the DT system to reflect changes in the physical system. In the DT hierarchy in Fig. 1, a strict one-to-many relationship is not enforced. This permits for different aggregation pathways to allow for different stakeholders' concerns to be better managed. For example, one aggregation can be based purely on physical proximity, irrespective of the type of asset, while another aggregation can consider only a specific type of asset.

The provision of software services through both the DT hierarchy and the SN allows the service to be hosted where it is the easiest to maintain. If the service only requires the data of a particular DT, then it is best to let that DT host the service. Otherwise, the service is placed in the SN to allow maximum flexibility. Similarly, keeping the functions that gather and prepare information for the users separate from the dashboard generation simplifies changing either the one or the other, as the needs of users or the users themselves change.

Another important feature to note is that each of the components (particularly the DTs and the services in the SN) operate as separate *holons* – autonomous and independent of the other components. These holons are loosely coupled and co-operative for the purpose of accomplishing the goal set out for the system by sending or responding to asynchronous messages containing data, information, requests, etc. The holonic behaviour also can be exploited by letting the different components of the DT system run as different processes (on the same or different computational hardware) or as different threads on the same hardware. The system components can therefore easily be distributed across computational resources.

This holonic nature simplifies the reconfiguration of the DT system to reflect changes to the physical reality or the users' needs. Changes to user needs will be catered for by providing new services in the SN, or by adding DT services to the DTs in the hierarchy. Physical infrastructural changes, such as the addition/removal of infrastructure, can easily be implemented by the addition/removing DTs from the system. Changes in data sources occur when, for example, a company is contracted to provide new data on the condition of a piece of infrastructure that already has a DT component in the system. Such changes primarily affect the WL and the DT configuration files.

Experience has shown (as in the case study) that, due to the modular nature of the architecture, many reconfigurations of the DT system can be accomplished by only changing configuration data or files, without any software code changes. This leads to minimal disruption and effort in maintaining an up-to-date digital representation of the physical infrastructure.

6   Conclusions and Future Work

This paper presents a software architecture for a DT system aimed at the integration of data silos, developed within the context of railway infrastructure. However, the application of this architecture is not limited to railway infrastructure, but can be used in other domains as well, since the nature of the problem of data silos is similar across most domains.

The system is designed to be holonic, and with the DT hierarchy created intuitively. The holonic nature of the architecture enables distributed operation, and system reconfiguration – both in terms of its reflection of the physical reality (such as with physical infrastructure or data source changes) and in terms of the services offered to the user.

The intuitive creation of the DT hierarchy removes the restrictions relating to DT hierarchy or behavioural differences between DTs, which were imposed by previous architectures. This allows for multiple, custom aggregation pathways, which can be used as a means to support the separation of stakeholder/user concerns.

Future work would be to compare different implementations of the internal architecture of the DTs – implementing the architecture proposed in [12] as the internal DT architecture, and comparing that to the implementation of the SLADT architecture as presented here. Furthermore, a detailed case study of this architecture, with multiple SN services and a more diverse range of data sources (such as static data silos, dynamic IoT sensors, digital human feedback, and paper-based data sources) should also be investigated.

Acknowledgements. This research was supported in part by the National Research Foundation of South Africa (Grant Number 129360) and the PRASA Research Chair at Stellenbosch University.

References

1. Human, C.: A Design Framework for Aggregation in a System of Digital Twins. Ph.D. thesis, Stellenbosch University, South Africa (2022)
2. Patel, J., Member, S.: Bridging data silos using big data integration. Int. J. Database Manag. Syst. **11** (2019). https://doi.org/10.5121/ijdms.2019.11301
3. Malakuti, S., Juhlin, P., Doppelhamer, J., Schmitt, J., Goldschmidt, T., Ciepal, A.: An architecture and information meta-model for back-end data access via digital twins. In: 26th IEEE Conference on Emerging Technologies and Factory Automation (2021). https://doi.org/10.1109/ETFA45728.2021.9613724
4. Morgado, J.F., et al.: Mechanical Testing Ontology for Digital-Twins: a roadmap based on EMMO. In: 2nd International Workshop on Semantic Digital Twins (2020). http://ceur-ws.org/Vol-2615/paper3.pdf
5. Goldbeck, G., Simperler, A.: Strategies for industry to engage in materials modelling. Zenodo (2019). https://doi.org/10.5281/zenodo.3564455
6. Johnson, A., et al.: Informing the information requirements of a digital twin: a rail industry case study. In: Proceedings of the Institution of Civil Engineers - Smart Infrastructure and Construction, pp. 1–13 (2021). https://doi.org/10.1680/jsmic.20.00017
7. Malakuti, S.: Emerging technical debt in digital twin systems. In: 2021 26th IEEE International Conference on Emerging Technologies and Factory Automation (ETFA) (2021). https://doi.org/10.1109/ETFA45728.2021.9613538
8. Sharma, A., Kosasih, E., Zhang, J., Brintrup, A., Calinescu, A.: Digital twins: state of the art theory and practice, challenges, and open research questions. J. Ind. Info. Int. 100383 (2020)
9. Kahlen, F.-J., Flumerfelt, S., Alves, A. (eds.): Transdisciplinary Perspectives on Complex Systems. Springer, Cham (2017). https://doi.org/10.1007/978-3-319-38756-7
10. Mazayev, A., Martins, J.A., Correia, N.: Interoperability in IoT through the semantic profiling of objects. IEEE Access, **6**, 19379–19385 (2017). https://doi.org/10.1109/ACCESS.2017.2763425
11. Brodt, A., Schiller, O., Sathish, S., Mitschang, B.: A mobile data management architecture for interoperability of resource and context data. In: 12th IEEE International Conference on Mobile Data Management (2011). https://doi.org/10.1109/MDM.2011.81
12. Malakuti, S., Borrison, R., Kotriwala, A., Kloepper, B., Nordlund, E., Ronnberg, K.: An integrated platform for multi-model digital twins. In: IoT 2021: 11th International Conference on the Internet of Things (2021). https://doi.org/10.1145/3494322.3494324
13. Redelinghuys, A.J.H., Kruger, K., Basson, A.: A six-layer architecture for digital twins with aggregation. In: Borangiu, T., Trentesaux, D., Leitão, P., GiretBoggino, A., Botti, V. (eds.) SOHOMA 2019. SCI, vol. 853, pp. 171–182. Springer, Cham (2020). https://doi.org/10.1007/978-3-030-27477-1_13
14. Kruger, K., Human, C., Basson, A.: Towards the integration of digital twins and service-oriented architectures. In: Proceedings of SOHOMA 2021, Studies in Computational Intelligence, vol. 1034, pp. 131–143 (2022). https://doi.org/10.1007/978-3-030-99108-1_10

Factory – Product Lifecycle Value Stream for Industry 4.0

Digital Twin Lifecycle: Core Challenges and Open Issues

Farah Abdoune, Maroua Nouiri, Olivier Cardin$^{(\boxtimes)}$, and Pierre Castagna

Nantes Université, École Centrale Nantes, CNRS, LS2N, UMR 6004, 44000 Nantes, France
olivier.cardin@univ-nantes.fr

Abstract. While the notion of a Digital Twin has been around for almost two decades, it is still evolving as it spreads into new sectors and uses cases. As a result, there is an ever-increasing diversity of definitions. However, numerous significant challenges must be overcome to increase its practical viability. In this regard, this study summarized the key features of the DT from the literature, defined explicitly the DT lifecycle stages, and identified the core challenges that should be highlighted when implementing a DT in practice during each stage of its lifecycle. This paper's findings can serve as a roadmap for academics as they address the future directions of DT research and deployment.

Keywords: Digital twin · Lifecycle · Design · Validation · Implementation · Service · Maintenance

1 Introduction

The Digital Twin paradigm is at the forefront of Industry 4.0 due to its potentialities, enabling state monitoring [1], diagnostics [2], prognostics [3], asset management [4], decision-support [5] in manufacturing and also different services in aerospace [6], healthcare [7] and smart cities [8] in conjunction with sensory data collecting, big data analytics, and Artificial Intelligence [9]. The persistent and continuous applications of the Digital Twin (DT) by various stakeholders in different sectors raise a blooming of definitions. Nonetheless, the diverse DT perspectives are united under a common umbrella in the definition of Rosen et al. [10], described as "a very realistic model of the current state of the process and their behaviour in interaction with their environment in the real world".

From this definition, it can be summed up that the main role of the DT is to ensure the real-data acquisition and synchronization with the physical system to update the virtual counterpart, reflecting an accurate and reliable representation of the monitored system during its whole lifecycle [11]. Michael Grieves introduced the DT first in his course of Product Lifecycle Management. However, despite the growing adoption of the concept, the main attention is mostly drawn to the physical entity that the DT is reflecting through its lifecycle regardless of the scope of the entity (system, machine, component, or product) and the services that the DT can offer to the latter. Only few research works are focusing on the DT lifecycle and different requirements that this concept needs to satisfy to be as reliable as possible throughout its whole life phases. Indeed, there are

© The Author(s), under exclusive license to Springer Nature Switzerland AG 2023
T. Borangiu et al. (Eds.): SOHOMA 2022, SCI 1083, pp. 157–167, 2023.
https://doi.org/10.1007/978-3-031-24291-5_13

several core challenges that the DT needs to tackle from its cradle to its grave to ensure its role including modelling errors, parameter tuning, and synchronization. Several types of research have been conducted to discuss major challenges like standardization or interoperability, but there is still a need to point out some issues so that the DT can exhibit more of its potential. For this reason, in this paper, we discuss in detail the lifecycle stages of the DT from design to maintenance, and explore different encountered challenges.

The remainder of this paper is structured as follows. Section 2 discusses Digital Twin features. Section 3 presents the lifecycle of Digital Twins. Section 4 analyses challenges and open issues during DT's lifecycle and Sect. 5 formulates conclusions.

2   Digital Twin Features

When the DT concept was first initiated by Grieves [11], it was characterized by three components - the physical space, the virtual part, and the connection between the two. Since then, the interest in Digital has increased across different sectors, leading to a wide range of definitions and features that have enriched Grieves' first definition. For example, Tao et al. [12] proposed a five-dimension Digital Twin architecture, adding to the three initial components the service of the DT and data circulating between different entities. Another perspective based on the integration level, according to Kritzinger et al. [13], allows for a complete understanding of how a DT should be defined. In reality, a DT can be strongly or weakly connected with its physical counterpart, resulting in three degrees of digital replication:

– A digital model is a depiction of a physical thing that already exists in the real world without the need for any automated data transfer between the two;
– A digital model with an extra automated one-way flow between the state of an existing physical item and the virtual twin is known as a "Digital Shadow" (DS); as a result, if the physical object's state changes, the digital object's state also automatically changes, but not the other way around;
– The term "Digital Twin" (DT) refers to an object with the data flow from the physical to the digital object is fully integrated into both directions - from physical to digital and vice versa; consequently, the digital object may also act as a controlling instance of the physical object. As a result, a change in the state of the physical object causes a change in the state of the digital object and vice versa.

Based on these contributions, the following crucial features require additional focus when using DT in real applications [14] inspired by the multi-dimension digital twin model of Tao et al. [12]:

• **Model:** To begin with, a high-fidelity virtual model forms the basis of every DT. Understanding the physical world in its entirety is crucial to achieving this goal. The degree of reality abstraction used for the virtual representation directly affects the model quality. Different modelling opportunities can be offered: A geometric model, which uses the necessary data formats for computer information translation and processing, defines a physical thing in terms of its geometric shape, embodiment, and appearance. A behavioural model explains different actions that a physical entity can take to carry out its functions, react to changes, communicate with others, modify internal processes, maintain its health, etc. The rules derived from historical facts, specialized knowledge, and predetermined logic are described by asset of rules. The rules provide the virtual model the capacity to reason critically, make decisions, assess situations, and make predictions [15].

- **Data:** Data is the primary source of operation for all models and services. The data must go through several processes before it can be considered knowledge [16]. Each stage has to be reorganized by the DT's features. Additionally, DT is exceptional in that it not only analyses data from the physical world but also combines data produced by virtual models to increase the accuracy of the outcomes.
- **Service:** Service management and encapsulation are crucial components of DT. Value-adding services such as monitoring, simulation, verification, virtual experiment, optimization, digital education, etc. are DT's ultimate objective [9]. Additionally, DT may support several other services, including those related to resources, algorithms, knowledge, etc. As a result, integrating Artificial Intelligence models can enhance these services.
- **Connection:** Data, services, models, and the physical environment are not separated from one another. Through links among them, they are continually interacting with one another to further collective growth. The most important feature of the DT is the synchronization between the virtual model and the physical system [17]. In this sense, the frequency at which data is transferred between the physical system and the virtual model must be defined. Information is shared in both directions between the real and virtual environments. This update is frequently referred to as real-time in the literature [18] where changes in the physical state are updated in the virtual representation immediately. In addition to synchronization, data connectivity is a crucial feature of the DT. Data connectivity must be online and bidirectional to be qualified as a digital twin [19–21]. It is assumed that online means that data is automatically exchanged between the actual system and the virtual representation and vice versa [18]. The DT should be able to collect data from the physical system and then give feedback that serves a specific objective, which may or may not result in physical system changes. Assuming that the connection from the virtual to the physical system will necessarily result in system changes is deductive reasoning.

3 Digital Twin Lifecycle

Considering the aforementioned requirements, to ensure the deployment of an accurate DT it is important to assess different phases of its lifecycle and define the task performed at each stage to improve the accuracy and quality of the DT. Similarly to software development, we can distinguish the following stage of the DT's lifecycle:

- *Design phase*: implies the creation of the model chosen among the different models listed before, after having a good understanding of the physical system and knowledge of different data and parameters related to its process [22].
- *Validation phase*: the model needs to be evaluated against the physical system to close the gap between the virtual model and the real system [23].

- *Implementation phase*: to deploy the model to production, data connection and frequency update need to be detailed in this phase, as well as the choice of technological infrastructure to ensure scalability [24].
- *Service phase*: the DT is operating and achieving its desired objectives, from monitoring to diagnostics, in conjunction with artificial intelligence techniques among others for knowledge extraction [25].
- *Maintenance phase*: when the model is affected by performance degradation in time due to intrinsic or extrinsic conditions, it needs to be maintained and updated [26].

4 Digital Twin Core Challenges and Open Issues

The system's non-linear behaviour, intricate interrelationships of stochastic nature between its elements which makes it unpredictable [27], as well as multiple states of operation [28] make difficult the modelling of these systems. A system of this complexity cannot be accurately replicated in virtual reality. As a result, establishing and improving DT is a drawn-out process. The virtual representation of the real entity remains imperfect. High standardization, modularization, infrastructure, and robustness are attributes of a good virtual model [29], see Fig. 1.

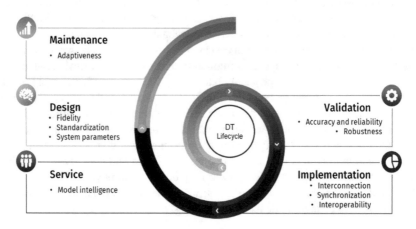

Fig. 1. The core challenges of the DT during its lifecycle stages

Encoding, interface, and communication protocol standardization are done to make information sharing and integration easier. Through the separation of different components, modularization promotes flexibility and scalability. Information transfer time and cost must be decreased. Additionally, the ability of models to handle varied uncertainty is essential. However, stating these attributes is easier said than done. Several challenges and open issues are still not tackled. For this reason, we are discussing different challenges encountered by the DT during its lifecycle.

Figure 1 sums up the DT lifecycle and the core challenges encountered by the DT analysed. Even if the design and validation phases can be thought of as an iterative process to correct the model, the DT development lifecycle can be seen from a spiral cycle angle where the DT continuously offers an increasingly complete and robust model due to its main dynamic feature (as Liu et al. [18] suggest giving the definition of a living model that continually adapts to changes in the environment).

4.1 Design Stage

Standardization
It is currently challenging to create an adequate model for a DT using standard approaches, because of the complexity of real-world systems. Digital Twins are created using a variety of technologies, interfaces, communication protocols, models, and data as there is not a clear reference design [30]. To specify reference architectural elements, information models and communication protocols, standard digital twin solutions should be created [31]. In this context, various existing standards and protocols can be used as a replacement [32]. For example, as part of the RAMI4.0, Plattform Industrie 4.0 includes the Asset Administration Shell (AAS) [33]. The ISO 23247 standard was recently published; it includes the principles and requirements for developing DTs in manufacturing [34].

Model Fidelity
Although it is often believed that a greater fidelity model guarantees a closer alignment between the physical system and virtual representation, this assumption may not be true if the data collected from the system is inappropriate for the level of model fidelity. The advantages of the higher fidelity model may not be realized due to observational and modelling errors brought on by inadequate data. Even if the required data could be gathered, a high-fidelity strategy may still face difficulties with data storage management, data transport restrictions, computer processing capacity, and decision support turnaround times. This lends more credence to the claim that the definition of a digital twin should not be based on the degree of model abstraction; the model's fidelity should rely on the use case [35].

System Parameters
System parameters can remain constant or change over time. The level of abstraction chosen for the virtual representation determines the parameters for the DT model. These parameters can be unknown because not all of the system states and parameters are easily swapped between the physical world and the virtual representation. This is because not all system inputs can be accurately measured due to either technological or financial constraints. Also, even when these parameters are known, due to external factors such as environmental conditions they can vary in time [36]. The DT should be able to adapt to the parameter's variability. Methods such as Parameter sensitivity analysis can be used to overcome this challenge. Other researchers prefer to use AI methods to tune these parameters in the next stage of the DT lifecycle.

4.2 Validation Stage

Accuracy and Reliability
DTs combine numerous models, with some based on physical principles and others

obtained empirically via machine learning methodologies. It is critical to fine-tune these models to correctly represent the behaviour of actual assets. Model tuning is an incredibly difficult process with no fixed principles or successful tuning techniques. Designers try to reduce the gap between model outputs and physical asset outputs while tweaking the models. This disparity, however, might be due to a malfunctioning sensor, data ambiguity, inaccurate parameters, model flaws, missing crucial model components, or a physical asset breakdown [37]. Regardless of the inherent difficulties associated with model tuning, the DT should be an accurate representation of the physical asset. Otherwise, the DT-derived insights are invalid.

The absence of approaches and procedures for validating virtual models makes the process time-consuming and difficult. Nevertheless, the authors of [38] reviewed the existing implicit approaches in the literature to reduce the gap between the virtual model and the real system through Machine Learning techniques, threshold setting, or even noise filters in automatic model generation.

Robustness

The main requirement of the DT is the ability to quantify, minimize, and convey the uncertainties that are present in the data and models, such as measurement noise and error propagation, modelling mistakes and approximations, and nuisance parameters in the model [15]. The challenge for the DT is to be insensitive to the above-mentioned uncertainties and be sensitive to the anomalies accruing in the physical system. This characteristic is qualified as robustness. Ensuring a robust DT is not always a trivial task. The main purpose of a DT is to monitor the real system by exhibiting the nominal behaviour that this system should have. The difference between the actual behaviour of the system and the nominal behaviour of the DT is then measured. Due to the uncertainties mentioned previously, the difference, also called residual in the ideal case, is equal to zero. However, this can never happen because a model as accurate as it is, is never a perfect replica of the system. In addition, the variability of the system is inherent, and different uncertain events can occur. For all these reasons, the DT should be able to separate between the non-zero residuals due to its intrinsic uncertainties and the variability of the physical system. In this context, data-driven methods are often used.

4.3   Implementation Stage

Synchronization

Because successful DT execution needs real-time access to created data, real-time connection and synchronization are a significant difficulty. Connecting and integrating the physical and virtual worlds is extremely difficult, owing to the physical environment's unique characteristics such as variability, uncertainty, and fuzziness, the different scales of the physical and virtual spaces, and the data generated continuously by various entities [27]. Real-time engagement will be easier to sustain if these difficulties can be overcome. According to [28], the synchronization between the virtual and physical models is critical, but it requires a breakthrough in high-speed transmission technology to stop being a problem. Other studies rely on probability distribution to model the stochastic nature of dynamic systems to quantify their uncertainty and variability and hence enable a better synchronization like Hidden Markov Models [39].

Data Interconnection

According to [40], the categorization of different DT integration levels by Kritzinger et al. [13] reveals that the majority of articles in the literature which assume to have developed DT fall under the digital model [41, 42] and digital shadow [43, 44]. This is because the majority of the claimed DT in literature does not present bilateral communication of information from the physical to the digital worlds and vice versa. Requiring all DTs to meet such a criterion is not feasible. As a result, a primary open research question consists of investigating why the automated bi-directional connection is not attained and understanding the challenges behind it, also examining the necessity of doing it in different use cases and the feasibility with current technologies.

The authors of [15] attempted to answer these questions. According to them, the requirement of an automated virtual-to-physical link can ignore the numerous offline human-in-the-loop (HITL) interactions with the physical system, such as data collection techniques like physical inspection, maintenance logs, etc., and system maintenance actions that could be used to update the virtual representation of system states. In other scenarios, the decision-making process does not necessitate the instantaneous transmission of information or requires the presence of an operator considering some contextual elements from the physical system.

Data Interoperability

While data is the foundation of DT, much existing field data does not adhere to a consistent data standard. Unstructured data (e.g., portable document format), semi-structured data (e.g., log files from an operator's integrity management software), and structured data are all possible (e.g., comma separated files, excel spreadsheets). Different suppliers' data integration systems also use different standards and methodologies to convey their data. Furthermore, existing data is frequently not connected to a single database and is housed in various locations. These characteristics make difficult integrating all existing and real-time data into a single data analysis module. As a result, an intermediate interpreter is needed to translate data from both proprietary and open access data sources into a common format that the DT can understand [37].

4.4   Service Stage

Model Intelligence

In the service stage, if a manufacturing process is examined a digital twin will aim at process monitoring and control, which will lead to error prediction and machine health tracking. The issue, however, is in enhancing the intelligence of these models with acquired data to continuously enhance the physical counterpart. A digital twin created for a machine, for example, will gradually adapt to the numerous scenarios that arise during the process. It also learns the machine's behaviour in such scenarios. As a result, it must specify its feedback. Over time, this learning process must be strengthened. For this aim, Machine Learning methods are widely used to learn the system's behaviour pattern enabling the DT to be self-evolving. However, the accuracy of such methods depends upon the quality and quantity of data used for training, yet the availability of data is not always trivial. The main challenge is the capacity for generalization enabling the learning process only with few data and the accurate response to new data [45].

4.5 Maintenance Stage

Model Update and Adaptiveness

When the DT is operating, complex software tools, hardware infrastructure, sensors, and asset life cycle data (e.g., measurements, simulations, models, asset status, anomalies, corrective measures, and operational parameters) are also running to meet the desired set objectives of the physical system. The physical system is in most cases known as a complex and dynamic system; this implies that changes in the statistical properties can occur, random external events can happen, and environmental conditions can vary over time. Hence, the DT must be maintained and updated to remain accurate and synchronized with the real system. To this aim, methods like change point detection or concept drift detection to explicitly identify the changes in the data distribution can be useful.

5   Conclusion

While the concept of a Digital Twin has been around for almost two decades, it is continually growing as it expands into new industries and uses cases. As a result, definitions are becoming increasingly diverse. However, multiple fundamental challenges must be addressed before increasing its viability including modelling mistakes, parameter uncertainties, real-time synchronization, robustness, adaptiveness against system changes, and so on. Several studies have been undertaken to examine important issues such as standardization and interoperability, but there is still a need to highlight some concerns so that the DT can demonstrate its full potential. The objective is to ensure DT reliability during its lifecycle stages.

As a result, in this article we intend to cover in depth the lifecycle stages of the DT, namely from design to maintenance and indicated the significant problems that should be addressed when implementing a DT in practice at each phase. The outcomes of this work can be used as a guideline for future researchers as they address the future paths of DT research and implementation.

References

1. Hu, S., Wang, S., Su, N., Li, X., Zhang, Q.: Digital twin based reference architecture for petrochemical monitoring and fault diagnosis. Oil Gas Sci. Technol. Rev. IFP Energies Nouv. **76**, 9 (2021). https://doi.org/10.2516/ogst/2020095
2. Castellani, A., Schmitt, S., Squartini, S.: Real-world anomaly detection by using digital twin systems and weakly-supervised learning. IEEE Trans. Ind. Inf. **17**(7), 4733–4742 (2021). https://doi.org/10.1109/TII.2020.3019788

3. Classens, K., Heemels, W.P.M.H.M., Oomen, T.: Digital twins in mechatronics: from model-based control to predictive maintenance. In: 2021 IEEE 1st International Conference on Digital Twins and Parallel Intelligence (DTPI), Beijing, China, pp. 336–339 (2021). https://doi.org/10.1109/DTPI52967.2021.9540144

4. Macchi, M., Roda, I., Negri, E., Fumagalli, L.: Exploring the role of digital twin for asset lifecycle management. IFAC-PapersOnLine **51**(11), 790–795 (2018). https://doi.org/10.1016/j.ifacol.2018.08.415

5. Fathy, Y., Jaber, M., Nadeem, Z.: Digital twin-driven decision making and planning for energy consumption. JSAN **10**(2), 37 (2021). https://doi.org/10.3390/jsan10020037

6. Tuegel, E.J., Ingraffea, A.R., Eason, T.G., Spottswood, S.M.: Reengineering aircraft structural life prediction using a digital twin. Int. J. Aerosp. Eng. **2011**, e154798 (2011). https://doi.org/10.1155/2011/154798

7. Liu, Y., et al.: A novel cloud-based framework for the elderly healthcare services using digital twin. IEEE Access **7**, 49088–49101 (2019). https://doi.org/10.1109/ACCESS.2019.2909828

8. Petrova-Antonova, D., Ilieva, S.: Digital twin modeling of smart cities. In: Ahram, T., Taiar, R., Langlois, K., Choplin, A. (eds.) IHIET 2020. AISC, vol. 1253, pp. 384–390. Springer, Cham (2021). https://doi.org/10.1007/978-3-030-55307-4_58

9. Cai, Y., Starly, B., Cohen, P., Lee, Y.-S.: Sensor data and information fusion to construct digital-twins virtual machine tools for cyber-physical manufacturing. Procedia Manuf. **10**, 1031–1042 (2017). https://doi.org/10.1016/j.promfg.2017.07.094

10. Rosen, R., von Wichert, G., Lo, G., Bettenhausen, K.D.: About the importance of autonomy and digital twins for the future of manufacturing. IFAC-PapersOnLine **48**(3), 567–572 (2015). https://doi.org/10.1016/j.ifacol.2015.06.141

11. Grieves, M.: Digital twin: manufacturing excellence through virtual factory replication, White paper, vol. 1, pp. 1–7 (2014)

12. Tao, F. , et al.: Five-dimension digital twin model and its ten applications, J. Jicheng Z. Xitong. Comput. Integr. Manuf. Syst. CIMS **25**, 1–18 (2019). https://doi.org/10.13196/j.cims.2019.01.001

13. Kritzinger, W., Karner, M., Traar, G., Henjes, J., Sihn, W.: Digital twin in manufacturing: a categorical literature review and classification. IFAC-PapersOnLine **51**(11), 1016–1022 (2018). https://doi.org/10.1016/j.ifacol.2018.08.474

14. Qi, Q., et al.: Enabling technologies and tools for digital twin. J. Manuf. Syst. **58**, 3–21 (2021). https://doi.org/10.1016/j.jmsy.2019.10.001

15. Van Der Horn, E., Mahadevan, S.: Digital twin: generalization, characterization and implementation. Decis. Supp. Syst. **145**, 113524 (2021). https://doi.org/10.1016/j.dss.2021.113524

16. Tao, F., Qi, Q., Liu, A., Kusiak, A.: Data-driven smart manufacturing. J. Manuf. Syst. **48**, 157–169 (2018). https://doi.org/10.1016/j.jmsy.2018.01.006

17. Angrish, A., Starly, B., Lee, Y.-S., Cohen, P.H.: A flexible data schema and system architecture for the virtualization of manufacturing machines (VMM). J. Manuf. Syst. **45**, 236–247 (2017). https://doi.org/10.1016/j.jmsy.2017.10.003

18. Liu, Z., Meyendorf, N., Mrad, N.: The role of data fusion in predictive maintenance using digital twin. AIP Conf. Proc. **1949**(1), 020023 (2018). https://doi.org/10.1063/1.5031520

19. Abramovici, M., Göbel, J.C., Savarino, P.: Reconfiguration of smart products during their use phase based on virtual product twins. CIRP Ann. **66**(1), 165–168 (2017). https://doi.org/10.1016/j.cirp.2017.04.042

20. Demkovich, N., Yablochnikov, E., Abaev, G.: Multiscale modeling and simulation for industrial cyber-physical systems. In: 2018 IEEE Industrial Cyber-Physical Systems (ICPS), pp. 291–296 (2018). https://doi.org/10.1109/ICPHYS.2018.8387674

21. Schleich, B., Anwer, N., Mathieu, L., Wartzack, S.: Shaping the digital twin for design and production engineering. CIRP Ann. **66**(1), 141–144 (2017). https://doi.org/10.1016/j.cirp. 2017.04.040

22. Zhang, L., Zhou, L., Horn, B.K.P.: Building a right digital twin with model engineering. J. Manuf. Syst. **59**, 151–164 (2021). https://doi.org/10.1016/j.jmsy.2021.02.009

23. Khan, A., Dahl, M., Falkman, P., Fabian, M.: Digital twin for legacy systems: simulation model testing and validation. In: 2018 IEEE 14th International Conference on Automation Science and Engineering (CASE), pp. 421–426 (2018). https://doi.org/10.1109/COASE.2018. 8560338

24. Redelinghuys, A.J.H., Basson, A.H., Kruger, K.: A six-layer architecture for the digital twin: a manufacturing case study implementation. J. Intell. Manuf. **31**(6), 1383–1402 (2019). https:// doi.org/10.1007/s10845-019-01516-6

25. Qi, Q., Tao, F., Zuo, Y., Zhao, D.: Digital twin service towards smart manufacturing. Procedia CIRP **72**, 237–242 (2018). https://doi.org/10.1016/j.procir.2018.03.103

26. Yu, J., Song, Y., Tang, D., Dai, J.: A digital twin approach based on nonparametric Bayesian network for complex system health monitoring. J. Manuf. Syst. **58**, 293–304 (2021). https:// doi.org/10.1016/j.jmsy.2020.07.005

27. Efthymiou, K., Pagoropoulos, A., Papakostas, N., Mourtzis, D., Chryssolouris, G.: Manufacturing systems complexity review: challenges and outlook. Procedia CIRP **3**, 644–649 (2012). https://doi.org/10.1016/j.procir.2012.07.110

28. Saez, M., Maturana, F., Barton, K., Tilbury, D.: Anomaly detection and productivity analysis for cyber-physical systems in manufacturing. In: 2017 13th IEEE Conference on Automation Science and Engineering (CASE), pp. 23–29 (2017). https://doi.org/10.1109/COASE.2017. 8256070

29. Tao, F., Zhang, M., Nee, A.Y.C.: Applications of digital twins, in digital twin driven smart manufacturing, pp. 29–62. Elsevier (2019). https://doi.org/10.1016/B978-0-12-817630-6.000 02-3

30. Semeraro, C., Lezoche, M., Panetto, H., Dassisti, M.: Digital twin paradigm: a systematic literature review. Comput. Ind. **130**, 103469 (2021). https://doi.org/10.1016/j.compind.2021. 103469

31. Lu, Q., Xie, X., Parlikad, A.K., Schooling, J.M.: Digital twin-enabled anomaly detection for built asset monitoring in operation and maintenance. Autom. Constr. **118**, 103277 (2020). https://doi.org/10.1016/j.autcon.2020.103277

32. Zheng, X., Lu, J., Kiritsis, D.: The emergence of cognitive digital twin: vision, challenges and opportunities. Int. J. Prod. Res., 1–23 (2021). https://doi.org/10.1080/00207543.2021. 2014591

33. Schweichhart, K.: Reference architectural model industrie 4.0 (rami 4.0). An Introduction. https://www.plattform-i40. Plattform Industrie 4.0 (2016)

34. Shao, G., Helu, M.: Framework for a digital twin in manufacturing: scope and requirements. Manuf. Lett. **24**, 105–107 (2020). https://doi.org/10.1016/j.mfglet.2020.04.004

35. Aivaliotis, P., Georgoulias, K., Arkouli, Z., Makris, S.: Methodology for enabling digital twin using advanced physics-based modelling in predictive maintenance. Procedia CIRP **81**, 417–422 (2019). https://doi.org/10.1016/j.procir.2019.03.072

36. Escobet, T., Bregon, A., Pulido, B., Puig, V. (eds.): Fault Diagnosis of Dynamic Systems: Quantitative and Qualitative Approaches. Springer, Cham (2019). https://doi.org/10.1007/ 978-3-030-17728-7

37. Wanasinghe, T.R., et al.: Digital twin for the oil and gas industry: overview, research trends, opportunities, and challenges. IEEE Access **8**, 104175–104197 (2020). https://doi.org/10. 1109/ACCESS.2020.2998723

38. Abdoune, F., Cardin, O., Nouiri, M., Castagna, P.: About perfection of digital twin models, in service oriented, holonic and multi-agent manufacturing systems for industry of the future. Stud. Comput. Syst., 91–101 (2022). https://doi.org/10.1007/978-3-030-99108-1_7

39. Ghosh, A.K., Ullah, A.S., Kubo, A.: Hidden Markov model-based digital twin construction for futuristic manufacturing systems. AIEDAM **33**(03), 317–331 (2019). https://doi.org/10.1017/S089006041900012X

40. Negri, E., Berardi, S., Fumagalli, L., Macchi, M.: MES-integrated digital twin frameworks. J. Manuf. Syst. **56**, 58–71 (2020). https://doi.org/10.1016/j.jmsy.2020.05.007

41. Mohd Salleh, N.A., Kasolang, S., Mustakim, M.A., Kuzaiman, N.A.: The study on optimization of streamlined process flow based on delmia quest simulation in an automotive production system. Procedia Comput. Sci. **105**, 191–196 (2017). https://doi.org/10.1016/j.procs.2017.01.206

42. Kaylani, H., Atieh, A.M.: Simulation approach to enhance production scheduling procedures at a pharmaceutical company with large product mix. Procedia CIRP **41**, 411–416 (2016). https://doi.org/10.1016/j.procir.2015.12.072

43. Vachálek, J., Bartalský, L., Rovný, O., Šišmišová, D., Morháč, M., Lokšík, M.: The digital twin of an industrial production line within the industry 4.0 concept. In: 2017 21st International Conference on Process Control (PC), pp. 258–262 (2017). https://doi.org/10.1109/PC.2017.7976223

44. Weyer, S., Meyer, T., Ohmer, M., Gorecky, D., Zühlke, D.: Future modeling and simulation of CPS-based factories: an example from the automotive industry. IFAC-PapersOnLine **49**(31), 97–102 (2016). https://doi.org/10.1016/j.ifacol.2016.12.168

45. Kaur, M.J., Mishra, V.P., Maheshwari, P.: The convergence of digital twin, IoT, and machine learning: transforming data into action. In: Farsi, M., Daneshkhah, A., Hosseinian-Far, A., Jahankhani, H. (eds.) Digital Twin Technologies and Smart Cities. IT, pp. 3–17. Springer, Cham (2020). https://doi.org/10.1007/978-3-030-18732-3_1

Distributed Control Architecture for Managing Internal Risks in Hazardous Industries

Auwal Shehu Tijjani[1]([✉]), Eddy Bajic[2], Thierry Berger[1], Michael Defoort[1], Yves Sallez[1], Mohamed Djemai[1,4], Clement Rup[2], and Kais Mekki[3]

[1] LAMIH CNRS UMR 8201, Polytechnic University Hauts-de-France, Valenciennes, France
{AuwalTijjani.Shehu,Thierry.Berger,Michael.Defoort,Yves.Sallez, Mohamed.Djemai}@uphf.fr
[2] CRAN CNRS UMR 7039, University of Lorraine, Nancy, France
eddy.bajic@univ-lorraine.fr
[3] OKKO SAS, Rémering-les-Puttelange, France
kais.mekki@okko-france.com
[4] QUARTZ EA 7393, ENSEA, Cergy-Pontoise, France
mohamed.djemai@ensea.fr

Abstract. This paper proposes a new distributed control architecture for the risk management of hazardous industrial facilities. The objective is to guarantee a high level of risk management in the operating conditions of an industrial facility by means of anticipation, prevention, avoidance and warning of any critical situation that could affect goods and people. Such a situation could appear by any abnormal positioning, handling and storage conditions of physical assets (e.g., pallets with potentially dangerous products, ...) as well as their interactions with other assets or moving entities (e.g., forklift, autonomous ground vehicle, ...), including human operators. The proposed methodology includes three levels of risk management defined according to different decision time horizons. At the reactive behaviour level, static and moving assets (considered as agents) are monitored in order to check that no risky situation occurs. At the proactive behaviour layer, a potentially risky situation is avoided by a mobile asset by adjusting its trajectory with only local knowledge of its neighbouring assets (e.g., two autonomous forklifts carrying incompatible assets that move to a crossing area) in a distributed way. At the higher level of forecasting behaviour, by taking into account the experience from history and past situations, the warehouse management system can modify the storage locations of certain physical assets.

Keywords: Internet of Things (IoT) · Risk management · Monitoring system · Cyber physical system · Industrial warehouse · Multi-agent system (MAS) · Decentralized planning

© The Author(s), under exclusive license to Springer Nature Switzerland AG 2023
T. Borangiu et al. (Eds.): SOHOMA 2022, SCI 1083, pp. 168–180, 2023.
https://doi.org/10.1007/978-3-031-24291-5_14

1   Introduction

Context and Motivation: The safety of people and goods is an uncompromising aspect of all modern-day and near-future industries [1]. Indeed, several accidents have occurred in the last decade (e.g., Beirut port 2020, Iqoxe 2020, Lubrizol 2019, BASF 2016, . . .) and reaffirm that safety is essential for any industrial setting [2]. For instance, a fatal accident occurred at the Lubrizol site on September 26, 2019. The main cause of this accident can be traced back due to the inappropriate positioning of different dangerous products in large quantities. Hence, the conventional techniques of storing two or more dangerous products in industries, under less supervision from experts may lead to fire outbreaks or explosions. Additionally, this can also result in potentially adverse effects not only on the health of humans and industrial facilities, but also on the large perimeter of the industrial setting. Therefore, handling, storing, transporting and disposing of hazardous substances give rise to specific constraints, typically in the manual or automated management of warehouse activities. Besides working pressure, other factors such as collisions between forklift trucks and foot workers in a congested area containing industrial materials (within close proximity) can often lead to an increase in catastrophic accidents during warehouse management.

State of the Art: Based on the above issues, a study of industrial accidents has been conducted by Dutch warehouses from 2007 to 2011 in [3]. The concept of "safety consciousness" has been proposed, in this work, to guarantee safety performance in warehouses. This method helps to avoid the consequences of harmful accidents by improving the individuals' awareness both at cognitive as well as behavioural levels. In the same vein, a behaviour-based safety management system deployed in Hong Kong construction industries has been investigated by [4]. Also, in [5], a sensitizing model has been proposed to support empirical research for studying different high-risk incidents relating to industries. A similar investigation by [6] has revealed that a vast number of hazardous accidents are mainly caused by the temporary storage and transportation of dangerous products, especially at the port areas and marshaling yards. In this work, many case studies concerning accidents in typical transportation interfaces have been analyzed. Such an analysis is still relevant in modern-day organizations, including industrial facilities. For this reason, it can be directly extended to the cooperation of products, people and resources having different conflicting interests towards achieving individual goals with a satisfactory level of risk management. Another real-life scenario happened in 2001 when the AZF chemical fertilizer plant (in France) was blasted by an explosion of tons of ammonium nitrate. This fatal incident was caused by human error resulting from improper positioning of the products. Recently, both Beirut (in Lebanon) and Dhaka (in Bangladesh) have become devastated due to explosions. In the case of Beirut, the problem was caused by ammonium nitrate stored at the port, while for the latter city, it could be, as well, due to inappropriate storage of chemical products. Hence, the development of innovative tools and methods to reduce these risks is of utmost importance within industrial facilities.

Recent evidence shows that for typical risk management of an industrial warehouse, accurate and effective localization, information gathering and monitoring are the main components for designing tools and services using either centralized, distributed, or hybrid architecture. What we know about risk management, from this evidence is largely based upon different operations of industrial warehouses, ranging from receiving, storing, picking and shipping operations. For instance, in [7], some methodologies have been proposed and discussed concerning the assessment of risks. In contrast to earlier fatal accidents, the authors focus only on addressing deliberate acts like terrorist attacks. Also, [8] proposed to deploy an analytic tool based on the well-known big data scheme to develop a policy for operational risk management in the energy industry. However, this strategy may be prone to cyber-attacks. Turning now to the project-based studies conducted by research institutes (National Research and Safety Institute for the Prevention of Occupational Accidents and Diseases, INRS, France; French national institute for industrial environment and risks, INERIS, France; Occupational Safety and Health Administration, OSHA, USA), several recommendations have been proposed to completely avoid or at least minimize the effects of chemical risk [9]. Within these recommendations, some research works attempt to perform a risk management operation autonomously, using autonomous multi-agent systems (MAS) configured in either centralized or distributed architecture [10].

Focusing on the centralized approach, this architecture ensures that all the agents are fully connected to a central system, where each agent can only transmit its information (e.g., states) to the central control unit [11]. The controller evaluates this information and then sends the necessary action/decision to the corresponding agent through the same communication link. Although extensive research has been carried out on centralized strategies, there are serious drawbacks associated with this architecture. For instance, any failure of the control unit relating to its hardware or the communication link results in the total breaking down of the whole MAS. Besides this issue, scaling up this architecture to include more agents requires a higher communication bandwidth, due to an increase in the amount of data to be exchanged between the agents and the central controller. From the practical point of view, this leads to more constraints such as a high financial budget accompanied by upgrading the communication equipment, thereby making the scalability of this strategy a challenging problem. Conversely, a distributed scheme does not have any common control unit. Instead, agents exchange their information only with their neighbours [12]. Therefore, each agent uses received information from its neighbour/s to compute its own decision. Hence, information is shared locally in this architecture. This scheme, as well as the paradigm of intelligent product (IP), have been exploited by the scientific community to improve the smart operation of industrial warehouses [12]. Compared to the classical centralized scheme, the distributed IP technology tackles high uncertain/unpredictable events, changes and disruptions that may occur during operation to avoid the inherent risks of centralization.

Another critical aspect of the industrial warehouse is represented by the storage operations. Many research works have studied the product allocation problem with compatibility constraints (i.e., slots and shelves capacities, floating location size, storing time, products compatibility, ...). One of these works, within the context of hazardous industrial warehouses of chemical products as well as nuclear reactors, includes a systematic investigation [13], where incompatible products should not be allocated as neighbours. Recently, the IoT, like Radio Frequency Identification (RFID [14]) and smart object technologies, helped in providing an efficient tool for risk management of dangerous products in warehouses [15]. Besides making dynamic warehouse supervision possible, the concept of IoT enhances augmented service delivery even at the top management level [16]. Furthermore, information is seamlessly available at each point of the process in the IoT infrastructure when every relevant component–potentially anything acting in a hazardous setting–is turned into a communicating object. This communicating object paradigm connects different physical assets to the IoT infrastructure; thus, it can help to provide full control and efficient monitoring of these physical assets.

Objectives of the Work: This paper introduces a distributed control architecture using the MAS paradigm for the risk management of physical assets. The objective is to guarantee a high level of risk management in different operating conditions of an industrial facility by means of anticipation, prevention, avoidance and warning of any critical situation that could affect goods and people. Indeed, the present technological trends make our environment more and more uncertain and more and more susceptible to these critical problems. Also, the paper exploits the concept of negotiation mechanisms, based on IoT devices and infrastructure linked to MAS monitoring architecture. The proposed methodology includes three levels of risk management defined according to different decision time horizons. At the reactive behaviour level, static and moving assets are monitored in order to check that no risky situation occurs. At the proactive behaviour layer, a potentially risky situation is avoided by a mobile asset by adjusting its trajectory with only local knowledge of its neighbouring assets in a distributed way. At the higher level of forecasting behaviour, by taking into account the experience from the history and past situations, the warehouse management system can modify the storage locations of certain physical assets.

The remaining parts of this paper are structured as follows. Section 2 presents and discusses the main problems to be tackled for the risk management of industrial warehouses. The three levels of risk management proposed in this paper, are detailed and discussed in Sect. 3. Section 4 (resp. Section 5) describes the reactive (resp. proactive) behaviour layers. The last Section finalizes this paper with some concluding remarks and future perspectives.

2   Problem Setup

Several risky situations (as shown in Fig. 1) could appear due to any abnormal positioning, handling and storing conditions of the physical assets as well as their interactions within themselves or with other resources including human operators in an industrial warehouse. These situations can be considered generic as they are relevant to classical situations encountered in many enterprises dealing with risky and hazardous products in the chemical industry and many other domains (e.g., handling of goods in ports, operating of health facilities in hospitals, processing of dangerous goods in logistic warehouses, etc.). From these generic use-cases, it is clear that any critical and relevant physical asset and autonomous moving entity should be monitored according to its specifications and critical characteristics, in order to control that an asset is:

- At the right place in the industrial facility,
- Stored under acceptable ambient conditions related to its characteristics,
- In place for a limited and monitored time duration,
- In the neighbourhood of compatible other assets,
- Interacting with the authorized operators,
- Handled/transported by authorized apparatus,
- Transported safely by automatic transportation resources in a cooperative way.

Hence, this paper deals with these problems by designing IoT devices (equipped with communication functionality) and developing methods in a cooperative Cyber-Physical System (CPS) infrastructure for real-time risk management of a typical industrial warehouse. A three-level control architecture is introduced and discussed in the next Section.

3   Proposed Control Architecture

The deployment of a mechanism by exploiting the concept of distributed and cooperative MAS, supported by an IoT-based communicating object scheme, is an interesting solution for resolving the risk management within the context of hazardous industrial facilities. Besides dealing with high uncertainty/unpredictable events, changes, and disruptions, this technique guarantees communication and decision capabilities close to the physical world. Thus, it allows reducing the delay time in decision-making and to enhance the local dynamic disruptions while avoiding the inherent risks of centralizing all the information. Following this concept, we propose to improve the critical issue of collective risk awareness. In the first phase, we begin by equipping each physical asset with built-in monitoring, decision making and alerting capabilities, at the same time connected to a higher layer of risk monitoring system at the facility level. Therefore, the physical assets and the autonomous entities are controlled based on their specifications and critical characteristics, which prevent any critical situations, as previously illustrated in Fig. 1. The proposed architecture comprising three levels of risk management is designed with respect to the different decision time horizons, as depicted in Fig. 2. The functionalities of each of these three levels, embedded in the proposed control structure, are as follows:

- A lower "Reactive behaviour" level is developed to monitor each physical asset according to ambient conditions, surrounding operators and other assets in

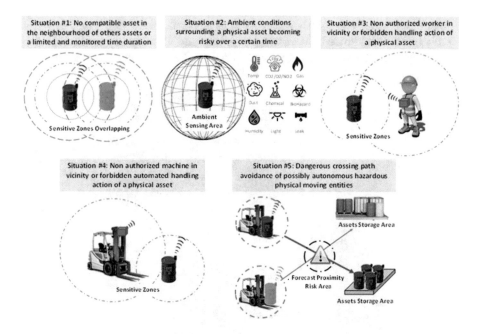

Fig. 1. Considered use-cases for the risk management of hazardous industrial facilities.

the vicinity. Even though the trajectory of each mobile asset is determined at this level, a signal is sent to the warehouse's top management system when other problems arise.

- An intermediate "Proactive behaviour" level is dedicated to a real-time analysis of potential risks relating to the trajectories of the physical assets. At this level, the problem of path crossing of non-compatible physical assets is resolved.

- At the higher level of "forecasting behaviour", the warehouse management system modifies the storage locations of certain physical assets based on previous experiences like history and past situations. The management system can also point out dangerous planned paths. Then, the system changes the trajectories of the corresponding mobile assets. For example, unidirectional corridors are designed at this layer to restrict the crossing of physical assets in areas with high activity of mobile agents (such as forklifts). This leads to a new strategy for modern warehouse reorganization.

Considering the three levels defined above, we propose the control architecture depicted in Fig. 3., which exhibits the following qualities:

1. The proposed scheme transforms the physical assets into risk management agents, enabling them to react in real-time. We have achieved this reactive behaviour by mounting an independent powered electronic device on each agent. The embedded device is fully autonomous with decision-making and

communication capabilities. Hence, the physical assets can play an active role in risk avoidance by detecting, avoiding and alerting any critical incident through our proposed intelligent and decentralized control architecture.

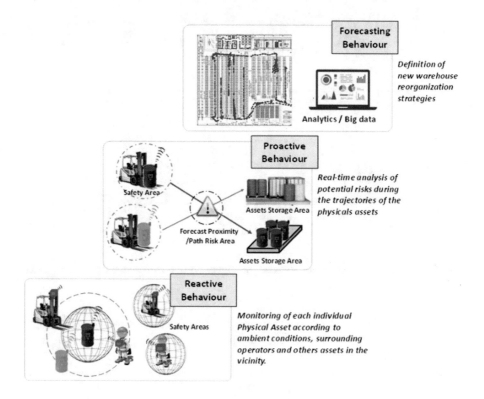

Fig. 2. The three levels of risk management of the proposed control architecture.

2. It can preventively react to deal with any forthcoming critical situations through anticipation of the assets' trajectories and storage planning. This proactive risk management layer has a short-term horizon. We have designed this layer following the notion of the high-level multi-agent model architecture.

3. The proposed architecture manages physical interactions and proximity effects between physical assets and operating resources, including human operators, according to the same risk management architecture in Fig. 3.

In summary, the proposed control architecture (Fig. 3) is developed from studies of holonomic systems (see [17] for more details), where a recursive decomposition of type "controller-controlled" technique has been used. Note that the last layer of our proposed control scheme provides a collaborative space, which facilitates the collaboration between the physical agents.

Remark 1. It is worth emphasizing that hereafter we only focus on the "Reactive" and "Proactive" behaviours.

4   Reactive Behaviour Level

Before describing this layer in our proposed control architecture, we begin by briefly explaining the basic principle of the design of an autonomous, reactive, and decision-making IoT device within the framework of this research. The overview of the proposed device named Safety Monitoring Communicative Agent (SaMoCa) figures out a smart entity and is illustrated in Fig. 4. It is worth mentioning that from Fig. 4, the designed innovative IoT device transforms any physical asset into SaMoCa, which is an autonomous agent with communication and autonomous monitoring and decision-making functionalities. Hence, each SaMoCa can send messages to other smart entities in a fully decentralized and autonomous approach. The received messages along with ambient conditions trigger the smart entity to make decisions and run algorithms to take the necessary actions for risk assessment. Furthermore, these actions mainly depend on a Real-Time Safety Agent Management Process (RTSAMP) exploiting adaptive protocols, measurements, and embedded decision-making algorithms. In summary, the functions of each SaMoCa include the following.

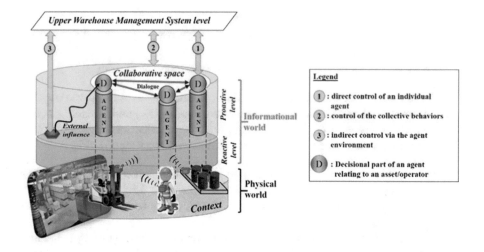

Fig. 3. Proposed control architecture.

1. Measure and monitor ambient conditions using customized onboard sensors.
2. Broadcast messages through a proper communication channel to the surrounding assets.
3. Identify the characteristics of the remaining physical assets which are near or in a distant vicinity, and then localize those nearby.
4. Analyze and identify any risks as well as critical issues on a real-time basis.

Fig. 4. Transformation from a physical asset to a SaMoCa.

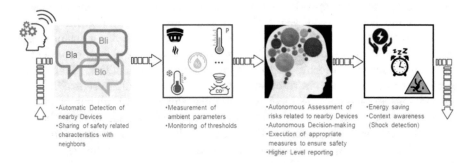

Fig. 5. Internal process autonomously operated by a SaMoCa when associated with a physical asset.

5. Take decisions on the security actions by anticipating and reporting any critical issue in the neighbourhood.
6. Report the incidence to the top-level risk management system.

At the reactive behaviour layer, the SaMoCa agents share the same physical and informational worlds in their vicinity. For monitoring and alarming activities, the assets are equipped with sensors and communication devices allowing them to continuously check that no risky situation occurs considering:

1. Their own status or location (leakage, movement detection, shock, indoor positioning system for autonomous/non autonomous forklifts, ...),
2. Their ambient and environmental conditions (e.g., temperature, pressure, humidity, gas concentration, ...),
3. Their compatibility constraints with other surrounding assets (e.g., chemical compatibility according to REACH regulation EC 1907/2006, assets volume or density in the vicinity, distance threshold, authorized operator, qualified handling resource, ...).

Based on such estimates for each SaMoCa, inference rules can be applied to detect the incompatible presence of other assets for a limited and monitored time duration (implying a risk for the concerned asset) or a risk due to the ambient conditions (fire, leakage, ...). If a risky situation is detected, a security alarm is sent to the community. Figure 5 details the internal process (RTSAMP) autonomously operated by a SaMoCa when associated with a physical asset, under energy constraints and context awareness capability.

5   Proactive Behaviour Level

A major part of the autonomy of autonomous mobile assets holds on the capacity of planning feasible trajectory in a particular environment. This environment may contain some areas in which the agent is not allowed to enter. Such forbidden areas are given thanks to the RTSAMP embedded in the IoT devices and the compatibility between assets.

Traditional motion planners consist in generating a collision-free trajectory from the initial to the final desired configurations. Depending on the distance that the agents have to travel, the computation of complete trajectories from start until finish may be computationally expensive. Moreover, the detection range of the physical sensors is limited. Therefore, the trajectories have to be computed online while the mission unfolds. It can be accomplished using an online receding horizon scheduler [18], in which partial trajectories from an initial state toward the goal are computed by solving an optimal control problem over a limited horizon.

In the proactive behaviour level, a potentially risky situation is avoided by a mobile asset by adjusting its trajectory with only local knowledge of its neighbouring assets in a distributed way. For example, two autonomous assets (autonomous ground vehicles) carrying incompatible products should appropriately navigate in order to avoid collisions. According to the exchanged information between two autonomous agents, their planned trajectories should be modified to avoid any unanticipated risky situation.

Here, we use a distributed scheme based on coordination by adjustment where two processes are running in parallel (see Fig. 6). By implementing the proposed proactive layer in simulation the result for a fleet of five mobile agents which moves from a linear configuration to a triangular one in an environment with forbidden areas (denoted as obstacles) is shown in Fig. 7. Based on a receding horizon planner, each robot computes its own desired trajectory which guarantees coordination between cooperative moving entities and avoidance of forbidden areas (e.g., obstacles, assets with non-compatibility, etc.). Then, the trajectory tracking is achieved using a reactive navigation controller in order to take into account unexpected entities. The receding horizon planner computes the optimal collision-free trajectory as follows. First, each cooperative moving agent computes its intuitive trajectory (i.e., a trajectory without collision with static obstacles and incompatible assets), as depicted in Fig. 7. Then, neighbouring moving entities exchange their intuitive trajectory. Using exchanged information, each moving agent locally modifies its intuitive trajectory to avoid collision between themselves, as shown in Fig. 7. The reactive navigation controller uses artificial potential fields to track the planned trajectory while avoiding collision with unexpected entities.

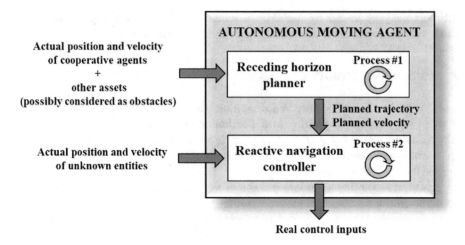

Fig. 6. Proposed proactive behaviour mechanism.

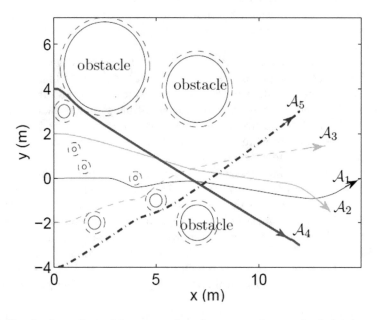

Fig. 7. Results for a fleet of 5 agents using the proposed proactive behaviour mechanism.

6   Conclusion

This paper introduced a distributed control architecture using the MAS paradigm for the risk management of physical assets in order to guarantee a high level of risk management in different operating conditions of an industrial facility by means of anticipation, prevention, avoidance and warning of any critical situation that could affect goods and people. The concept of negotiation mechanisms is investigated, based on IoT devices and infrastructure linked to MAS monitoring architecture. A three-level architecture has been proposed allowing real-time risk management of an industrial warehouse. At the reactive behaviour level, static and moving assets are monitored in order to check that no risky situation occurs. At the proactive behaviour layer, a potentially risky situation is avoided by a mobile asset by adjusting its trajectory with only local knowledge of its neighbouring assets in a distributed way.

It is worth mentioning that this work is part of the preliminary results of the I2RM project, which is just at its beginning, and several works will be performed in the future. It is planned for instance to be implemented on a customized industrial IoT device and architecture built by CRAN and the OKKO company, and also in a lab demonstrator at LAMIH (e.g., industrial mobile robots MiR100). The development of a digital twin to assist the facility risk management will also be proposed.

Acknowledgements. This research work is supported by the Hauts-de-France region and the ANR (French National Research Agency) under project ANR I2RM (Interactive and Intelligent physical assets control system for the Risks Management of hazardous industrial facilities).

References

1. Olive, C., O'Connor, T.M., Mannan, M.S.: Relationship of safety culture and process safety. J. Hazard. Mater. **130**(1–2), 133–140 (2006)
2. Reniers, G., Khakzad, N., Cozzani, V., Khan, F.: The impact of nature on chemical industrial facilities: Dealing with challenges for creating resilient chemical industrial parks. J. Loss Prev. Process Ind. **56**, 378–385 (2018)
3. De Koster, R.B., Stam, D., Balk, B.M.: Accidents happen: the influence of safety-specific transformational leadership, safety consciousness, and hazard reducing systems on warehouse accidents. J. Oper. Manag. **29**(7–8), 753–765 (2011)
4. Lingard, H., Rowlinson, S.: Behavior-based safety management in Hong Kong's construction industry. J. Safety Res. **28**(4), 243–256 (1997)
5. Le Coze, J.C.: Outlines of a sensitising model for industrial safety assessment. Saf. Sci. **51**(1), 187–201 (2013)
6. Christou, M.D.: Analysis and control of major accidents from the intermediate temporary storage of dangerous substances in marshalling yards and port areas. J. Loss Prev. Process Ind. **12**(1), 109–119 (1999)
7. Matteini, A., Argenti, F., Salzano, E., Cozzani, V.: A comparative analysis of security risk assessment methodologies for the chemical industry. Reliab. Eng. Syst. Saf. **191**, 106083 (2019)
8. Goel, P., Datta, A., Mannan, M.S.: Application of big data analytics in process safety and risk management. In: 2017 IEEE International Conference on Big Data (Big Data), pp. 1143-1152. IEEE (2017)
9. Hollnagel, E.: Safety-I and Safety-II: The Past and Future of Safety Management. CRC Press, Boca Raton (2018)
10. Ozcan, T., Celebi, N., Esnaf, S.: Comparative analysis of multi-criteria decision making methodologies and implementation of a warehouse location selection problem. Expert Syst. Appl. **38**(8), 9773–9779 (2011)

11. Xuan, P., Lesser, V.: Multi-agent policies: from centralized ones to decentralized ones. In: Proceedings of the First International Joint Conference on Autonomous Agents and Multiagent Systems: part 3, pp. 1098-1105 (2002)

12. He, W., Xu, W., Ge, X., Han, Q.L., Du, W., Qian, F.: Secure control of multi-agent systems against malicious attacks: a brief survey. IEEE Trans. Ind. Inform. **18**, 3595–3608 (2021)

13. Gaci, O., Mathieu, H.: A dynamic risk management in chemical substances warehouses by an interaction network approach. New Frontiers in Graph Theory, 451-465 (2012)

14. Rahman, F., Bhuiyan, M.Z.A., Ahamed, S.I.: A privacy preserving framework for RFID based healthcare systems. Futur. Gener. Comput. Syst. **72**, 339–352 (2017)

15. Trab, S., et al.: A communicating object's approach for smart logistics and safety issues in warehouses. Concurr. Eng. **25**(1), 53–67 (2017)

16. Liu, Z., Xie, K., Li, L., Chen, Y.: A paradigm of safety management in industry 4.0. Syst. Res. Behav. Sci. **37**(4), 632–645 (2020)

17. Sallez, Y., Berger, T., Trentesaux, D.: Open-control: a new concept for integrated product-driven manufacturing control. IFAC Proc. Volumes **42**(4), 2065–2070 (2009)

18. Defoort, M., Kokosy, A., Floquet, T., Perruquetti, W., Palos, J.: Motion planning for cooperative unicycle-type mobile robots with limited sensing ranges: a distributed receding horizon approach. Robot. Auton. Syst. **57**(11), 1094–1106 (2009)

Blockchain Adoption for Autonomous Train: Opportunities and Challenges

Melissa Hassoun[(✉)], Yassine Idel Mahjoub, and Damien Trentesaux

Université Polytechnique Hauts-de-France, LAMIH-UMR CNRS 8201, Mont Houy,
Valenciennes, 59313 Valenciennes, Hauts-de-France, France
{melissa.hassoun,yassine.idelmahjoub,damien.trentesaux}@uphf.fr

Abstract. Autonomous Train is increasingly attracting the interest of researchers and industrials due to its numerous advantages regarding security and safety, service improvement, and costs reduction. Nonetheless, the Autonomous Train still faces many challenges including safety and security, fleet coordination, interoperability, and traceability. In this paper, the disruptive technology Blockchain is explored for a potential adoption for Autonomous Train deployment in order to overcome some of the Autonomous Train challenges. Finally, an illustrative example of a potential Blockchain application in the Autonomous Train domain is presented.

Keywords: Blockchain · Smart contracts · Autonomous train · Decentralisation · Traceability · Security · Automation

1 Introduction

The interest given to autonomous transportation systems is expanding significantly. Industrialists and fleet operators are becoming competitors in this field. In our research, we focus on autonomous train (hereafter abbreviated AT). A train is said to be autonomous (and not automatic) when it is capable to make decisions enabling it to adapt to different situations, and can function in an open environment (i.e., a portion of a track with neither fully secured nor controlled access) under unstructured and dynamic conditions [62]. It is obvious that safety and security are the most important issues to meet when developing autonomous systems, especially when they are directly related to human life or very important material losses, as in the case of ATs. Moreover, a full deployment of ATs requires the automation of many railway tasks such as interlocking system, signalling control, communication, maintenance, billing etc.

Currently, several works are under development to achieve the deployment of the AT using advanced communication and internet technologies such as the Internet of Things (IoT), Artificial Intelligence (AI), 5G etc. that allow a centralised control of the train [56]. Nevertheless, in recent years, a disruptive technology has emerged called Blockchain. Its first implementation by Satoshi

© The Author(s), under exclusive license to Springer Nature Switzerland AG 2023
T. Borangiu et al. (Eds.): SOHOMA 2022, SCI 1083, pp. 181–195, 2023.
https://doi.org/10.1007/978-3-031-24291-5_15

Nakamoto in 2009 gave birth to the cryptocurrency Bitcoin [45]. This technology was initially used in the financial sector but it quickly attracted attention in many other areas: supply chain, healthcare, IoT, manufacturing, transportation, and energy to name few [3], especially after the integration of Smart Contracts (hereafter SC), a concept first coined by Nick Szabo in 1994 [58] into the Blockchain for the very first time in Ethereum [47]. A Smart Contract adds a logical layer to the Blockchain by its ability to execute automatically if predefined conditions are met [5].

The purpose of this paper is to explore the opportunities and challenges of Blockchain and SCs adoption for ATs deployment.

The reminder of this paper is organised as follows: Sect. 2 presents some backgrounds of Blockchain and SCs. Section 3 makes a literature review of Blockchain applications. Section 4 presents AT expectations and challenges. Blockchain opportunities and challenges for ATs are presented in Sect. 5. An illustrative example is given in Sect. 6 Finally, Sect. 7 provides a short discussion and Sect. 8 concludes the paper.

2   Backgrounds

This section presents some preliminary information about technologies we are going to leverage in our work, namely Blockchain and SCs.

2.1   Blockchain

Blockchain is a particular kind of the so called Distributed Ledger Technology (DLT) that enables the storage of decentralised, trusted, secure, and chronological data [73]. Blockchain encompasses a series of blocks linked by cryptographic hashes, where each block contains a list of timestamped transactions [71] as illustrated by the example in Fig. 1.

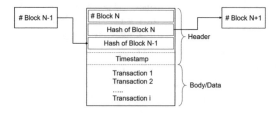

Fig. 1. Blockchain structure

Transactions are handled by nodes which can be ordinary computers connected through a peer-to-peer (P2P) network [40]. Although there is no central authority that manages transactions in a Blockchain, nodes use their appropriate public key to be authenticated and their private key to sign and unlock the transaction [9]. Table 1 summarises some of the Blockchain's characteristics.

Table 1. Blockchain characteristics

Characteristic	Explanation
Security	By the means of cryptographic hashes
Decentralisation	A peer-to-peer network without central authority
Immutability	An append-only store of transactions
Transparency	All the transactions are visible to everyone
Anonymity	Using cryptographic public-private key pair
Fault tolerant	The failure of one component does not involve the failure of the whole system

After the transaction is proposed by an entity in the network, it has to be approved by the majority of the nodes in order to be registered onto the Blockchain; this is the consensus mechanism. Various consensus mechanisms have been developed: Proof of Work (PoW), Proof of Stake (PoS), Delegated Proof of Stake (DPoS), Proof of Elapsed Time (PoET) etc., to maintain unanimity, authenticity, and integrity in distributed and decentralised networks [53]. The mentioned algorithms are typically used in public Blockchains which are permissionless Blockchains where every network participant can read, write, and send transactions without any restriction (e.g., Bitcoin and Ethereum) [72]. In contrast, there are permissioned Blockchains where the nodes have to be accepted in the network to perform operations. When the access is controlled by a single entity it is called a private Blockchain, as Hyperledger Fabric [22], and when it is controlled by a group of bodies it is called a consortium Blockchain which is a hybrid of public and private Blockchains. There are consensus algorithms dedicated to permissioned Blockchains such as Practical Byzantine Fault Tolerant (PBFT), Proof of Authority (PoA), and Tendermint [20]. The selection of Blockchain type to adopt depends on the use case requirements.

2.2 Smart Contracts

As mentioned earlier, Szabo [58] introduced a concept in 1994 that permitted to translate legal contracts into codes called Smart Contracts, and embedding them in hardware or software assets that can execute them without needing trusted third party. In a Blockchain-based system, SCs are self-executing deterministic scripts recorded in the Blockchain that run when predefined conditions are satisfied (by following an *if-else-if* statement) so that a logical and computational layer is added to the trust infrastructure supported by Blockchain [19].

Once a SC is agreed and signed by all parties, it is added to the Blockchain as a program code. Then, it is shared among all the nodes in the Blockchain throughout the P2P network [67] so that every participant (e.g., a user or a client application) can trigger it by sending a transaction recorded in an immutable manner in the Blockchain [75]. This new generation of Blockchain provides accurate, secure, faster, more efficient, transparent, and low cost transactions [7]. In addition, it can rely either on the information on the Blockchain (On-Chain) or on an external data source (Off-Chain) by the means of oracles [67].

Different Blockchain platforms are deploying SCs. Each platform adapts SCs to its specific characteristics (execution environment, language, data-model, consensus etc.) [75]. For instance, in Ethereum (the first that implemented SCs), Solidity is the most high level language used to write SC's code which is then converted into bytecode to be compiled in the Ethereum Virtual Machine (EVM) [69]. In contrast, many conventional languages are used in Hyperledger Fabric to write the SC (known as chaincode) such as Golang, Java, or JavaScript, which are executed in a Docker container [12]. The Codra platform also supports SCs written in Java or Kotlin and executed across the Java Virtual Machine (JVA) [15]. Several other platforms supporting SCs are detailed in the literature across multiple comparisons [14,63,75].

3 Blockchain Applications: Short Overview

This section presents some works in the literature that are related to Blockchain and SC applications in several areas.

Blockchain and SCs are getting increasingly popular they are attracting interest in different fields including IoT, energy, healthcare, finance, government, supply chain, manufacturing, agriculture, and transportation [3,6].

In the manufacturing domain, the work efficiency is improved using Blockchain that enables the sharing of data on the ledger (e.g., spare parts tracking) [6]. Authors in [27] applied a directed acyclic graph structured Blockchain to address the Single Point of Failure (SPOF) and malicious attacks problems in Industrial IoT systems. A decentralised approach *FabRec* is developed in [8] in order to make information available to all manufacturing organizations on the P2P network. Blockchain is also applied to enhance security and privacy in smart factory [64]. A Blockchain-based distributed framework is proposed in [54] for the automotive industry in smart cities.

Table 2 summarizes some of Blockchain applications in different other fields.

Table 2. Blockchain applications

Field	Applications	Ref.
IoT	Management, control, and security of IoT devices	[30]
	Blockchain-IoT combination for sharing services and resources and automating existing time-consuming workflows	[19]
	Dataset sharing and Blockchain-based compromised firmware detection and self-healing for security	[13]
	Blockchain-based scheme of firmware update	[33]
	Blockchain-based decentralised system for identification and authentication of IoT devices	[26]
Energy	Secure transaction of energy production and consumption information between different stakeholders	[4]
	Machine-to-Machine (M2M) electricity market with anonymity	[36,55]
	Energy grid security enhancement through a digitized market	[44]
	Green energy assistance through a decentralised model in order to minimize the environmental impact	[28]
Healthcare	Design of a Blockchain-based architecture FHIRChain for clinical data sharing	[74]
	A decentralised record management system handling electronic medical records called MedRec for auditability, interoperability, and accessibility of data	[10]
	Blockchain-based system MeDShare that tackles the problem of medical data sharing in a trustless environment	[70]
Supply Chain	A Blockchain-based agri-food supply chain traceability system using RFID	[60]
	Scalable Blockchain framework ProductChain	[39]
	Hyperledger Fabric-based agri-food supply chain traceability system	[40]
	Agricultural food supply chain traceability based on SCs	[65]
	Product traceability system in the supply chain based on SCs	[66]
Transportation	Review of potential Blockchain applications in railway	[46,52]
	Preliminary study of Blockchain-based intelligent transportation systems with a case study on realtime ride-sharing services	[73]
	Blockchain-based prototype implementation for railway control	[32]
	Blockchain-based architecture of a state channel for Smart Mobility Systems	[48]
	Blockchain-based digital railway ticketing using Hyperledger architecture	[51]
	Blockchain-based model sharing and calculation method for urban rail intelligent driving systems using SCs to implement the management of the entire federal reinforcement learning	[34]
	Robust and distributed level crossing control architecture for accident prevention in automated trains using Blockchain and SCs	[43]

4   Autonomous Train: Expectations and Challenges

As indicated previously, the advances in technology conducted several researches to reach the deployment of a fully AT. In this section, some of advantages and challenges of AT are presented.

First, ATs are supposed to improve safety and security as the human errors are eliminated and the computerized systems provide more precise movements than humans, enhanced perception capabilities, and fast reactions in dangerous situations [68]. Second, they will increase the capacity and utilization of rail lines by reducing the gaps between consecutive ATs without affecting the safety, thus increasing the service frequency in stations, reducing waiting time, and improving the alighting and boarding process [16] which will further improve the overall service reliability, availability, and flexibility. Third, operational costs can be limited due to the reduction in the train crew size and the associated management, training, and labour costs [68]. Furthermore, ATs are meant to consume less energy as acceleration, traction, and braking procedures along with air-conditioning system are optimized. Moreover, ATs could control their own energy consumption, save energy, and decrease energy consumption peaks [59]. Finally, an efficient fleet management can be performed by deploying more ATs during peak hours to meet the demand and removing unused ATs during off-peak hours in order to facilitate timely maintenance [56].

Despite the mentioned advantages, AT operation faces several challenges and that current studies attempt to overcome that have been classified differently in the literature [56, 62, 68]. In the following, the most relevant ones are summarized.

Safety and security issues are of big importance, since AT is a safety-critical system. Ensuring a safe behaviour of the AT is essential to avoid railway incidents. In addition, ATs rely on computer-based technologies which can lead to cyber-attacks; therefore, cyber-security measures are vital to develop. Communication and surveillance technologies have to be upgraded to mitigate the risk of intrusion of people or objects on rail tracks.

Learning, informational and decision-making issues have to be addressed taking into account physical and operational attributes. A critical element is the perception function that determines the train's decision quality by providing an appropriate representation of the internal and external environment including speed and position measurements, track detection, signals gathering, health status of the train etc. Another element is about analysis and decision-making that requires a certain degree of autonomy along with storage of data, events, decisions, and performed tasks.

Fleet coordination and interoperability problems are not to be overlooked. On one side, an appropriate cooperation between trains (autonomous and with driver) sharing the same infrastructure and energy suppliers is required in order to improve service quality and minimize energy costs. On the other side, interoperability of the AT with different elements on which it relies such as information systems, train stations, emergency units, maintenance centres, passengers, and onboard employees have to be considered.

Another considerable class of issues is related to standardization, ethics, human acceptance and interaction, and employment. Last but not least, a fully AT deployment requires the automation of some railway tasks such as interlocking, signalling, billing, maintenance etc.

Besides these listed issues, some others could be considered such as SPOF problem related to centralized control that could be damaging if the control centre fails. The consequences of such an incident could be harmful and costly. Moreover, the train's traceability is needed for maintenance and scheduling requirements.

To achieve a fully AT with a high degree of security, researchers used many innovative technologies including IoT [1,24], AI [21,50], and big data [29,31]. Nevertheless, each technology has its limits such as IoT that faces different challenges including the availability of internet everywhere and at no cost, security issues, low-cost smart sensing system development, scalability, fault tolerance, power consumption, energy etc. [17,42]. Despite its success, AI still has many challenges regarding societal and legal issues, ethical issues, and technical issues such as the limits of recognition approaches [25,49,57,61]. In order to overcome some of the mentioned AT challenges, we propose to adopt the previously introduced disruptive technology Blockchain. The next section explores a potential Blockchain adoption for AT deployment.

5   Blockchain Adoption for AT

5.1   Blockchain Opportunities

Blockchain provides several benefits and opportunities. This section discusses how Blockchain can be beneficial in addressing several issues related to AT.

Security: To improve security in AT, cybersecurity could be enhanced by the means of cryptography and consensus algorithms, since Blockchain is in its core a secure technology. In addition, the SPOF problem could be addressed thanks to replicated, shared, and synchronized data among the Blockchain so that the failure of a component will not involve the failure of the entire system. Moreover, data confidentiality, availability, and integrity is guaranteed.

Decentralisation: As Blockchain is a decentralised technology that operates on a P2P network without a central authority sharing the data among all the nodes, existing centralization problem in AT could be eliminated so that trains and infrastructure elements could be able to make collaborative decisions via consensus algorithms.

Task automation: When deploying ATs, Blockchain could be beneficial to automate some railway tasks that surround the AT such as interlocking, signalling, billing, maintenance etc. using SCs.

Digitisation: By the means of digitization techniques, trains and infrastructure elements could be considered as Non-Fungible Tokens (NFTs), which are virtual assets stored on the Blockchain and not interchangeable with other digital assets [18], to leverage their benefits including the proof of ownership so that they cannot be counterfeited, and to ensure their visibility and traceability.

Traceability: Blockchain provides traceability through data immutability along with tokenization using NFTs that contain metadata of each asset. This could be exploited for train's position and health, spare parts, and maintenance traceability.

Figure 2 illustrates a potential Blockchain adoption in a railway system that deployings ATs enabling the communication, decentralisation, tasks automation, and traceability. A more detailed illustrative example is presented in the next section.

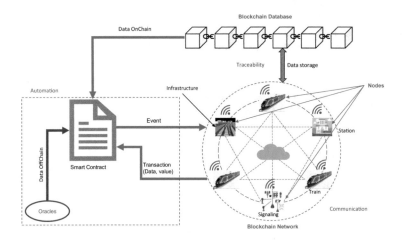

Fig. 2. Blockchain adoption in a railway system

5.2   Blockchain Challenges

Blockchain technology is still immature and faces several challenges. In this section, some of the Blockchain challenges are discussed.

The main issues in Blockchain adoption are technical ones including scalability, throughput and latency [37]. Public Blockchains (e.g., Bitcoin and Ethereum) are affected by these problems and cannot be used for AT deployment. However, there exist private Blockchains such as Hyperledger that could be more appropriate for such application [2]. Moreover, Blockchain may lead to significant environmental costs due to the high energy consumption while executing consensus or mining algorithms [11]. Usability issues are also to be considered as Blockchain is a new and complex technology which can generate fear and doubt among users. Another critical issue is related to regulation and policies because Blockchain is still novel and not regulated yet [35]. Furthermore, interoperability between different Blockchain ecosystems can represent a challenge since each

Blockchain has its own characteristics (type, consensus algorithm, coding language etc.) [38]. Last but not least, Blockchain privacy issues (such as the leakage of user identity and transaction data) must not be ignored to protect users [23].

6   Illustrative Example

In this section, an illustration of Blockchain adoption for AT is presented. This example is expected to address some of aforementioned AT deployment challenges including security, SPOF problem, tasks automation, and traceability.

For instance, Hyperledger Fabric could be chosen for implementation as it offers the possibility of permissioned ecosystems which is suitable for our application. Figure 3 illustrates a standard transaction flow regarding this platform. A user proposes a transaction to the peers which could be servers in the network that execute the SC and endorse a first response to the user. After that, the transaction is submitted to the ordering service that is responsible of ordering transactions in blocks before broadcasting them among the peers. Then, the transaction is validated by the peers. Finally, the user is notified and the Blockchain database (ledger) is updated. All these steps are labelled from 1 to 9 on the figure.

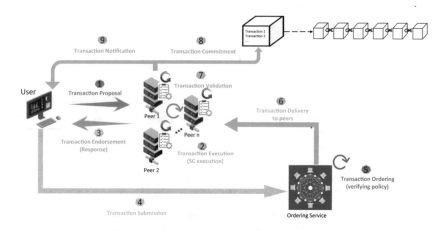

Fig. 3. Transaction flow in Hyperledger Fabric

The global idea is to apply the above transaction flow to a railway system deploying ATs. In the following, a potential example is introduced and highlighted by Fig. 4. Hereafter, only the steps performed by the user and the peers are described in details.

Infrastructure elements and ATs could be considered as peers in the network that are responsible of the SC execution and transaction validation. In addition, ATs could be assumed as assets and represented as NFTs incorporating a specific metadata that allows train's traceability, visibility, and authenticity. The ordering service could be any railway organization responsible of railway management. The transaction could be proposed either by the train or an infrastructure element (e.g., blocks) and will have as purpose the safe movement of the train on the tracks. This way, the interlocking system along with signalling could be automated by the means of SCs that are executed when all safety rules are satisfied. Decentralisation is also satisfied since the decision is made by all the railway systems, and the train remains traceable through tokenization.

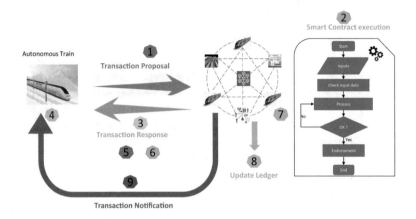

Fig. 4. An illustrative example of Blockchain adoption for AT

For instance, once the train arrives at a specific distance from a station, a transaction is triggered to request permission to enter the station. The transaction proposal is made by the train and sent to different servers acting as peers, which can be the station, other trains, and the entities responsible for the organisation of trains in the stations. These peers execute the smart contract in charge of authorising the train's entry, the SC being a code that contains the necessary conditions for a train to enter the station in compliance with safety measures. Once the SC is executed, all the peers propose the same answer to the train and the transaction is then sent to an ordering service which in turn integrates the transaction into the Blockchain. Finally, a notification is sent to the train allowing it to enter the station and the ledger is updated. Thus, the security of the movement and data is verified, the traceability of all trains entering the stations is ensured, as well as all the other advantages of the technology mentioned previously, namely decentralisation, authenticity, and transparency.

7   Short Discussion

The application of Blockchain technology in industrial engineering is still in its infancy but presents huge advantages considering several recurring (i.e., known for years) safety, accuracy, traceability and security related issues. From our perspective, this field of research is promising.

One important point is that this approach is not specific for AT in fact. Even if our work is focused on AT field, some of the ideas presented in this paper could be customized to be applied in other areas. For instance, in supply chain, the train could be substituted by a product moving through the supply chain elements. In flexible manufacturing cells, trains could be replaced by pallets or Automated Guided Vehicles (AGV) and the infrastructure by the production system or the path that the AGVs and pallets travel on. Blockchain could be used to overcome data security issues related to computer-based technologies used in AGVs such as Bluetooth and to ensure the traceability of the products or the pallets, as suggested by [41]. For instance, each AGV can act as a user proposing a transaction whose aim is to ensure that a given task is carried out safely, and the servers acting as peers will be able to execute the smart contract containing all the conditions for performing the task. After the ordering process, the AGV can execute its task and the whole network of AGVs together with the different parts of the system will have participated in the decision making ensuring a decentralisation of the system. Events are then recorded in the distributed ledger as proof of all the production activities done. Aside transportation, other similarities can be found in the service domains, typically in healthcare to ensure the safety and unicity of drug doses using NFTs, to ensure the traceability of patient journey etc.

8 Conclusion

This paper explored the Blockchain opportunities for AT deployment. First, some backgrounds of Blockchain and SCs are outlined. Moreover, a short overview of Blockchain applications is presented. Further, some of the AT expectations and challenges are listed, and a potential Blockchain adoption to overcome these challenges is highlighted. Finally, a practical example is presented to illustrate Blockchain adoption for AT deployment.

Although Blockchain is still in its infancy, its salient features make it a promising technology for AT deployment in order to address security, decentralisation, automation, and traceability issues. Nevertheless, Blockchain is still immature and needs more advancements to be enhanced and fully implemented. Even so, Blockchain is already applied in several industrial fields especially IoT, supply chain, manufacturing etc.

This article remains a prospective one and will certainly be followed by more practical papers dealing with case studies such as tasks automation.

References

1. IoT proving its worth to rail industry at a time of crisis. https://www.railtech. com/digitalisation/2020/04/14
2. Pros and cons of hyperledger fabric for blockchain networks. https:// www.devteam.space/blog/pros-and-cons-of-hyperledger-fabric-for-blockchain- networks/

3. Abou Jaoude, J., Saade, R.G.: Blockchain applications-usage in different domains. IEEE Access **7**, 45360–45381 (2019)
4. Ahl, A., Yarime, M., Tanaka, K., Sagawa, D.: Review of blockchain-based distributed energy: implications for institutional development. Renew. Sustain. Energy Rev. **107**, 200–211 (2019)
5. Alharby, M., Van Moorsel, A.: Blockchain-based smart contracts: a systematic mapping study. arXiv preprint arXiv:1710.06372 (2017)
6. Alladi, T., Chamola, V., Parizi, R.M., Choo, K.K.R.: Blockchain applications for industry 4.0 and industrial IoT: a review. IEEE Access **7**, 176935–176951 (2019)
7. Allam, Z.: On smart contracts and organisational performance: a review of smart contracts through the blockchain technology. Rev. Econ. Bus. Stud. **11**(2), 137–156 (2018)
8. Angrish, A., Craver, B., Hasan, M., Starly, B.: A case study for blockchain in manufacturing:"fabrec": a prototype for peer-to-peer network of manufacturing nodes. Procedia Manuf. **26**, 1180–1192 (2018)
9. Aydar, M., Cetin, S.C., Ayvaz, S., Aygun, B.: Private key encryption and recovery in blockchain. arXiv preprint arXiv:1907.04156 (2019)
10. Azaria, A., Ekblaw, A., Vieira, T., Lippman, A.: MedRec: using blockchain for medical data access and permission management. In: 2016 2nd International Conference on Open and Big Data (OBD), pp. 25–30. IEEE (2016)
11. Badea, L., Mungiu-Pupăzan, M.C.: The economic and environmental impact of bitcoin. IEEE Access **9**, 48091–48104 (2021)
12. Baliga, A., Solanki, N., Verekar, S., Pednekar, A., Kamat, P., Chatterjee, S.: Performance characterization of hyperledger fabric. In: 2018 Crypto Valley Conference on Blockchain Technology (CVCBT), pp. 65–74. IEEE (2018)
13. Banerjee, M., Lee, J., Choo, K.K.R.: A blockchain future for internet of things security: a position paper. Digital Commun. Netw. **4**(3), 149–160 (2018)
14. Benahmed, S., et al.: A comparative analysis of distributed ledger technologies for smart contract development. In: 2019 IEEE 30th Annual International Symposium on Personal, Indoor and Mobile Radio Communications (PIMRC), pp. 1–6. IEEE (2019)
15. Brown, R.G., Carlyle, J., Grigg, I., Hearn, M.: Corda: an introduction. R3 CEV, August **1**(15), 14 (2016)
16. Castells, R.M., Graham, I., Andrade, C., Churchill, G., Cox, C.: Automated metro operation: greater capacity and safer, more efficient transport. PTI **60**, 15–21 (2011)
17. Chen, Y.K.: Challenges and opportunities of internet of things. In: 17th Asia and South Pacific design automation conference, pp. 383–388. IEEE (2012)
18. Chohan, U.W.: Non-fungible tokens: blockchains, scarcity, and value. Critical Blockchain Research Initiative (CBRI) Working Papers (2021)
19. Christidis, K., Devetsikiotis, M.: Blockchains and smart contracts for the internet of things. IEEE Access **4**, 2292–2303 (2016)
20. Dib, O., Brousmiche, K.L., Durand, A., Thea, E., Hamida, E.B.: Consortium blockchains: overview, applications and challenges. Int. J. Adv. Telecommun. **11**(1&2), 51–64 (2018)
21. Etxeberria-Garcia, M., Labayen, M., Zamalloa, M., Arana-Arexolaleiba, N.: Application of computer vision and deep learning in the railway domain for autonomous train stop operation. In: 2020 IEEE/SICE International Symposium on System Integration (SII), pp. 943–948. IEEE (2020)

22. Falazi, G., Hahn, M., Breitenbücher, U., Leymann, F., Yussupov, V.: Process-based composition of permissioned and permissionless blockchain smart contracts. In: 2019 IEEE 23rd International Enterprise Distributed Object Computing Conference (EDOC), pp. 77–87. IEEE (2019)

23. Feng, Q., He, D., Zeadally, S., Khan, M.K., Kumar, N.: A survey on privacy protection in blockchain system. J. Netw. Comput. Appl. **126**, 45–58 (2019)

24. Fraga-Lamas, P., Fernández-Caramés, T.M., Castedo, L.: Towards the internet of smart trains: a review on industrial IoT-connected railways. Sensors **17**(6), 1457 (2017)

25. Gerke, S., Minssen, T., Cohen, G.: Ethical and legal challenges of artificial intelligence-driven healthcare. In: Artificial Intelligence in Healthcare, pp. 295–336. Elsevier (2020)

26. Hammi, M.T., Hammi, B., Bellot, P., Serhrouchni, A.: Bubbles of trust: a decentralized blockchain-based authentication system for IoT. Comput. Secur. **78**, 126–142 (2018)

27. Huang, J., Kong, L., Chen, G., Wu, M.Y., Liu, X., Zeng, P.: Towards secure industrial IoT: blockchain system with credit-based consensus mechanism. IEEE Trans. Industr. Inf. **15**(6), 3680–3689 (2019)

28. Imbault, F., Swiatek, M., De Beaufort, R., Plana, R.: The green blockchain: managing decentralized energy production and consumption. In: 2017 IEEE International Conference on Environment and Electrical Engineering and 2017 IEEE Industrial and Commercial Power Systems Europe (EEEIC/I&CPS Europe), pp. 1–5. IEEE (2017)

29. Johnson, E., Nica, E.: Connected vehicle technologies, autonomous driving perception algorithms, and smart sustainable urban mobility behaviors in networked transport systems. Contemp. Read. Law Soc. Justice **13**(2), 37–50 (2021)

30. Khan, M.A., Salah, K.: IoT security: review, blockchain solutions, and open challenges. Futur. Gener. Comput. Syst. **82**, 395–411 (2018)

31. Konecny, V., Barnett, C., Poliak, M.: Sensing and computing technologies, intelligent vehicular networks, and big data-driven algorithmic decision-making in smart sustainable urbanism. Contemp. Read. Law Soc. Justice **13**(1), 30–39 (2021)

32. Kuperberg, M., Kindler, D., Jeschke, S.: Are smart contracts and blockchains suitable for decentralized railway control? arXiv preprint arXiv:1901.06236 (2019)

33. Lee, B., Lee, J.H.: Blockchain-based secure firmware update for embedded devices in an internet of things environment. J. Supercomput. **73**(3), 1152–1167 (2017)

34. Liang, H., Zhang, Y., Xiong, H.: A blockchain-based model sharing and calculation method for urban rail intelligent driving systems. In: 2020 IEEE 23rd International Conference on Intelligent Transportation Systems (ITSC), pp. 1–5. IEEE (2020)

35. Lu, Y.: The blockchain: state-of-the-art and research challenges. J. Ind. Inf. Integr. **15**, 80–90 (2019)

36. Lundqvist, T., De Blanche, A., Andersson, H.R.H.: Thing-to-thing electricity micro payments using blockchain technology. In: 2017 Global Internet of Things Summit (GIoTS), pp. 1–6. IEEE (2017)

37. Mahjoub, Y.I., Chargui, T., Bekrar, A., Trentesaux, D.: Supply chain application of blockchain-based solutions for cyber-physical systems: Review and prospects. In: Borangiu, T., Trentesaux, D., Leitão, P., Cardin, O., Joblot, L. (eds.) SOHOMA 2021. Studies in Computational Intelligence, vol. 1034, pp. 545–558. Springer, Cham (2022). https://doi.org/10.1007/978-3-030-99108-1_39

38. Mahjoub, Y.I., Hassoun, M., Trentesaux, D.: Blockchain adoption for SMEs: opportunities and challenges

39. Malik, S., Kanhere, S.S., Jurdak, R.: Productchain: scalable blockchain framework to support provenance in supply chains. In: 2018 IEEE 17th International Symposium on Network Computing and Applications (NCA), pp. 1–10. IEEE (2018)
40. Marchese, A., Tomarchio, O.: An agri-food supply chain traceability management system based on hyperledger fabric blockchain. In: Proceedings of the 23rd International Conference on Enterprise Information Systems (ICEIS2021), vol. 2, pp. 648–658 (2021)
41. Mrabet, H., Alhomoud, A., Jemai, A., Trentesaux, D.: A secured industrial internet-of-things architecture based on blockchain technology and machine learning for sensor access control systems in smart manufacturing. Appl. Sci. **12**(9), 4641 (2022)
42. Mukhopadhyay, S.C., Suryadevara, N.K.: Internet of things: challenges and opportunities. Internet of Things **9**, 1–17 (2014)
43. Muniandi, G.: Blockchain-based robust and distributed level crossing control architecture for accident prevention in automated trains (2021)
44. Mylrea, M., Gourisetti, S.N.G.: Blockchain for smart grid resilience: exchanging distributed energy at speed, scale and security. In: 2017 Resilience Week (RWS), pp. 18–23. IEEE (2017)
45. Nakamoto, S.: Re: Bitcoin p2p e-cash paper. The Cryptography Mailing List (2008)
46. Naser, F.: The potential use of blockchain technology in railway applications: an introduction of a mobility and speech recognition prototype. In: 2018 IEEE International Conference on Big Data (Big Data), pp. 4516–4524. IEEE (2018)
47. Omohundro, S.: Cryptocurrencies, smart contracts, and artificial intelligence. AI Matters **1**(2), 19–21 (2014)
48. Pedrosa, A.R., Pau, G.: ChargeItUP: on blockchain-based technologies for autonomous vehicles. In: Proceedings of the 1st Workshop on Cryptocurrencies and Blockchains for Distributed Systems, pp. 87–92 (2018)
49. Perc, M., Ozer, M., Hojnik, J.: Social and juristic challenges of artificial intelligence. Palgrave Commun. **5**(1), 1–7 (2019)
50. Plissonneau, A., Trentesaux, D., Ben-Messaoud, W., Bekrar, A.: AI-based speed control models for the autonomous train: a literature review. In: 2021 Third International Conference on Transportation and Smart Technologies (TST), pp. 9–15. IEEE (2021)
51. Preece, J., Easton, J.: Blockchain technology as a mechanism for digital railway ticketing. In: 2019 IEEE International Conference on Big Data (Big Data), pp. 3599–3606. IEEE (2019)
52. Preece, J., Easton, J., Preece, J., Easton, J.: A review of prospective applications of blockchain technology in the railway industry. Preprint submitted Int. J. Railw. Technol. 1–22 (2019)
53. Sankar, L.S., Sindhu, M., Sethumadhavan, M.: Survey of consensus protocols on blockchain applications. In: 2017 4th international conference on advanced computing and communication systems (ICACCS), pp. 1–5. IEEE (2017)
54. Sharma, P.K., Kumar, N., Park, J.H.: Blockchain-based distributed framework for automotive industry in a smart city. IEEE Trans. Industr. Inf. **15**(7), 4197–4205 (2018)
55. Sikorski, J.J., Haughton, J., Kraft, M.: Blockchain technology in the chemical industry: machine-to-machine electricity market. Appl. Energy **195**, 234–246 (2017)
56. Singh, P., Dulebenets, M.A., Pasha, J., Gonzalez, E.D.S., Lau, Y.Y., Kampmann, R.: Deployment of autonomous trains in rail transportation: current trends and existing challenges. IEEE Access **9**, 91427–91461 (2021)

57. Symeonidis, G., Groumpos, P.P., Dermatas, E.: Traffic light detection and recognition using image processing and convolution neural networks. In: Kravets, A.G., Groumpos, P.P., Shcherbakov, M., Kultsova, M. (eds.) CIT&DS 2019. CCIS, vol. 1084, pp. 181–190. Springer, Cham (2019). https://doi.org/10.1007/978-3-030-29750-3_14

58. Szabo, N.: Formalizing and securing relationships on public networks. First monday (1997)

59. Thong, M., Cheong, A.: Energy efficiency in Singapore's rapid transit system. Journeys **I**, 38–47 (2012)

60. Tian, F.: An agri-food supply chain traceability system for china based on RFID & blockchain technology. In: 2016 13th International Conference on Service Systems and Service Management (ICSSSM), pp. 1–6. IEEE (2016)

61. Tizhoosh, H.R., Pantanowitz, L.: Artificial intelligence and digital pathology: challenges and opportunities. J. Pathol. Inform. **9**(1), 38 (2018)

62. Trentesaux, D., et al.: The autonomous train. In: 2018 13th Annual Conference on System of Systems Engineering (SoSE), pp. 514–520. IEEE (2018)

63. Valenta, M., Sandner, P.: Comparison of ethereum, hyperledger fabric and corda. Frankfurt Sch. Blockchain Center **8**, 1–8 (2017)

64. Wan, J., Li, J., Imran, M., Li, D., et al.: A blockchain-based solution for enhancing security and privacy in smart factory. IEEE Trans. Industr. Inf. **15**(6), 3652–3660 (2019)

65. Wang, L., Xu, L., Zheng, Z., Liu, S., Li, X., Cao, L., Li, J., Sun, C.: Smart contract-based agricultural food supply chain traceability. IEEE Access **9**, 9296–9307 (2021)

66. Wang, S., Li, D., Zhang, Y., Chen, J.: Smart contract-based product traceability system in the supply chain scenario. IEEE Access **7**, 115122–115133 (2019)

67. Wang, S., Yuan, Y., Wang, X., Li, J., Qin, R., Wang, F.Y.: An overview of smart contract: architecture, applications, and future trends. In: 2018 IEEE Intelligent Vehicles Symposium (IV), pp. 108–113. IEEE (2018)

68. Wang, Y., Zhang, M., Ma, J., Zhou, X.: Survey on driverless train operation for urban rail transit systems. Urban Rail Transit **2**(3), 106–113 (2016)

69. Wohrer, M., Zdun, U.: Smart contracts: security patterns in the ethereum ecosystem and solidity. In: 2018 International Workshop on Blockchain Oriented Software Engineering (IWBOSE), pp. 2–8. IEEE (2018)

70. Xia, Q., Sifah, E.B., Asamoah, K.O., Gao, J., Du, X., Guizani, M.: MeDShare: trust-less medical data sharing among cloud service providers via blockchain. IEEE Access **5**, 14757–14767 (2017)

71. Xu, X., Weber, I., Staples, M.: Architecture for Blockchain Applications. Springer, Cham (2019)

72. Yang, R., et al.: Public and private blockchain in construction business process and information integration. Autom. Constr. **118**, 103276 (2020)

73. Yuan, Y., Wang, F.Y.: Towards blockchain-based intelligent transportation systems. In: 2016 IEEE 19th International Conference on Intelligent Transportation Systems (ITSC), pp. 2663–2668. IEEE (2016)

74. Zhang, P., White, J., Schmidt, D.C., Lenz, G., Rosenbloom, S.T.: FHIRChain: applying blockchain to securely and scalably share clinical data. Comput. Struct. Biotechnol. J. **16**, 267–278 (2018)

75. Zheng, Z., et al.: An overview on smart contracts: challenges, advances and platforms. Futur. Gener. Comput. Syst. **105**, 475–491 (2020)

Solving a Job Shop Scheduling Problem Using Q-Learning Algorithm

Manal Abir Belmamoune[1](✉), Latéfa Ghomri[1], and Zakaria Yahouni[2]

[1] Manufacturing Engineering Laboratory of Tlemcen (MELT), University of Tlemcen, Tlemcen, Algeria
{manalabir.belmamoune,latefa.ghomri}@univ-tlemcen.dz
[2] Univ. Grenoble Alpes, CNRS, Grenoble INP G-SCOP, 38000 Grenoble, France
zakaria.yahouni@grenoble-inp.fr

Abstract. Job Shop Scheduling Problem (JSSP) is among the combinatorial optimization and Non-Deterministic Polynomial-time (NP) problems. Researchers have contributed to this area using several methods. Among them, we cite machine learning algorithms, more precisely Reinforcement Learning (RL). This algorithm is suitable for the discussed problem as agents can learn decisions and optimize them according to the environment's state. Once the learning process is efficient, RL can be used in real-time to cope with changes. This paper deals with the JSSP using the RL algorithm, more specifically a Q-learning algorithm. We propose a new representation of the state of the environment based on machine loads and the agent is evaluated twice using two different methods. The actions selected by the agent are the dispatching rules. Our QL approach with such representation is compared with the results obtained by the literature.

Keywords: Job shop scheduling problem · Reinforcement learning · Dispatching rules · Q-learning

1 Introduction

Production planning ensures that the company can develop, manufacture and finalize products efficiently and within predefined deadlines. Scheduling is an important step when planning a production process. Manufacturing companies can allocate tasks to needed resources efficiently by using production scheduling. Job Shop Scheduling Problems (JSSP) consist of scheduling jobs with different rooting on different machines. An example of such a problem is assembling different products on different machines; each product needs a specific set of machines. These problems are currently one of the most relevant issues in manufacturing, especially for systems that require a high degree of flexibility to meet customer needs. As manufacturers adapt to this demand, they will make changes to increase system flexibility, which will inevitably increase the complexity of the overall system [1]. JSSP are combinatorial optimization problems and Non-Deterministic Polynomial-time (NP) hard problems.

One resolution method consists in using dispatching rules due to their simplicity and ease of use, such as Shortest Processing Time (SPT), First in First Out (FIFO), Longest

© The Author(s), under exclusive license to Springer Nature Switzerland AG 2023
T. Borangiu et al. (Eds.): SOHOMA 2022, SCI 1083, pp. 196–209, 2023.
https://doi.org/10.1007/978-3-031-24291-5_16

Processing Time (LPT), Last in First Out (LIFO), and Earliest Due Date (EDD). The problem is that dispatching rules may be much more appropriate for one problem and not for another. Moreover, due to the large-scale combinatorial optimization problems, it is complicated to use exact methods; this is the reason solutions related to Artificial Intelligence may be more appropriate in terms of time computation. Generally, meta-heuristics such as Genetic Algorithms (GA) are used [2], or Machine Learning such as Supervised Learning [3], Unsupervised Learning [4], and RL methods [5].

In this paper, we are interested in JSSP resolution using Machine Learning (ML), in particular, Reinforcement Learning (RL). RL consists of autonomous agents (software programs) that learn to choose appropriate actions/decisions in a state (e.g., workshop configuration at a specific time) to achieve their goals by interacting with the environment [6]. The reason for this choice is the dynamic environment that places the agent in a situation of taking a real-time decision and evaluating these decisions (actions). There are many RL algorithms, such as SARSA [7], actor-critic algorithm [8], and Q-learning algorithm [9]. In this research, we focus on the Q-learning algorithm (QL) which is a model-free RL algorithm in which the agent learns from real-world experience examples from the environment and never uses the generated next-state predictions. Such an algorithm is also easy to implement compared to other more complicated ones. Based on the state of the environment (e.g., machine availability), an action such as assembling a component is considered. This action is evaluated using what is called a reward. The reward (for example the production time, machine utilization, etc.) is a function that helps the agent evaluate its decisions and avoid taking the same incorrect actions in future states. Actions of the same problem are taken in a loop called an *episode* until the learning process is finished. After this learning phase (many episodes), the agent is capable of evaluating a next action and taking the appropriate decision. In this paper, we propose a new representation of the state, actions and rewards. The state represents the machine's queue for jobs that need to be ordered. An action taken by the agent consists of selecting the most appropriate dispatching rule according to which it will schedule the jobs. This action is evaluated twice; the first time during the job's assignment, and the second time after obtaining the final schedule, hence knowing the makespan (the max. Completion time of jobs, C_{max}). The results of our approach are compared with those of the dispatching rules (FIFO, LIFO, SPT, and LPT), of the meta-heuristics GA [10], Hybrid GA [11], and of Deep reinforcement learning (DRL) [12].

This paper is organized as follows: Sect. 1 presents work related to JSSP using the RL algorithms. Section 2 describes the JSSP. Section 3 introduces the RL algorithm that was used and its implementation. Section 4 analyses and compares the results obtained. Section 5 formulates conclusions and enounces future research perspectives.

2   Related Works

Job shop-type workshops are the most complicated in terms of assigning tasks to machines, where each job has its specific routing as shown in Fig. 1, and the processing time of tasks in the same machine is not necessarily identical. A real example of JSSP can be the assembling of different types of products using a set of machines.

RL can be used to solve JSSP and is defined as a technique that learns what an agent does to interact with a given environment to maximize rewards [21]. A learning agent must be able to sense to some extent the state of its environment and take actions that affect the state. The agent must also have one or more targets related to the state of the environment [22]. In recent years, several papers related to JSSP have used different types of RL algorithms, with different objective functions, state representations and reward functions. Makespan was used as objective function in [13, 14], and [15]; the authors chose the QL algorithm, as this is one of the most used algorithms in JSSP.

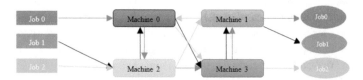

Fig. 1. Job shop example

Wang et al. [12] proposed a dynamic scheduling method based on deep reinforcement learning (DRL). They adopted the proximal policy optimization (PPO) to find the optimal policy for scheduling to handle the large state scale. To represent the state, they used three matrices: a machine matrix, a processing time matrix for each job which does not change over time, and a job processing status matrix. The function of the reward is related to the utilization of the machine and the obtained makespan. Liu et al. [8] proposed deep reinforcement learning to deal effectively with the JSSP. The proposed network comprises an actor network and a critical network. They represent the state of the environment in three matrices: a process time matrix, a boolean matrix of the job assigned to each machine and a boolean matrix of completed jobs; the reward is evaluated by three different functions: processing time of the selected job, remaining processing time of the job and the comparison of the smallest makespan.

Jiménez et al. [16] developed a tool for JSSP which works with the Q-learning algorithm and can be adapted to different scheduling scenarios such as Flexible JSSP and Parallel Machines problems. Zhou et al. [17] used a DRL; they built two networks - the prediction and target network. Waschneck et al. [18] proposed cooperative DQN agents, which utilize deep neural networks trained with user-defined objectives to optimize scheduling. They applied their study in a small factory simulation of an abstracted frontend-of-line semiconductor production facility. Tassel et al. [19] represented a DRL environment for Job-Shop Scheduling; they proposed a meaningful state representation using a matrix in which each row describes information about the job such as the allocation of operations. They designed a dense reward function based on the scheduled area. After each action, they computed the difference between the duration of the allocated operations and the idle time of a machine. Samsonov et al. [20] proposed a new form of reward and designed a new space for action. They represented the state by six characteristics: the machine states, the sum of all operations' processing times currently in the queue of each machine, the sum of all operations' processing times for each job, the duration of the next operation for each job, the index of the next required machine for each job, and finally the time already elapsed until a given time moment. The reward of the agent is evaluated at the end of the episode.

All authors who contributed to this field used different approaches to define the states and evaluate the agent's actions. They choose to represent the state by giving all information about the environment to the agent. However, this representation of the state may not be necessary; the reason is that the agent that selects the set of jobs on the machine does not need to have information about other jobs on the other machines. Action selection in most works is based on the selection of jobs to be executed on a machine. This puts the agent in a situation where multiple actions representing all orders are proposed on the same machine. We can find this phenomenon in large workshops. In our study, we have a different representation of the state, and the action/reward consists of selecting the most appropriate dispatching rule depending on the current situation. Before defining this representation, we introduce the problem statement in the next section.

3   Problem Description

In a job shop there is a set of n jobs $\{J_1, J_2, J_3, \ldots, J_n\}$, each job having its operations: $J_i = \{O_{i1}, O_{i2}, O_{i3}, \ldots O_{ij} \ldots, O_{im}\}$ with $i \in \{1, n\}, j \in \{1, m\}$ and m is the number of operations executed on m machines: $\{M_1, M_2, M_3, \ldots, M_m\}$. All the tasks or the operations O_{ij} are defined with a start time t_{ij}, a processing time p_{ij}, and a completion time C_{ij}.

The completion time of operations on a machine is referred to as C_j. The maximum load $maxLoad_j$ of a machine M_j is the sum of the processing times of all operations to be executed on this machine. The instantaneous load $instLoad_j$ represents the load of the machine after the affectation of an operation (it is the ratio between executed operations and $maxLoad_j$). $batch_j$ and seq_j are the current queues of M_j and the current sequence of tasks of this machine, respectively.

Important constraints should be respected to avoid overlapping during the allocation of operations to the machines. The precedence constraint is presented in Fig. 2.a, where a new operation O_{ij+1} of J_i cannot be started until the previous one O_{ij} is finished.

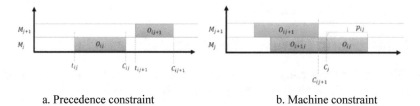

a. Precedence constraint   b. Machine constraint

Fig. 2. Constraints presentation

The other constraint related to the completion time of machine C_j states that we cannot start an operation on a machine without finishing the previous one, as shown in Fig. 2.b. To calculate it, we must know the max. Value of the next operation's completion time C_{ij+1} on a machine M_{j+1} and the completion time of the machine, which is:

$$C_j = p_{ij} + \max(C_j, C_{ij+1})$$

The objective of this study is to minimize the total completion time of all jobs in all machines, $C_{max}(makespan) = \min(C_{\max}, withC_{max} = \max(C_j))$. The RL technique can be used to schedule jobs (operations) optimally.

From different RL algorithms, we selected Q-learning which represents a form of model-free RL algorithm. The core of the algorithm is a simple value iteration update; each state-action pair (s, a) has a Q-value associated. When action a is selected by the agent located in state s_t, the return of the environment after applying an action is represented as reward r_t. The Q-value for that state-action pair is updated based on the reward received when selecting that action, and the best Q-value for the subsequent state s_{t+1} [23]. The Q-values are saved in a Q-table, making the agent access them rapidly. Bellman updates the Q-values is as follows:

$$Q^{new}(s_t, a_t) = (1 - \alpha)Q^{old}(s_t, a_t) + \alpha(r_t + \gamma(\max(Q(s_{t+1}, a)))) \tag{1}$$

where α represents the learning rate; it is the acceptance probability of the target value. $(1 - \alpha)$ is the probability of keeping the old Q-value and γ is a discount factor, used to balance the immediate and the future reward. $(\max(Q(s_{t+1}, a)))$ represents the maximum Q-value of the next state, and this is the reason why the Q-learning is an off-policy algorithm - the policy used during the evaluation stage can differ from the one used in the improvement stage, which leads to more exploration at the expense of convergence speed [24]. In other words, it is not necessarily that the action a_t chosen in the state s_t is the same one as a in the target $(r_t + \gamma(\max(Q(s_{t+1}, a))))$.

The goal is to find the optimal policy in the long term, to reach this objective. The agent needs a process to explore the environment by taking actions randomly and interacting with the environment; it exploits by maximizing its gains when it finds an optimal policy. The ε-greedy policy allows the agent to explore with a probability of ε and to exploit with a probability of $1 - \varepsilon$. The use of Q learning in this field has given interesting results. It represents the most adaptable algorithm to real-time decision-making situations, which is the case in industrial systems in general. The implementation of this algorithm requires a well-defined representation of its elements, which we define in detail in the next section. This helps the agent avoid situations that do not represent reality and may interfere with the learning phase.

4   Reinforcement Learning Implementation

We will present in this section the basic elements of the algorithm in terms of the JSSP, then the learning phase and finally the application phase of the algorithm.

4.1   State Representation

Our objective is to minimize the total completion time C_{max}. Compared to literature, we have represented the state as a machine with at least two tasks in the queue. With only this information, an agent must choose an action that represents the selection of one prioritized task in the queue. We have chosen to define the state in this way because the necessary information for the agent specifies which task must be allocated at this moment. There is no decision to take for the machines that have only one job in their queue. Therefore, the state in Eq. (1) is $s_t = batch_j\ when|batch_j| \geq 2$.

4.2 Action Selection

In our study, we tried to choose the action differently; the agent will choose one of the popular dispatching rules for each decision. There are two for this selection. Firstly, it is essential to schedule using a rule in a state; each rule may perform better in some states. Secondly, the agent will choose only four actions which are the dispatching rules SPT, LPT, FIFO and LIFO instead of choosing all the possible Jobs. Therefore, the actions in Eq. (1) are:

$$a_t = \{SPT, LPT, FIFO, LIFO\}$$

QL is executed using many episodes/iterations; for each one different decision are taken. For each decision, the agent will apply an action for one machine or many actions for different ones, depending on the availability of the machines in this episode. The episode is executed and repeated until the end of the allocation of all the jobs in all the machines and the final makespan is obtained.

4.3 Reward Function

The reward function represents an important step in the implementation of the algorithm. The reward is most often proportional to the optimization target or a value that highly correlates with it (e.g., makespan and average utilization) [24]. In our study, the reward is related to the machine loads ($instLoad_j$) and the completion time of the job. Loading machines in a balanced manner over time provides system flexibility by increasing the machine's utilization rate and availability when needed; this leads to minimal completion time, but it does not guarantee optimality. When the agent is in a state s_t and it must take a decision on M_j, it calculates the load of each machine ($instLoad_j$), then it calculates the average loads of all the machines except M_j. Then, the current load of $M_j(instLoad_j)$ is compared with the average to see if the machine is balanced with the others or not. In this step, we will evaluate the action chosen by the agent, and whether it deserves a reward or a penalty. For instance, if $instLoad_j = 20\%$ and the average load of the other machines is 60%, this means that the machines are not well balanced. In this case, the correct action is to choose the rule that prioritizes the job with the longest processing time. If the chosen rule does not select first the job with this characteristic, a penalty is applied.

The proposed reward function is calculated using the resulting completion time (after applying the action chosen by the agent) C_j and the sum of all completion times that can be obtained with the other actions (if they were chosen). This process evaluates the policy chosen by the agent, assumes cooperation between the dispatching rules in the same episode and optimizes the policy:

$$r_t = \begin{cases} (1 - C_j)/ \sum_{k=1}^{4} C_j \ (\textbf{all actions}), & \textit{if chosen ction balances the loads} \\ -(1 - C_j)/ \sum_{k=1}^{4} C_j \ (\textbf{all actions}), & \textit{otherwise} \end{cases}$$

Another simple method of representing the reward is proposed:

$$r_t = \begin{cases} +1, & \textit{if chosen action balances the loads} \\ -2, & \textit{otherwise} \end{cases}$$

The agent will be evaluated twice, the first time during the learning process in an episode, i.e.; when assigning jobs to machines, using loads and completion times as we mentioned before; the second time at the end of the episode where it obtains the C_{max}. In this case, the evaluation is done using the C_{max} value. If this is the smallest value obtained until that moment, the rewards of all the actions taken in this episode will be maximized; otherwise, nothing will be added. This policy helps the agent to maximize the gain only for the right decisions. The evaluation at the end of the episode is represented as follows:

$$r_t = ((C_{max}(maximal) - C_{max})/C_{max}(maximal)) \times episode,$$
$$C_{max} \leq C_{max}(maximal)$$

4.4 Learning Phase

In this phase, we will use the Q-learning algorithm adapted to the different characteristics of the JSSP. During the learning phase, the agent will try to maximize the rewards of a good solution. We have defined the reward specifying how it will maximize each time it finds a good solution to its exploration path. It must also increase the exploitation rate each time it finds a good solution; therefore, a mechanism was added to help the agent to exploit more solutions each time it finds a good solution. This is conducted by increasing the value of ε each time until ε reaches the value 0.999. In this way, the agent does not explore too much, which leads to slow learning and does not converge quickly to the best solution. Table 1 resumes the learning phase.

4.5 Final Model

After the end of the learning phase, the Q-table contains all the states experienced by the agent, as well as the Q-values of each action applied in each state. Actions are considered as good decisions with maximum Q-values or bad decisions with minimum Q-values, since the learning strategy leads to maximizing the gain for good decisions. In this phase, the agent's optimal policy π^* is used to choose actions with maximum Q-value. This brings us to the solution approximated by the learning algorithm:

$$\pi^*(s_t) = \max(Q(s_t, a))$$

Table 1. Learning phase algorithm

1.	initialize the Q − table empty
2.	calculate the $maxLoad_j$ and $instLoad_j$ of all the machines
3.	$\varepsilon = 0.2,\ \alpha = 0.9,\ \gamma = 0.75$
4.	**For several episodes:**

- **While the end of the allocation of all jobs is not done:**
    - a. **Fill the machines queue**
    - b. **If $|batch_j| \geq 2$:**
        - Add a row Q − table with 4 columns with null values
        - Choose an action using an ε − greedy policy
        - Calculate the reward using the first evaluation
        - Update the Q − value ($Q^{old}(s_t, a_t) \leftarrow Q^{new}(s_t, a_t)$) in the Q − table
        - Add the job to the machine sequence
        - Update the state of the batch ($s_t \leftarrow s_{t+1}$)
        - If there is another machine $[m + 1].batch \geq 2$ return to (b)
- **Compare the C_{max} of this episode with the smallest one obtained and compute the reward using the second evaluation**
- **Increase the ε value**
- **If $\varepsilon > 0.9$:**
    - $\varepsilon = 0.999$

5   Results and Discussion

QL01 represents the results when using the reward function related to the completion time of actions, and QL02 when the reward is +1 and −2. We have applied this experience to 46 benchmark instances from the OR Library [25] and compared the results given by QL01 and QL02 with the dispatching rules (FIFO, SPT, LPT), GA [10], HGA [11], DRL [12] and the optimal solution as well. The algorithms are executed under 1000 episodes for small instances (ft06, ft10, la01 to la15, orb01 to orb03 and orb07) and 10000 episodes for mean and large scales (la16 to la40) in a reasonable time. The results are shown in Table. 2. Best results of each instance are highlighted.

Table 2. Results of QL01 and QL02, dispatching rules (SPT, LPT, FIFO), meta-heuristics (GA, HGA), and DRL.

Workshop	SPT	LPT	FIFO	OPT	QL 01		QL 02		GA [10]		HGA [11]		DRL [17]	
					C_{max}	Error	C_{max}	Error	C_{max}	Error	C_{max}	Error	C_{max}	Error
ft06 (06 × 06)	88	77	65	55	57	−2	57	−2	**55**	0	**55**	0	57	−2
ft10 (10 × 10)	1074	1295	1184	930	1017	−87	999	−69	994	−64	**938**	−8	1033	−103

(continued)

Table 2. (*continued*)

Workshop	SPT	LPT	FIFO	OPT	QL 01		QL 02		GA [10]		HGA [11]		DRL [17]	
					C_{max}	Error	C_{max}	Error	C_{max}	Error	C_{max}	Error	C_{max}	Error
la01 (05 × 10)	751	822	772	666	**666**	0	**666**	0	667	−1	**666**	0	**666**	0
la02 (05 × 10)	821	990	830	655	685	−30	685	−30	676	−21	**655**	0	715	−60
la03 (05 × 10)	672	825	755	597	623	−26	619	−22	627	−30	**597**	0	634	−37
la04 (05 × 10)	711	818	695	590	620	−30	620	−30	608	−18	**590**	0	665	−75
la05 (05 × 10)	610	693	610	593	**593**	0	**593**	0	**593**	0	**593**	0	**593**	0
la06 (05 × 15)	1200	1125	926	926	**926**	0	**926**	0	**926**	0	**926**	0	**926**	0
la07 (05 × 15)	1034	1069	1088	890	**890**	0	967	−77	891	−1	**890**	0	894	−4
la08 (05 × 15)	942	1035	980	863	**863**	0	876	−13	**863**	0	**863**	0	**863**	0
la09 (05 × 15)	1045	1183	1018	951	**951**	0	**951**	0	**951**	0	**951**	0	**951**	0
la10 (05 × 15)	1049	1132	1006	958	**958**	0	**958**	0	**958**	0	**958**	0	**958**	0
la11 (05 × 20)	1473	1467	1272	1222	**1222**	0	**1222**	0	**1222**	0	**1222**	0	**1222**	0
la12 (05 × 20)	1203	1240	1039	1039	**1039**	0	**1039**	0	**1039**	0	**1039**	0	**1039**	0
la13 (05 × 20)	1275	1230	1199	1150	**1150**	0	**1150**	0	**1150**	0	**1150**	0	**1150**	0
la14 (05 × 20)	1427	1434	1292	1292	**1292**	0	**1292**	0	**1292**	0	**1292**	0	**1292**	0
la15 (05 × 20)	1339	1612	1587	1207	**1207**	0	1302	−95	1256	−49	**1207**	0	1212	−5
la16 (10 × 10)	1156	1229	1180	945	983	−38	995	−50	993	−48	**945**	0	/	/
la17 (10 × 10)	924	940	943	784	800	−16	800	−16	804	−20	**784**	0	/	/
la18 (10 × 10)	981	1114	1049	848	873	−25	861	−13	874	−26	**848**	0	/	/
la19 (10 × 10)	940	1062	983	842	875	−33	875	−33	895	−53	**844**	−2	/	/
la20 (10 × 10)	1000	1272	1272	902	939	−37	941	−39	942	−40	**911**	−9	/	/
la21 (10 × 15)	1324	1451	1265	1046	1107	−61	1126	−80	1180	−134	**1046**	0	/	/

(*continued*)

Table 2. (*continued*)

Workshop	SPT	LPT	FIFO	OPT	QL 01		QL 02		GA [10]		HGA [11]		DRL [17]	
					C_{max}	Error	C_{max}	Error	C_{max}	Error	C_{max}	Error	C_{max}	Error
la22 (10 × 15)	1180	1315	1369	927	1022	−95	1026	−99	1103	−176	**935**	−8	/	/
la23 (10 × 15)	1162	1302	1354	1032	1038	−6	1053	−21	1100	−68	**1032**	0	/	/
la24 (10 × 15)	1203	1245	1141	935	1000	−65	1021	−86	1077	−142	**953**	−18	/	/
la25 (10 × 15)	1449	1374	1283	977	1074	−97	1059	−82	1116	−139	**984**	−7	/	/
la26 (10 × 20)	1499	1564	1372	1218	1302	−84	1327	−109	1433	−215	**1218**	0	/	/
la27 (10 × 20)	1784	1700	1644	1252	1364	−112	1361	−109	1469	−217	**1256**	−4	/	/
la28 (10 × 20)	1610	1844	1532	1273	1358	−85	1363	−90	1408	−135	1225	48	/	/
la29 (10 × 20)	1556	1720	1540	1238	1363	−125	1348	−110	1439	−201	1196	42	/	/
la30 (10 × 20)	1792	1866	1664	1355	1390	−35	1440	−85	1546	−191	**1355**	0	/	/
la31 (10 × 30)	1954	2340	1918	1784	1857	−73	1802	−18	1906	−122	**1784**	0	/	/
la32 (10 × 30)	2165	2513	2110	1850	1914	−64	1883	−33	2002	−152	**1850**	0	/	/
la33 (10 × 30)	1901	2306	1873	1719	1817	−98	1802	−83	1838	−119	**1719**	0	/	/
la34 (10 × 30)	2005	2324	1925	1721	1828	−107	1794	−73	1934	−213	**1721**	0	/	/
la35 (10 × 30)	2118	2421	2142	1888	1985	−97	1925	−37	2106	−218	**1888**	0	/	/
la36 (15 × 15)	1854	1946	1516	1268	1415	−147	1372	−104	1480	−212	**1287**	−19	/	/
la37 (15 × 15)	1632	1944	1873	1397	1533	−136	1561	−164	1606	−209	**1408**	−11	/	/
la38 (15 × 15)	1395	1732	1475	1196	1334	−138	1351	−155	1435	−239	**1219**	−23	/	/
la39 (15 × 15)	1540	1822	1532	1233	1358	−125	1360	−127	1456	−223	**1245**	−12	/	/
la40 (15 × 15)	1493	1822	1604	1222	1311	−89	1315	−93	1492	−270	**1241**	−19	/	/
orb01 (10 × 10)	1478	1410	1368	1059	**1119**	−60	1177	−118	1212	−153	/	/	1131	−72

(*continued*)

Table 2. (*continued*)

Workshop	SPT	LPT	FIFO	OPT	QL 01		QL 02		GA [10]		HGA [11]		DRL [17]	
					C_{max}	Error	C_{max}	Error	C_{max}	Error	C_{max}	Error	C_{max}	Error
orb02 (10 × 10)	1175	1293	1007	888	**918**	−30	922	−34	960	−72	/	/	993	−105
orb03 (10 × 10)	1179	1430	1405	1005	**1071**	−66	1075	−70	1162	−157	/	/	1092	−87
orb07 (10 × 10)	475	470	504	397	**418**	−21	423	−26	408	−11	/	/	432	−35

Our approaches gave good results compared with the dispatching rules. In all instances, the proposed algorithms produced solutions much closer to the optimal solution compared to the minimal solution among the dispatching rules. The reason is that it is better to use a mix of rules in a problem rather than using only one rule. By comparing the difference between the GA and optimal solution and the difference between the QL01and optimal solution, 89,13% of the 46 instances give best results than GA, especially the instances with a large scale. For QL02, for 80,43% of the 46 instances QL01 gives a better solution compared with GA in large scales and some average scales instances except la07, la08, la15, and la16. By comparing with HGA, our algorithm QL01 gives a good result for 27,90% of the compared instances, and 20,93% for QL02.

The reason our approaches do not give better results compared with HGA was caused by limiting the possible actions. The agent chooses four actions instead of all the possible Jobs that can be allocated. For DRL, QL01 gives a good solution for all instances (100%), and 80,95% compared with QL02. Figure 3 shows the convergence of the learning algorithm. At the beginning of learning, the agent explores the environment. Once it finds a better solution, it exploits and keeps a probability of exploring.

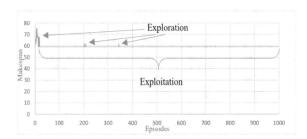

Fig. 3. Convergence of solution for instance ft06 (06 × 06) QL01

As shown in the picks in Fig. 4, in case the agent finds a bad solution, it goes back to the good solution found and maximizes it; in case it finds another good solution, it will maximize the rewards for the new one.

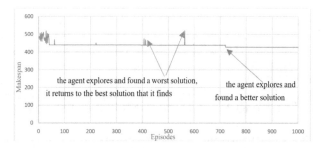

Fig. 4. Convergence of solution for instance orb07 (10 × 10) QL02

The large size instances do not restrict the agent to search the solution space and find a solution that approaches the optimal solution, see Fig. 5. The exploration time of the environment increases every time we have a larger instance. This is due to the expansion of the solution space from a small sized instance to an instance of large size.

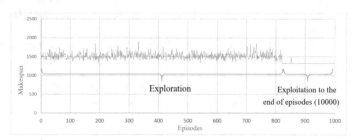

Fig. 5. Convergence of solution for la40 (15 × 15) QL02

6 Conclusion

In our article, we discussed the JSSP problem of production planning process in a manufacturing system. We used Q-Learning as solver because it is best suited for this type of problem. A new representation of the environment's state is proposed, that only affects the machines on which the agent will make decisions. Two new reward functions are introduced based on machine loads. The agent's decision is then twice evaluated, first during the assignment of jobs and second at the end of the schedule.

The application of the algorithm is compared with work described in the literature. The results of the algorithm outperformed the genetic algorithm and dispatching rules. The HGA gave a good result compared to our algorithm because the agent that we programmed has the choice between only four actions as well as a single agent in charge of assigning jobs to machines. Future perspectives consist of implementing our algorithm in a disturbed environment where real-time decisions are taken. A comparative analysis with constraint programming methods in terms of time can be made for small instances.

References s

1. Kardos, C., Laflamme, C., Gallina, V., Sihn, W.: Dynamic scheduling in a job-shop production system with reinforcement learning. Procedia CIRP **97**, 104–109 (2020). https://doi.org/10.1016/j.procir.2020.05.210

2. Chryssolouris, G., Subramaniam, V.: Dynamic scheduling of manufacturing job shops using genetic algorithms. J. Intell. Manuf. **12**, 281–293 (2001). https://doi.org/10.1023/A:1011253011638

3. Schmidt, J., Stober, S.: Approaching scheduling problems via a deep hybrid greedy model and supervised learning. In: Proceedings of the 17th IFAC Symposium on Information Control Problems in Manufactur-ing Budapest (2021). IFAC-PapersOnLine, Vol. 54, Issue 1, pp. 805-810. https://doi.org/10.1016/j.ifacol.2021.08.095

4. Cheng, C.-Y., Pourhejazya, P., Ying, K.-C., Lin, C.-F.: Unsupervised learning-based Artificial Bee Colony for minimizing non-value-adding operations. J. Appl. Soft Comput., 107280 (2021). https://doi.org/10.1016/j.asoc.2021.107280

5. Lang, S., Kuetgens, M., Reichardt, P., Reggelin, T.: Modeling production scheduling problems as reinforcement learning environments based on discrete-event simulation and OpenAI Gym. In: Proceedings of the 17th IFAC Symposium on Information Control Problems in Manufactur-ing Budapest (2021). IFAC-PapersOnLine, Vol. 54, Issue 1, pp. 793-798. https://doi.org/10.1016/j.ifacol.2021.08.093

6. Wang, Y.-C., Usher, J.M.: Application of reinforcement learning for agent-based production scheduling. Eng. Appl. Artif. Intell. **18**(1), 73–82 (2004). https://doi.org/10.1016/j.engappai.2004.08.018

7. Aissani, N., Trentesaux, D.: Efficient and effective reactive scheduling of manufacturing system using Sarsa-multi-objective agents. In: Proceedings of the 7th International Conference on MOSIM, Paris, pp. 698–707 (2008). file:///C:/Users/BT/Downloads/MOSIM08_aissani_etal_finalx.pdf

8. Liu, C.-L., Chang, C.-C., Tseng, C.-J.: Actor-critic deep reinforcement learning for solving job shop scheduling problems. IEEE Access **8**, 71752–71762 (2020). https://doi.org/10.1109/ACCESS.2020.2987820

9. Wei, Y., Zhao, M.: Composite rules selection using reinforcement learning for dynamic job-shop scheduling. In: IEEE Conference on Robotics, Automation and Mechatronics, vol. 2, pp. 1083–1088 (2004). https://doi.org/10.1109/RAMECH.2004.1438070

10. Ombuki, B.M., Ventresca, M.: Local search genetic algorithms for the job shop scheduling problem. Appl. Intell. **21**, 99–109 (2004). https://doi.org/10.1023/B:APIN.0000027769.48098.91

11. Qing, R., Wang, Y.: A new hybrid genetic algorithm for job shop scheduling problem. Comput. Oper. Res. **39**(10), 2291–2299 (2012). https://doi.org/10.1016/j.cor.2011.12.005

12. Wang, L., et al.: Dynamic job-shop scheduling in smart manufacturing using deep reinforcement learning. Comput. Netw. **190**, 107969 (2021). https://doi.org/10.1016/j.comnet.2021.107969

13. Gabel, T., Riedmiller, M.: On a successful application of multi-agent reinforcement learning to operations research benchmarks. In: IEEE International Symposium on Approximate Dynamic Programming and Reinforcement Learning, pp. 68–75 (2007). https://doi.org/10.1109/ADPRL.2007.368171

14. Yingzi, W., Xinli, J., Pingbo, H., Kanfeng G.: Pattern driven dynamic scheduling approach using reinforcement learning. In: Proceedings of the IEEE International Conference on Automation and Logistics, Shenyang, pp. 514–519 (2009). https://doi.org/10.1109/ICAL.2009.5262867

15. Wang, Y.-F.: Adaptive job shop scheduling strategy based on weighted Q-learning algorithm. J. Intell. Manuf. **31**(2), 417–432 (2018). https://doi.org/10.1007/s10845-018-1454-3

16. Martínez Jiménez, Y., Coto Palacio, J., Nowé, A.: Multi-agent reinforcement learning tool for job shop scheduling problems. In: Dorronsoro, B., Ruiz, P., de la Torre, J.C., Urda, D., Talbi, E.-G. (eds.) OLA 2020. CCIS, vol. 1173, pp. 3–12. Springer, Cham (2020). https://doi.org/10.1007/978-3-030-41913-4_1

17. Zhou, L., Zhang, L., Horn, B.K.P.: Deep reinforcement learning-based dynamic scheduling in smart manufacturing. Procedia CIRP **93**, 383–388 (2020). https://doi.org/10.1016/j.procir.2020.05.163

18. Waschneck, B., et al.: Optimization of global production scheduling with deep reinforcement learning. Procedia CIRP **72**, 1264–1269 (2018). https://doi.org/10.1016/j.procir.2018.03.212

19. Tassel, P., Gebser, M., Schekotihin, K.: A reinforcement learning environment for job-shop scheduling. arXiv preprint arXiv:2104.03760 (2021)

20. Samsonov, V., et al.: Manufacturing control in job shop environments with reinforcement learning. In: Proceedings of the 13th International Conference on Agents and Artificial Intelligence (ICAART 2021), pp. 589–597 (2021). https://doi.org/10.5220/0010202405890597

21. Usuga Cadavid, J.P., Lamouri, S., Grabot, B., Pellerin, R., Fortin, A.: Machine learning applied in production planning and control: a state-of-the-art in the era of industry 4.0. J. Intell. Manuf. **31**(6), 1531–1558 (2020). https://doi.org/10.1007/s10845-019-01531-7

22. Palacio, J.C., Jiménez, Y.M., Schietgat, L., Van Doninck, B., Nowé, A.: A Q-learning algorithm for flexible job shop scheduling in a real-world manufacturing scenario. Procedia CIRP **106**, 227–232 (2022). https://doi.org/10.1016/j.procir.2022.02.183

23. Rinciog, A., Meyer, A.: Towards standardizing reinforcement learning approaches for stochastic production scheduling. arXiv preprint arXiv:2104.08196 (2021)

24. OR - Library. http://people.brunel.ac.uk/~mastjjb/jeb/orlib/files/jobshop1.txt

Semantic Approach to Formalize Knowledge from Building Renovation Domain: Application to the IsoBIM Project

Haya Naanaa[(✉)], Hind Bril El Haouzi, William Derigent, and Mario Lezoche

CRAN CNRS, UMR 7039, Université de Lorraine, 54506 Vandœuvre-lès-Nancy, France
{haya.naanaa,hind.el-haouzi,william.derigent,
mario.lezoche}@univ-lorraine.fr

Abstract. The introduction of semantic technologies to the construction field is a concept that has been explored many times over the last few years. However, building renovation tends to be overlooked when talking about advancements in construction. Even if renovation projects consists 57% of construction activities, the renovation field is still the least developed. Information exchange between the project's different actors is still a persistent problem. Therefore, this paper proposes a semantic approach that aims to formalize knowledge from the renovation domain, using BIM as the information source, to design a dynamic digital representation of a renovation project (construction digital twin). In this context, and to highlight the remaining scientific questions, a case study of a renovation project is used to assess the current state of the renovation process and to test the ontology under development.

Keywords: Semantic web · Building renovation · Digital twin

1 Introduction

Buildings in Europe are responsible for about 36% of all CO_2 emissions, 55% of the electricity consumption and 40% of the total energy consumption [1]. Thus, the building renovation[1] rate in Europe is expected to increase in the near future in an attempt to soften the impact on the environment. Till today, the renovation field relies heavily on manual labour, which is reflecting in the current renovation rate, since less than 2.5% of existing buildings are renovated each year[2].. Information technology is expected to play a considerable part in enhancing the progress of renovation projects. However, the construction sector suffers from specificities preventing its introduction. Indeed, since more than 93% of the construction companies regroups less than 10 employees, most of these companies do not have the resources required to adopt and deploy computer

---

[1] This paper deals with the renovation of the exteriors of buildings. Hence, the term "building renovation" refers to "renovation of the external envelope of buildings using external insulation.".

[2] www.europarl.europa.eu

© The Author(s), under exclusive license to Springer Nature Switzerland AG 2023
T. Borangiu et al. (Eds.): SOHOMA 2022, SCI 1083, pp. 210–221, 2023.
https://doi.org/10.1007/978-3-031-24291-5_17

tools to automate their processes. Moreover, the strong fragmentation of the sector leads to interoperability issues and therefore information exchange problems, that also need sufficient resources and expertise to be solved. Consequently, problems arise regarding collaboration between actors, planning updates, respect of deadlines, strongly disturbing the evolution of the construction project. This observation is even more correct for the renovation domain.

As a result, the ANR ISOBIM project[3], summarized in Fig. 1, aims to speed up the renovation process by providing low-cost and fully integrated computer tools to companies of the renovation sector. Based on the Lean and BIM paradigms, the IsoBIM platform ought to cover the entire process of renovation of buildings, from the identification of the constructive solution, through the development of configuration and layout models to the development of models for planning and monitoring construction projects. In this context, one of the objectives is to be able to ensure the management of all project information in a formal and integrated repository capable of covering the design, layout, planning and monitoring phases which this research will attempt to achieve.

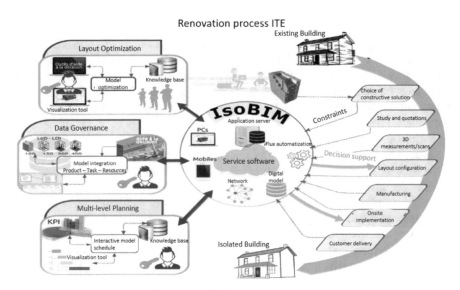

Fig. 1. IsoBIM project overview

The IsoBIM platform manages two different representations of the construction project: the first one is a static view of the renovated building, composed of the old building on which external insulation panels are plugged. The data involved here is data related to the building dimensions and geometry plus some information regarding the panels (sizes, materials, required handling resources, …). The second one is a dynamic view, related to the renovation process itself, that will be monitored and managed through the platform. Authors of [2] define the construction digital twin as a system able to

---

[3] ANR ISOBIM project website: http://isobim.cran.univ-lorraine.fr.

monitor a physical asset, improve its operational efficiency and enable predictive maintenance. This digital twin uses real-time data generated via sensors or manual entries to record and analyse the real-time structural and environmental parameters of a physical asset and to improve the building's interaction with the environment and with users. As a result, the IsoBIM platform can be considered as a digital mock-up of the renovated building coupled with a digital twin of the renovation project.

The question that we will attempt to answer in this paper is: How to formalize the information required for a renovation project? Indeed, this paper proposes a framework that enables the management of a renovation project data through a formal repository, using a dedicated renovation ontology. The paper is organized as follow: In Sect. 2, a quick look is done at the state of the art of semantic technologies in the construction industry. Section 3 presents the general envisaged framework and its current state of deployment. In Sect. 4, a case study featuring the ontology under development is presented. Finally, the last section is reserved for conclusions and future works.

2   Use of BIM and Semantic Technologies in the Construction Field

Conducting a review of the construction industry, authors of [3] conclude that till today this industry suffers from lack of research and development, and poor technology advancements. Building renovation seems to be even more neglected in this area. From the renovation perspective, introducing automation requires the availability of certain information related to building geometry, planning, scheduling and the implementation of on-site monitoring, in an ideal case where a BIM model would be accessible. According to the authors of [4], the digital twin paradigm aims to enhance existing construction processes and BIM models with their underpinning semantics within the context of a cyber-physical synchronicity, where digital models reflect the construction physical assets at any given moment in time. Furthermore, based on the review work [5] of the usage of semantic web technologies in AEC (Architecture, Engineering and Construction) industry, there is a noticeable interest in the use of semantics in the construction field. These technologies are mainly used as complementary to existing BIM models, in order to overcome interoperability issues. The goal is to be able to serialize a BIM model to a common data model so that it can automatically be parsed into a different data structure.

When talking about automation in the construction industry, two main approaches arise: data-driven and knowledge-driven approaches. The meaning of data-driven is the practice of collecting and analyzing data to derive insights and solutions. A data-driven approach facilitates predicting the future by using past and current information. However, data-driven approaches rely on extensive information which require the existence of a knowledge base. Without the availability of good quality data, the risk of making false assumptions increases [6]. Whereas knowledge-based approaches are less reliant on data, the importance of creating a knowledge environment is recognizes, which is necessary to enhance the quality of decision making through the development process, and to re-use and share the knowledge in order to address the different problems that may occur throughout a project.

Thus, a review of various methodologies of BIM usage and semantic web technologies in the AEC domain, specifically in renovation applications, is necessary for a better understanding of the construction data managing.

2.1 Data-Driven Approaches

In this area, the authors of [7] propose a novel approach combining Building Information Modelling (BIM), lean project production systems, automated data acquisition from construction sites and supply chains, and artificial intelligence to generate a data-driven planning and control workflow for the design and construction of buildings while benefiting from digital twin information systems.

The authors identify two types of information, *virtual* and *physical*. Physical information is found in building, components and their relationships, resources and their actions. These can be measured and monitored resulting in digital copies that represent the virtual information, used to build the digital twin which will represent both the construction product and process as shown in Fig. 2.

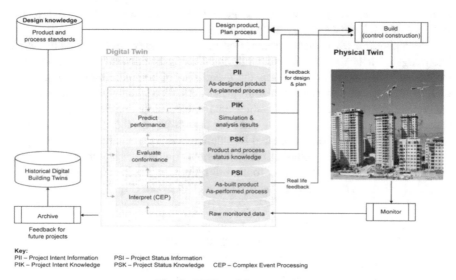

Fig. 2. DTC workflow process

The digital twin must contain all the virtual information. However, the information used to build the digital twin may change through time, hence two states are taken into consideration: the *future state* which is referred to as Project Intent Information (PII) and the *past state* defined as Project Status Information (PSI). At the start of a construction project, designers generate a planning which is used to create the PII. Once the construction phase begins, monitoring the physical building begins as well, the accumulated data generating the PSI. Contractors then use the PII to guide and control the construction process. A specialized function compares the PII and PSI to detect the degree of deviation and check if it is acceptable or requires an intervention. In the latter case, designers are set to interfere and update the planning which is added to the PII. The updated version of the PII guides the project and the cycle continues until completion. All the data gathered throughout the lifecycle of the project is then archived.

This approach is still mainly theoretical given that the implementation of DTC requires some advancements in areas such as: data fusion, data storage mechanisms, algorithms for maintaining consistency among digital twins, as well as data science methods and algorithms for monitoring, interpretation, simulation, and optimization.

Even though the implementation of digital twins in the construction industry is trending (Opoku et al. 2021), when reviewing the digital twin application in the construction industry in 2021, it is found that there is no research on implementing this technology in renovation applications.

2.2 Knowledge-Driven Approaches

When gathering different information from various sources, miscommunication and lack of shared understanding is bound to happen, which can lead to complications that can disrupt the work progress. The need to overcome the interoperability issues among software tools and to improve data exchange in general motivated the usage of semantic web technologies and especially ontologies [5].

The authors of [8] present a renovation project ontology which formally models knowledge from the renovation field, while considering the different requirements, constraints and also taking into accounts the installation of common renovation products such as windows and thermal insulation panels. At the beginning, data regarding the planning of the project and other complementary documents is collected, then verified based on the ontology requirements. Once the project begins, it is divided by RenovationProduct which has Component and InstallationActivity and is linked to a BuildingInterface. The installation activities for each renovation product require specific MaterialsAndResources, Workforce, and ToolsAndEquipment. The installation activities are subjected to certain constraints found in the class Constraint (Fig. 3).

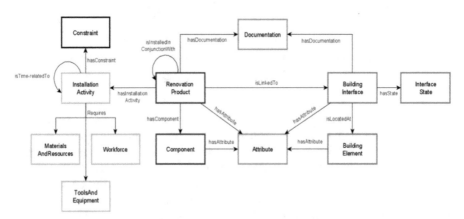

Fig. 3. Reno-Inst ontology overview

This approach is verified and validated by experts in the field through multiple workshops; then the model is implemented in a case study. However, this approach presents some limitations: only three renovation products are considered. The available ontological resources are not used in this approach, and no link to BIM or IFC is established.

Authors of [9] propose a semantic approach that aims to automate the safety checking of a construction project. The framework, shown in Fig. 4, consists of four modules.

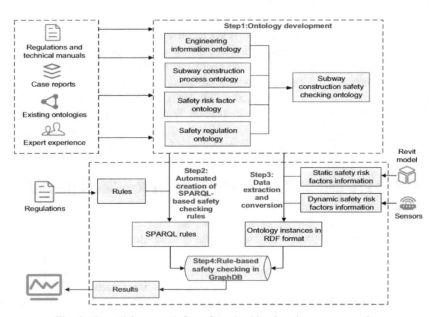

Fig. 4. Overall framework for safety checking in subway construction

The first of them is the ontology development module that creates a standardized format for the heterogenous information regarding the construction safety checking, by semantically linking four ontologies. The second module creates a set of rules for safety checking in which SPARQL-based safety checking rules are automated. The third module handles the safety risk factors related to data extraction from BIM models and sensors. The extracted data is then converted into ontology instances. The last module specifies where the rule-based safety checking takes place. Ontology instances that do not meet the constraints are filtered and the safety checking results are exported.

This approach is validated through a case study which proves the feasibility and the efficiency of the model. However, some limitations arise; the approach cannot extract information directly from BIM models or sensors.

2.3 Challenges and Problem Statement

Till today, no efficient approach was developed to support the renovation domain. The traditional ways of renovating buildings by relying on manual labour are incredibly time consuming and complicated. There is a need for information collection, and multiple studies are developed to assess the methodologies of building digitalization.

Moreover, semantic web technologies are leveraged in an attempt to formalize and organize all information required for renovation. Many approaches are developed; however, all of them have some limitations. Also, after reviewing the literature not enough studies are conducted regarding the renovation phase. Renovation is rarely considered, even though the construction industry is one of the largest industries and renovation projects represent 57% of all construction projects [8].

At present, no approach successfully offers the knowledge formalization of a renovation process that covers the lifecycle of the whole project from the design phase until the implementation phase. Most of the data-driven approaches are dedicated to construction in general and do not offer solutions to the renovation domain. Although some knowledge-driven approaches address renovation, the offered solutions are limited to static data management; the only approach found that uses an ontology for the dynamic management addresses issues on safety checking in a construction project. Therefore, the aim of this paper is to present a novel framework that aims to automate building renovation by formalizing the information gathered from BIM in a dedicated ontology.

3   A New Framework for Knowledge Formalization in the Building Renovation Sector

This section presents a framework that is currently under development. Given the complexity and individuality of renovation projects, having a fully automated process is unrealistic; however, once the process is decomposed into tasks, some of them can be automated. Concerning automation, having an information management system that includes a database for the project and ensures the information exchange between different parties is very useful. The benefit of having such a robust database is that it facilitates the successful implementation of digital twins in the renovation industry. The chosen approach for this repository is the ontological one. The aim of the ontology in development is to formalize the knowledge required for the building renovation domain. The reason to create an ontology for the building domain is to help project managers to organize the planning and the execution of renovation projects. The renovation ontology will be designed to represent all the concepts related to requirements, planning, scheduling, installation, and execution of a renovation project. Moreover, it will encompass concepts and properties regarding the tasks required for the installation of thermal insulation panels on buildings with various geometric complexities.

Normally, in a renovation project different actors are responsible for different parts of the project; usually information will be a side effect to the progression of their work. The interest behind this approach is the development of a common knowledge base that is able to organize the information generated throughout the entire project.

Ontologies are very powerful when it comes to mapping, representing and retrieving knowledge [10]; the proposed approach focusses on this aspect. The objective of the described project is to develop a collaborative and semi-automatic process for the renovation of buildings. A SaaS (Software as a Service) solution, also currently in development, is envisaged to bring together the multi-level layout, as well as a renovation ontology that organizes and collects information while offering a knowledge base that can be utilized at different stages of the project. To achieve the highest level of automation, the passage from a model generated by one of the actors of a renovation project to another model must be done via an automatic transformation method (Li et al., 2022). Such a solution establishes a semantic and procedural link between all actors.

The framework is presented in Fig. 5. The main interest in this framework is the development of a quadruple ontology handling different information.

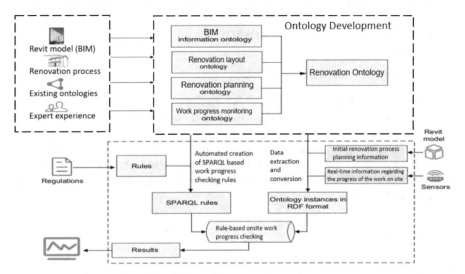

Fig. 5. Novel framework overview

In order to get a clear idea about the renovation process and to help the ontology development, multiple experts in the domain were interviewed. These experts belong to two main categories: construction companies that conduct renovation projects, and research laboratories specialized in renovation. Experts were asked to provide a clear idea of the renovation process and the challenges they face when handling complex projects. Special attention was given to their suggestions on how to improve the proposed approach based on real needs. They were also questioned about the execution phase: once the on-site implementation begins, what kind of information should be collected. After comparing the answers, a common ground between all experts was found.

When dealing with a renovation project, the building must be divided into facades that must be located using longitude, latitude and altitude data. Several elements can belong to a façade; the elements can be inside or outside the building: doors, windows, cross-walls and floor-ends. Each of these elements is assigned information about the type, location, material properties and dimensions.

In projects for the renovation of an existing building, usually an old building, there is usually no BIM model, and 2D drawings are not available. Hence, at the beginning of the project the first step is to conduct a series of 3D scans to generate the BIM model of the building. Once the BIM model is available, the BIM information ontology can be enriched with the information provided by the BIM model of the existing building.

The next step is the layout generation. A software that is under development will enable the automatic generation of the renovation layout. The output of the software is a JSON file offering the required information for the renovation layout ontology. This layout can easily serve as the basis for the preliminary project planning. The renovation planning ontology is built based on the initial planning generated from the renovation layout, and the fourth and final ontology is the work progress monitoring ontology.

One of the many benefits of using an ontology is the manageability of new information. The Revit model combined with the initial planning will be compared to the real state of the site collected by sensors in real time, so that any deviation from the planning will be detected automatically using SPARQL rules.

3.1 Ontology Development Methodology

There is no "best" way or methodology for developing ontologies. Many approaches can be followed to build an ontology, such as the iterative approach which is best fitted to build an ontology given the currently available resources. The iterative approach consists of beginning with a rough first pass at the ontology, then revising and refining the evolving ontology and filling in the details. To understand what kind of information needs to be provided by the experts, several meetings are organized with experts in the renovation field [11].

4   Case Study

The case study taken into consideration in this paper is an actual building, located in Nancy, France. The renovation process envisioned for this building consists of installing thermal insulation panels for an envelope of the whole building. Multiple visits were organized to collect as much data as possible. The following documents were transferred: technical files, energy bills and maintenance costs for the years 2018 to 2020, floor plans of the ground floor and a current floor of the building, facade plans.

Some scans of the building's outside were also conducted to retrieve the exterior geometry and, after eliminating the noise from the scans, the point clouds were obtained and the 3D model of the building was generated (Fig. 6), and enriched by the information regarding the interior design and thermal properties of the building.

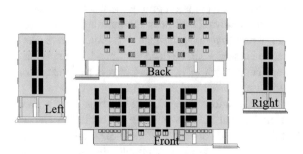

Fig. 6. 3D model of Jarville building

Thus, a first draft of the ontology shown in Fig. 7 could be built based on the infor-mation obtained. The main classes of the ontology define a general view of the aspects related to renovation projects, requirements, constraints, and the installation of renovation products. A class hierarchy is created with the main concepts identified during the conducted interviews. Class properties between the concepts were established to represent their relationships. The ontology has two main classes: facade and facade_components which define all the elements belonging to a building. The facade_components class consists of construction entities with a characteristic technical function, form or position. These entities possess certain data properties necessary to generate the renovation layout. Each façade_components element has a height, width, x_dim, y_dim and mechanical_resistance, which are crucial information for renovation.

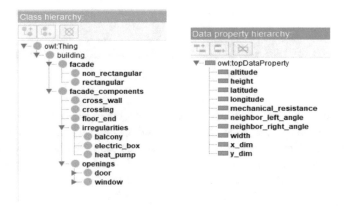

Fig. 7. Ontology hierarchy

A deeper look in the ontology is presented in Fig. 8.

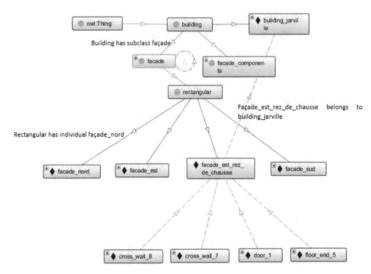

Fig. 8. Ontology overview

Each building will be linked to facade and facade_components that include architectural elements such as cross_walls, floor_ends and openings which can be door or window. As mentioned previously, each element has data properties describing the dimensions and the location relative to the facade to which it belongs. Other properties such as neighbor_right_angle are necessary to locate the facade with respect to neighbour elements. Up to now, building elements, their properties and their relationships with other elements have been defined in our project, but constraints not yet.

5   Conclusion and Perspectives

Over the past few years, the construction field started witnessing digitalization and automation; even semantics approaches were explored. Although renovation projects represent about 57% of all construction activities, they seem to be neglected; hence, the need for an ontology to facilitate and delegate the information management of the different phases of a renovation project.

A framework in development is presented in this paper, that ensures the mapping, representation and retrieval of information through a renovation ontology. A case study is used to test the ontology.

Because this approach is still in the development phase, it will be beneficial to do a systematic literature review. The idea is to enrich what already exists instead of redefining concepts, since renovation ontologies already exist.

References

1. European Commission. In Focus: Energy Efficiency in Buildings; European Commission: Brussels, Belgium (2020). https://ec.europa.eu/info/news/focus-energy-efficiency-buildings-2020-feb-17_en Accessed 17 Sep 2021

2. Khajavi, S.H., Motlagh, N.H., Jaribion, A., Werner, L.C., Holmstrom, J.: Digital twin: vision, benefits, boundaries, and creation for buildings. IEEE Access **7**, 147406–147419 (2019). https://doi.org/10.1109/ACCESS.2019.2946515

3. Opoku, D.G.J., Perera, S., Osei-Kyei, R., Rashidi, M.: Digital twin application in the construction industry: a literature review. J. Build. Eng. **40**, 102726 (2021). https://doi.org/10.1016/j.jobe.2021.102726

4. Tao, F., Cheng, J., Qi, Q., Zhang, M., Zhang, H., Sui, F.: Digital twin-driven product design, manufacturing and service with big data. Int. J. Adv. Manuf. Technol. **94**(9–12), 3563–3576 (2017). https://doi.org/10.1007/s00170-017-0233-1

5. Pauwels, P., Zhang, S., Lee, Y.C.: Semantic web technologies in AEC industry: a literature overview. Autom. Constr. **73**, 145–165 (2017). https://doi.org/10.1016/j.autcon.2016.10.003

6. Wu, L., Ji, W., Feng, B., Hermann, U., AbouRizk, S.: Intelligent data-driven approach for enhancing preliminary resource planning in industrial construction. Autom. Constr. **130** (2021). https://doi.org/10.1016/j.autcon.2021.103846

7. Sacks, R., Brilakis, I., Pikas, E., Xie, H. S., Girolami, M.: Construction with digital twin information systems. Data-Centric Eng. **1**(6) (2020). https://doi.org/10.1017/dce.2020.16

8. Amorocho, J.A.P., Hartmann, T.: Reno-Inst: An ontology to support renovation projects planning and renovation products installation. Adv. Eng. Inf. **50** (2021). https://doi.org/10.1016/j.aei.2021.101415

9. Li, X., Yang, D., Yuan, J., Donkers, A., Liu, X.: BIM-enabled semantic web for automated safety checks in subway construction. Autom. Constr. **141**, 104454 (2022). https://doi.org/10.1016/j.autcon.2022.104454

10. Gruber, T.: Circumscription-a form of non-monotonic reasoning. In J. F. Conceptual Structures, Information Processing in Mind and Machine, vol. 43, Issue 6 (1995). Addison Wesley. http://www.w3.org/TR/owl-features/, http://suo.ieee.org/

11. Noy, N.F., Mcguinness, D.L.: Ontology development 101: a guide to creating your first ontology (2001). https://www.researchgate.net/publication/243772462_Ontology_Development_101_A_Guide_to_Creating_Your_First_Ontology

12. www.europarl.europa.eu

Physical Internet Supply Chain: Short Literature Review and Research Perspective on Strategic and Operational Levels

Maroua Nouiri[1]([✉]), Tarik Chargui[2], Abdelghani Bekrar[2], and Damien Trentesaux[2]

[1] Nantes Université, École Centrále Nantes, CNRS, LS2N, UMR 6004, 44000 Nantes, France
maroua.nouiri@univ-nantes.fr
[2] Univ. Polytechnique Hauts-de-France, LAMIH, CNRS, UMR 8201,
59313 Valenciennes, France
{tarik.chargui,abdelghani.bekrar,damien.trentesaux}@uphf.fr

Abstract. Various technologies have been introduced recently to enhance the performance and sustainability of the global supply chain network. Physical Internet is one of the most studied paradigms that considers the whole supply chain instead of focusing only on the production sites and assembly lines of Industry 4.0 with Cyber Physical Systems. The Physical Internet is designed to improve the sustainability economically, socially and environmentally. Most of the reviewed papers focus on economic objectives. Sustainability in its social and environmental form is ignored in the literature, except for a few works that consider energy consumption and CO_2 emissions. In this paper, we present a short literature review on the applications and role of the Physical Internet in the improvement of the logistics global performance as well as its sustainability. The assignment problem is highlighted at both strategic and operational levels. At a strategic level, the Physical Internet (PI) network considered in this work is a multi-plant, multi-product supply chain. At operational level, the PI-hub is studied. The results of the analysis are discussed, and future research directions are proposed.

Keywords: Physical Internet · Multi-plant · Multi-product · PI-hub · Strategic · Operational · Supply chain

1 Introduction

Supply chain performance has largely evolved in recent decades. With automation and robotics, production lines became more efficient and effective [1]. The usage of new technologies such as ERP in combination with Internet of Things (IoT) helped with the flexibility and the adaptation of production and assembly lines [2].

Various concepts and paradigms have been introduced in recent years. The principal ones are Industry 4.0, Internet of Things, Cyber Physical Production Systems and Physical Internet [3]. The first three were more focused on the interconnectivity of production and assembly lines. However, Physical Internet includes the whole supply chain from suppliers, factories, distribution centres, to retailers and final customers [2].

© The Author(s), under exclusive license to Springer Nature Switzerland AG 2023
T. Borangiu et al. (Eds.): SOHOMA 2022, SCI 1083, pp. 222–230, 2023.
https://doi.org/10.1007/978-3-031-24291-5_18

The main idea of Physical Internet is to create a global supply chain network that can be used by all its elements. The concept is based on sharing the supply chain resources such as transportation vehicles (freight trains, trucks, ships, etc.), storage spaces (warehouses, cross-docking hubs, container terminals), and handling resources (robots, forklifts, etc.) [4].

The main goal of Physical Internet is to ensure sustainability of the supply chain. It consists of making the global supply chain sustainable socially, economically and environmentally. Social sustainability consists in providing a better working environment for human operators by decreasing, for example, the travelled distance of truck drivers. The economical aspect consists on reducing energy consumption in general (electrical energy, fuel, etc.). From the environmental point of view, the sustainability consists in reducing trash materials and gas such as CO_2 emissions.

For a proper functioning, Physical Internet needs three main elements: PI-containers, PI-nodes, and PI-vehicles. PI-containers are standardized with additional functionalities [5]. For example, the PI-containers can move automatically in the PI-hub sorting zones. Figure 1 illustrates the three main objectives of the PI concept.

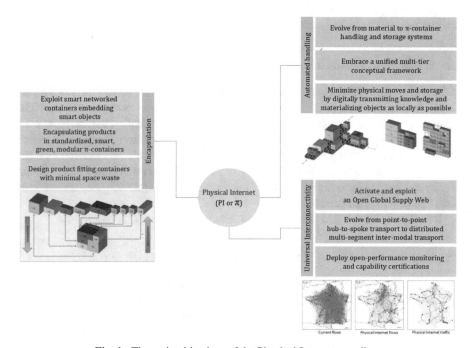

Fig. 1. The main objectives of the Physical Internet paradigm

The literature about Physical Internet addresses various optimization problems from strategic to operational levels. The strategic decisions concern mainly the assignment of PI-hubs and the routing of vehicles in the PI network. The operational ones are more focused on local optimizations decisions such as trucks assignment and routing of PI-containers. The objective of this papers is to provide a short literature review and research perspective on strategic and operational levels in the PI network.

The remaining of this paper is organized as follows. Section 2 and 3 present the strategic and operational decisions respectively. Section 4 provides a discussion while highlighting a relevant research directions. A conclusion and some perspectives are given in Sect. 5.

2   Strategic Decisions in Physical Internet Supply Chain Network

As described in the previous section, the Physical Internet supply chain network is composed by different PI-Nodes. The complexity of strategic decisions is related to the size of the supply chain network. In this section, a detailed description of the assignment problem in Single Plant Physical Internet Network (SP-PISCN) and Multi Plant Physical Internet Supply Chain Network (MP-PISCN) are given.

2.1   Single Plant Classical Supply Chain Network

In the single plant classical supply chain network (SP-CSCN), the product flow's direction is predetermined. The client demand is pre-assigned to its own point of sale. The replenishment orders are also pre-assigned to a specific distribution centre. How-ever, as the PISCN is based on full connectivity between PI-hubs, an assignment research problem is raised due to the complexity of replenishment node's selection. In fact, the replenishment orders from any given points of sale can be served by any PI-hub around the network and the hubs can be supplied by any other PI-hub or the only single connected plant that provides the product.

It is already known that the single plant inventory problem is NP-Hard [6].

2.2   Multiple Plant Classical Supply Chain Network

The assignment problem in the MP-PISCN is more complex than the single one based on the fact that a multi-plant assignment problem can be decomposed into m single plant inventory problems. This implies that the multi-plant assignment problem is NP-Hard. The main difference between both PISCN is the number of plants and the type of the produced product. Indeed, the source replenishment of PI-Hubs can be any other PI-hub or any connected plant that produces the requested type of the product [7].

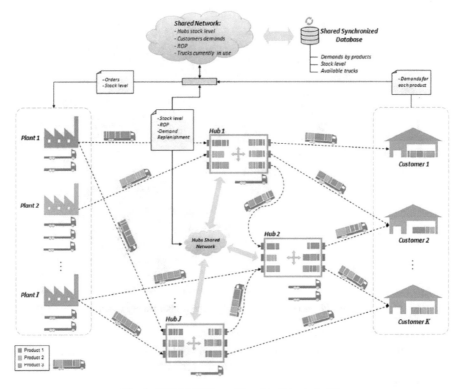

Fig. 2. MP-PISCN considered in this study [8].

Strategic decisions concern solving the external assignment problem of the global network. However, the operational decisions include solving the internal assignment problem inside the PI-Hub that consists, for example, of assigning trucks to the docks, routing of PI-containers inside the sorting area, and assigning the PI-containers to the outgoing train.

As presented in Fig. 2, plant #1 provides for example products for different PI-hubs [8]. Figure 3 is inspired from [9] and offers a closer look at the PI-hub operations inside the global supply chain. In the next section, operational decisions will be presented for both the classical cross-docks and PI-hubs.

Nouiri et al. proposed in [7] a multi-agent model for a multi-plant, multi-product supply chain network that supports an open network with n nodes (plants, retailers, etc.). Three replenishment policies are proposed with different criteria of selection. A simulation tool was used to implement the proposed multi-agent model. Sustainability in global supply chain have been considered by many researchers. Authors of [17] were interested in sustainable order allocation and a strategic design on the supply chain network using a multi-objective hybrid approach. The proposed approach was validated on an automobile industry case study. In the context of artificial intelligence, Kantasa-Ard et al., [18] used recurrent neural network to predict customer demand in the Physical Internet supply chain.

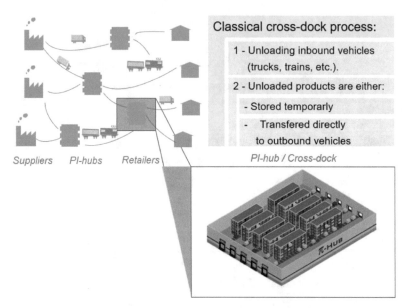

Fig. 3. PI-hub's operations inside the global supply chain

In order to minimize the CO_2 emission, authors of [19] developed a non-dominated genetic algorithm (NSGA-II) for a food supply chain taking into consideration the perishability of products.

3   Operational Decisions in Physical Internet Supply Chain

In the Physical Internet network, PI-hubs are the most demanding resources in the supply chain in terms of storage capacity and docks assignment. The operational level of decisions in the supply chain consists on scheduling local assignment, and sequencing of trucks and storage operations inside the PI-hubs.

3.1   Classical Hubs

Before addressing the Physical Internet hubs, we need to present the main assignment problems in the classical hubs or cross-docks. Various literature reviews have been done on the cross-docking problems. For a comprehensive state-of-the-art of cross-docking problems, the interested reader may refer to [10] who classifies the articles based on their decision level from strategic, tactical to operational problems. Authors of [11] show to what extent these operational problems addressed in the literature match the current needs and concerns of the cross-dock managers. Their review is followed up by [12]. All these articles focus (implicitly) on road-road cross-docks; for a review of decision problems for rail-rail and rail-road facilities, refer to [10].

Cross-docking network-related problems are mentioned by [10] as a special class of cross-docking problems, but authors in [13] observe that managing a cross-docking network implies decisions on all three decision levels, i.e., strategic, tactical, and operational. Therefore, they propose a new classification covering both cross-dock facility and cross-docking network management, on all three decision levels. They also address synchronization issues, i.e., problems in which it is important to have an information exchange between the facility and the network or between the different decision levels for an efficient management of the global system.

Because synchronization with the network is a key element in PI cross-docks, our work builds upon the classification proposed and summarized by [13]. The next section concern PI-hubs.

3.2 PI-Hubs

PI-hubs are one of the most impacting elements of the Physical Internet supply chain performance [14]. The optimization of PI-hubs scheduling and assignment operations has a major impact on the global PI performance. PI-hubs are classified into different categories depending on multi-modality on the vehicles, as presented in Fig. 4.

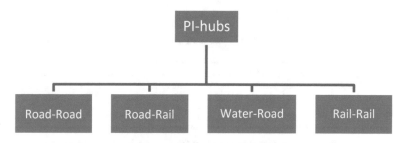

Fig. 4. Multi-modality of physical internet hubs

In the context of the PI, the truck assignment problem is amongst the most studied. Authors of [4] suggest that earliest and latest arrival and departure times might be contractually defined for each PI container; that could therefore be considered as a hard constraint.

Authors of [15] use a simulation-optimization method to solve a truck-to-door scheduling problem in a cross-docking platform. Although the authors claim their approach is a complement to the PI concept, how they consider the distinctive features of a PI cross-dock is unclear.

Several articles study the case of a train-truck PI hub, focusing on the train to truck operations only, when the train is unloaded and the PI containers are loaded into several trucks [20]. Walha et al. simultaneously model the scheduling of trucks (or trains) at the PI hub gates to minimize the routing distance, and the allocation of each incoming PI container to an outbound truck/train/barge to use as few of them as possible [16]. Chargui et al. [9] combined optimization with simulation to generate robust solutions that stay feasible in case of disruption in the sorting area of the two-way Road-Rail PI hubs.

Other research consider the routing of vehicles as well. For instance, after formulating the problem mathematically, [21] developed a co-evolutionary method for the integrated vehicle routing and scheduling with cross-docks. Most of the works above addressed the sustainability aspect only in the classical context. To the best of our knowledge, approaches for strategic decisions were not well studied in the PI context.

Figure 5 sums up the mapping of the assignment problem from operational level in-side the PI-hub to the strategic level inside the global network. The problem consists of assigning the tasks to the resources in order to optimize a certain criteria. The tasks in MP-PISCN are the client's demand and the replenishment order of the PI-hub.

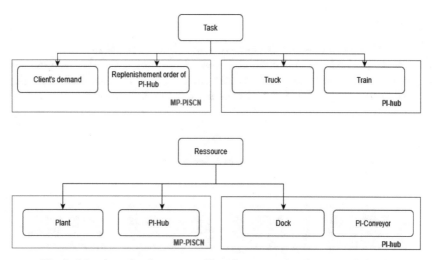

Fig. 5. Mapping of assignment problem from operational to strategic levels

Inbound/outbound vehicles (train and trucks) can be considered as tasks of the internal assignment problem in PI-hub. The resources used to solve assignment problems of MP-PISCN are plant and PI-hub, and dock and PI-conveyor for PI-hub scheduling problems. Solving such problems requires objectives to be optimized like resource usage, minimize tardiness, travelled distance, etc. For that reason, different solving approaches are used and explored by researchers. The next section provides a discussion about the different gaps in the literature at both strategic and operational levels. Some relevant research questions will be introduced.

4   Discussion

As reviewed in the literature, most of the papers dealing with the Physical Internet addressed the global supply chain with all its components (suppliers, hubs, retailers, etc.). Different problems were tackled, from scheduling vehicles routes to replenishment strategies. At the operational level, most of the works focus on the assignment of trucks to the docks using different approaches such as mathematical programming (mixed-integer programming models, multi-objective models), meta-heuristics (tabu search, simulated annealing genetic algorithms, etc.). Simulations have been also used to study various scenarios with statistical analysis. Moreover, most of the papers consider only a single plant that produces one or multiple types of products, and only few of them consider multi-plant with multiple products. To the best of our knowledge, there are no works that have combined assignment problem at both strategic and operational decisional levels in Physical Internet network, which rises the following research directions:

- A detailed systematic literature review that analyses the different solving approaches at all decisional levels in static environment.
- Studies that consider internal and external perturbations (accident, PI-hub failures, absenteeism of operators, etc.)
- More study about the investment cost of Physical Internet materials (PI-hub facilities, PI-container, PI-conveyor, etc.)
- More sustainable objectives need to be included in future studies other than CO_2 emission and energy consumption.
- The social sustainability objectives are ignored in current research and need to be considered.

5 Conclusion

This work deals with Physical Internet supply chain assignment problem at strategic and operational decisional levels. The Physical Internet network considered is composed by m plants, n PI-hubs and j retailers. A mapping of assignment problem from operational to strategic levels is highlighted. Strategic decisions in the multi-plant multi-product Physical Internet supply chain network concern the assignment of clients' demand to a PI-hub, replenishment order of PI-hub to other PI-hubs or other plants. Operational decisions concern the assignment of inbound/outbound PI-movers to dock or PI-conveyor.

Based on the analysed literature, we observe that different relevant research directions are not treated especially combining such a problem at both decisional levels and including perturbation events. Moreover, social sustainable goals are ignored in current research.

References

1. Pan, S., Trentesaux, D., McFarlane, D., Montreuil, B., Ballot, E., Huang, G.Q.: Digital interoperability in logistics and supply chain management: state-of-the-art and research avenues towards Physical Internet. Comput. Ind. **128**, 103435 (2021)
2. Yang, Y., Pan, S., Ballot, E.: Mitigating supply chain disruptions through interconnected logistics services in the physical internet. Int. J. Prod. Res. **55**(14), 3970–3983 (2017)
3. Nguyen, T., Duong, Q.H., Van Nguyen, T., Zhu, Y., Zhou, L.: Knowledge mapping of digital twin and Physical Internet in supply chain management: A systematic literature review. Int. J. Prod. Econ. **244**, 108381 (2022)
4. Montreuil, B., Meller, R. D., Ballot, E. (2013). Physical Internet foundations. In Proc. Service Orientation in Holonic and Multi-agent Manufacturing and Robotics, Studies in Computational Intelligence, pp. 151–166, Springer

5. Crainic, T.G., Montreuil, B.: Physical Internet enabled hyperconnected city logistics. Transportation Research Procedia **12**, 383–398 (2016)
6. Darvish, M., Larrain, H., Coelho, L.C.: A dynamic multi-plant lot-sizing and distribution problem. Int. J. Prod. Res. 6707–6717 (2016). https://doi.org/10.1080/00207543.2016.115 4623
7. Nouiri, M., Bekrar, A., Giret, A., Cardin, O., Trentesaux, D.: A multi-agent model for the multi-plant multi-product physical internet supply chain network. In: Borangiu, T., Trentesaux, D., Leitão, P., Cardin, O., Lamouri, S. (eds.) SOHOMA 2020. SCI, vol. 952, pp. 435–448. Springer, Cham (2021). https://doi.org/10.1007/978-3-030-69373-2_31
8. Nouiri, M., Bekrar, A., Trentesaux, D.: An energy-efficient scheduling and rescheduling method for production and logistics systems. Int. J. Prod. Res. **58**(11), 3263–3283 (2020)
9. Chargui, T., Bekrar, A., Reghioui, M., Trentesaux, D.: Proposal of a multi-agent model for the sustainable truck scheduling and containers grouping problem in a Road-Rail Physical Internet hub. Int. J. Prod. Res. **58**(18), 5477–5501 (2020)
10. Van Belle, J., Valckenaers, P., Cattrysse, D.: Cross-docking: State of the art. Omega **40**(6), 827–846 (2012)
11. Ladier, A.L., Alpan, G.: Cross-docking operations: Current research versus industry practice. Omega **62**, 145–162 (2016)
12. Theophilus, O., Dulebenets, M.A., Pasha, J., Abioye, O.F., Kavoosi, M.: Truck scheduling at cross-docking terminals: a follow-up state-of-the-art review. Sustainability **11**(19), 5245 (2019)
13. Buijs, P., Vis, I.F., Carlo, H.J.: Synchronization in cross-docking networks: A research classification and framework. Eur. J. Oper. Res. **239**(3), 593–608 (2014)
14. Chargui, T., Ladier, A.L., Bekrar, A., Pan, S., Trentesaux, D.: Towards designing and operating Physical Internet cross-docks: problem specifications and research perspectives. Omega **111**, 102641 (2022)
15. Pawlewski, P.: DES/ABS approach to simulate warehouse operations. In: Highlights of Practical Applications of Agents, Multi-Agent Systems, and Sustainability - The PAAMS Collection. PAAMS 2015. Communications in Computer and Information Science, vol. 524. Springer, Cham (2015). https://doi.org/10.1007/978-3-319-19033-4_10
16. Walha, F., Bekrar, A., Chaabane, S., Loukil, T.M.: A rail-road PI-hub allocation problem: Active and reactive approaches. Comput. Ind. **81**, 138–151 (2016)
17. Govindan, K., Jafarian, A., Nourbakhsh, V.: Bi-objective integrating sustainable order allocation and sustainable supply chain network strategic design with stochastic demand using a novel robust hybrid multi-objective metaheuristic. Comput. Oper. Res. **62**, 112–130 (2015)
18. Kantasa-Ard, A., Nouiri, M., Bekrar, A., Ait el Cadi, A., Sallez, Y.: Machine learning for demand forecasting in the Physical Internet: a case study of agricultural products in Thailand. Int. J. Prod. Res. **59**(24), 7491–7515 (2021)
19. Musavi, M., Bozorgi-Amiri, A.: A multi-objective sustainable hub location-scheduling problem for perishable food supply chain. Comput. Ind. Eng. **113**, 766–778 (2017)
20. Sallez, Y., Pan, S., Montreuil, B., Berger, T., Ballot, E.: On the activeness of intelligent Physical Internet containers. Comput. Ind. **81**, 96–104 (2016)
21. Yin, P.Y., Lyu, S.R., Chuang, Y.L.: Cooperative coevolutionary approach for integrated vehicle routing and scheduling using cross-dock buffering. Eng. Appl. Artif. Intell. **52**, 40–53 (2016)

Education for Industry 4.0

A Systems Engineering-Oriented Learning Factory for Industry 4.0

Theodor Borangiu$^{(\boxtimes)}$, Silviu Răileanu, Florin Anton, Iulia Iacob, and Silvia Anton

Department of Automation and Industrial Informatics, University Politehnica of Bucharest, Bucharest, Romania
{theodor.borangiu,silviu.raileanu,florin.anton}@upb.ro

Abstract. Industry 4.0 represents the new stage in the organization and control of the industrial value chain; its basis is represented by Cyber-Physical Systems that use intensively automation, bridge the physical and digital world through Industrial IoT, are controlled with distributed intelligence by smart products driving the production steps in closed-loop data models and controls, and operate in business models integrated in the cloud universal manufacturing space. With reference to the manufacturing models of the 4th industrial revolution, the foundational technologies and standards integrated in the reference architectural model Industry 4.0 from the perspective of production, digital, AI and cyber technologies are analysed. The paper describes a model of learning factory for Industry 4.0 based on an industrial real manufacturing cell controlled by a CPS, and on a 42-week didactic program with hands-on training and master course teaching.

Keywords: Industry 4.0 · Digital manufacturing · Robotics · Cyber-Physical System · Industrial IoT · Cloud service · AI · Education 4.0 · Hands-on training

1 Introduction

Industry 4.0 represents the vision for the Industry of the Future; the term 'Industry 4.0' defines a methodology to generate a transformation from machine dominant manufacturing to *digital manufacturing* [1]. The digital transformation of manufacturing advocated by Industry 4.0 is asked for by the economic and societal realities of the 21st century: markets are currently demanding customized, high-quality products with frequent characteristics changes, in highly variable batches with shorter delivery times, forcing companies to adapt their design, logistics and production processes in conjunction with multi-function, reconfigurable resource teams orchestrated in flexible, efficient, agile and reality-aware factories. New production and IC^{2}T (Information, Communication and Control technologies) are developed to sustain the digitalized production in the global vision of the fourth industrial revolution. Industry 4.0 points to digital production patterns in holistic and adaptive automation system architectures exposing cooperation between plant components, and product intelligence [2].

The production process has traversed in historical perspective several industrial revolutions, starting in 1784 with the first one - Industry 1.0, in which mechanical production

© The Author(s), under exclusive license to Springer Nature Switzerland AG 2023
T. Borangiu et al. (Eds.): SOHOMA 2022, SCI 1083, pp. 233–253, 2023.
https://doi.org/10.1007/978-3-031-24291-5_19

was based on water and steam powered engines. The second industrial revolution made possible since 1870 mass production in flow shop-type assembly lines using electrical energy and applying division of labour. Industry 3.0 marked the 'digital revolution' about 1970 with electronics, communication and information technologies that helped setting up shop floor devices for process automation: CNC, PLC, robots, and computing workstations. Then, in the second decade of the 21st century, the 4th industrial revolution, termed Industry 4.0, was established as a new level of interconnection in the organization and control of: i) the lifecycle of production assets and related production processes, and ii) the product lifecycle. The initiative 'Industry 4.0', announced at the Hannover fair in Germany in April 2011 [3], advocated the concept of global digitization based on pervasive instrumenting and real-time interconnection of resources and products allowing intelligent decision making for smart, safe and sustainable production [4]. This integrated vision promotes autonomy, self-behaviour, and cooperation of machines as approaches to solve adaptability to environment, agility to market and product changes, and increasingly personalized customer requirements [5].

In terms of manufacturing, several models are associated with the four industrial revolutions: a) *craft production* (began about 1850): it responded to specific customer orders, allowing high product variety and flexibility, each product treated as unique; this is a slow and expensive manufacturing model; b) *American system* (from 1913 to 1932): is related to the standardization of products' components; c) *mass production* (1932–1955): making of products at lower cost through large-scale manufacturing; this model allows limited variety of products, time and cost reduction; d) *lean production* (from 1955 to 1980): a multi-dimensional approach with a large variety of management practices (just in time production, quality systems, work teams, cellular manufacturing); originally derived from Toyota production system, this model is based on three important principles: delivering value as defined by the customer, eliminating waste, and continuous improvement; e) *mass customization* (started in 1980): combines business practices from mass- and craft production; this is a customer-centric model in which manufacturing systems are intelligent, faster, more flexible and interoperable.

Sustained by the continuous evolution of IC^{2}T, mass customization led to the emergence of three new manufacturing models that increased process automation, resource reality awareness, data-driven supply, logistics and production planning and control, and computer-based value chain optimization; these models and the technologies they use foreshadow the global universal manufacturing space defined by Industry 4.0: e1) *Computer Integrated Manufacturing* (CIM): the model describes the complete automation of a manufacturing plant, with all processes running under computer control and digital information linking them together [6]. CIM is a combination of different applications and technologies like computer aided-design (CAD), -engineering (CAE), -process planning (CAPP), -manufacturing (CAM), -quality control (CAQC), robotics, and enterprise planning and management (ERP). It is the integration of all enterprise operations that work with a common data repository; e2) *Intelligent* (or *smart*) *Manufacturing* (IM): a model-based, data driven concept of manufacturing performed by reconfigurable systems with distributed intelligence and resource sharing for reality-aware optimization of production. The IM model is based on intelligent science and technology and uses three classes of key enabling technologies (KET) that upgrade the design, production, management, and integration of the product's life cycle [7]:

- *Production technologies*: additive manufacturing, guidance vision in robotics, adaptive machining.
- *Digital* and *AI technologies*: smart sensors, edge computing, IoT, multi-agent (MAS) and holonic systems, product intelligence, high performance computing (HPC), machine learning (ML), prediction, augmented reality (AR), service oriented architectures (SOA).
- *Cyber technologies*: big data analytics, resource and product virtualization, fog computing, cloud services, digital twins, cyber-physical systems (CPS), software defined networks (SDN), cybersecurity, cognitive computing, human-robot interaction.

e3) *Cloud Manufacturing* (CMfg): a manufacturing model with support of cloud computing, virtualization and service-oriented technologies. This model transforms shop floor resources into services that can be comprehensively shared and circulated [8]. The model transposes pools of factory resources (robots, machines, controls) into *on-demand* making services; it also enables pervasive, *on-demand* network access to a shared pool of configurable HPC resources (servers, storage, applications) that can be rapidly provisioned and released as services to global MES workloads with minimal management. CMfg covers the whole product's life cycle from design, production and testing to after-sales services, and is hence considered as a parallel, networked, and intelligent manufacturing system. Manufacturing resources and capacities are virtualized, encapsulated, networked into services that can be accessed, invoked, and implemented [9].

Industry 4.0 (I4.0) describes a fundamental process of innovation and transformation based on data driving industrial value creation, generating new business, and working models in global digital ecosystems; flexible, highly dynamic and globally networked value-added manufacturing chains will replace rigid, explicit ones. This process is characterised by a new level of sociotechnical interaction between all the resources and participants involved in manufacturing networks (equipment, production facilities and systems) that are autonomous, reality-aware, capable of self-control in reaction to particular events, self-configuring, knowledge-based, instrumented, spatially distributed and incorporating optimized planning and management systems [10]. I4.0 permits customer specifications to be considered in the product design, engineering, order planning, manufacture and inspection phases, and frequently changed.

The 2030 Vision for Industry 4.0, formulated in a holistic approach by Plattform Industrie 4.0 [11], promotes the global design of open digital ecosystems by assuming three strategic fields of action: 1) *Autonomy*: self-determination and free scope for action guarantee competitiveness in digital business models by help of technology development, security and digital infrastructure; 2) *Interoperability*: cooperation in open ecosystems eases plurality and flexibility through regulatory framework (security data handling), standards and integration, AI and distribution of intelligence; 3) *Sustainability*: modern industrial value creation ensuring high standard of living by decent work and education, social inclusion, climate change mitigation and the circular economy.

Industry 4.0 involves the technical integration of Cyber-Physical Systems (CPS) into manufacturing and logistics activities and the use of the Internet of Things and

Services (IoT, IoS) in industrial processes. The shift is to networked manufacturing, self-organising adaptive logistics and customer-integrated engineering requiring business models that will be implemented by a highly dynamic network of businesses rather than by a single company [12]. A key component of the I4.0 vision is the smart factory embedded into inter-company value networks, characterised by end-to-end engineering that encompasses both the manufacturing process and the manufactured product, achieving seamless convergence of the digital and physical worlds.

Three new manufacturing models are associated with the 4th industrial revolution: f) *Manufacturing as a Service* (MaaS): the future cloud networked manufacturing is based on the MaaS concept, i.e., the shared use of networked manufacturing infrastructure to produce goods; companies use the Internet and the Cloud to share manufacturing equipment, capacity and services for agility to market demand, collaboration with the customer for product design, and proper asset management [13]; g) *Open manufacturing*: a new model of socioeconomic production in which goods are produced in an open, collaborative and distributed way, based on open design and open source methods; such a system may engage in making products designed and owned by other companies [14]; h) *Universal manufacturing*: an open manufacturing model with higher degree of standardization and formal representation of digital enterprise models in the cloud [15]; this model implements the many-to-many configuration between products and production systems and needs an increased presence of manufacturing enterprises, from small to large, in a cloud [16].

In the I4.0 framework, CPS take advantage from the integration of Cloud-based and Service-Oriented Architectures to deploy end-to-end support along both *product lifecycle* (design, engineering, manufacturing, after-sales services, de-manufacturing) and *factory lifecycle*. In a factory lifecycle perspective, CPS are able to interact with all the hierarchical layers of the automation pyramid (ERP, MES, SCADA, PLC, field level) and to empower the exchange of information across all the phases, resulting in a better product-service development (in terms of efficiency, timing, quality, others).

At the same time, new markets for CPS technologies and products will be created by implementing the dual I4.0 factory-product strategy:

- Horizontal integration through value networks: development of inter-company value chains and networks.
- End-to-end digital integration of engineering across the entire value chain of both the product and the associated manufacturing system.
- Vertical integration of flexible and reconfigurable manufacturing systems within businesses.

Industry 4.0 makes use of the Industrial Internet of Things (IIoT) and Services framework in which the CPS technology makes virtualized resources, products and orders to interact for the intelligent achievement of a production goal. The main focus of the I4.0 platform is realizing strongly coupled, smart manufacturing and logistics systems. CPSs pair a physical layer managed by IIoT technologies and a virtual layer managed by cloud computing technologies [17].

This paper proposes a Systems Engineering-oriented higher education (HE) program devoted to the knowledge and skills needed in order to put in practice the Industry of the Future (IoF) transformation set up by the 4th industrial revolution. The foundational technologies of Industry 4.0 that transform industrial production, the factory levels and the architecture layers on which they are related in RAMI 4.0 Reference Architectural Model Industry 4.0 for product development and production are described in Sect. 2 of the paper. They will be mapped in the courseware of the HE program; Sect. 3 discusses 'related work', i.e., worldwide programs aligned to the paradigm of the fourth educational revolution (Education 4.0) that exploits the potential of digital technologies and personalised knowledge and data. The new master program SELFI4.0 with hands-on training in Industry 4.0 technologies, CPS and IoT solution development is introduced in Sect. 4. Conclusions about the economic and technical feasibility of the educational program and partnership with industry needed are formulated in Sect. 5.

2 Foundational Technologies and Standards Integrated in the Reference Architectural Model Industry 4.0

The digital transformation of manufacturing promoted by Industry 4.0 is driven by nine foundational technology advances. Some of them are already used in manufacturing, although not in a generalised way; with Industry 4.0, they will transform production, interconnecting separated, locally optimized shop floor units (machines, cells) in fully integrated, globally optimized, agile and reality-aware production flows, changing thus conventional, fixed production rapports between suppliers, producers, and customers, and also between humans and machines:

- *Additive manufacturing*: a new key enabling production technology, component of the Direct Digital Manufacturing (DDM) concept that includes novel 3D printing, digital shape reconstruction and modelling technologies (the need for tooling and setup is reduced by producing parts directly based on digital models). Additive manufacturing is closely related to the CMfg model that moves, in the Industry 4.0 context, from production-oriented manufacturing processes to customer- and service-oriented ones; it is also associated to the Design as a Service (DaaS) concept when the enterprise uses the Internet and the Cloud to share DDM services with the customer for customized product design and production in small batches.
- *Autonomous robots*: in the IoF perspective, industrial robots and automated guided vehicles (AGV), already used in the last 60 years, will become more autonomous, flexible and collaborative. They will be able to interact with one another and with humans and learn from them to be more dexterous, e.g., by emulating manipulative gestures of expert workers.
- *Augmented reality*: AR-based systems will be used in the IoF to provide workers, via specialised devices, with real-time information to assist and improve work procedures; this is an Artificial Intelligence KET that will be intensively used for virtual training of human operators (managing emergencies, interacting with machi-nes) and support for decision making.
- *Simulation*: will use real-time data to mirror the physical world in an extended digital model, which can include production entities: resources, products, orders, processes, and humans. The *Digital Twin* (DT) is an AI-based KET that embodies this extended

model of digital 3D shape representation, operating mode, execution context, history of behaviour, time evolution and status; it consists of a virtual representation of a production asset (e.g., a robot) or system (a robot team) that is able to run on different simulation strategies and synchronizes the virtual and real system, thanks to data sensed by smart devices (e.g., IoT gateways), mathematical models and real time data processing in IoT aggregation nodes. DT-based simulation is used for resource health monitoring, anomaly detection, predicting unexpected events, and optimizing production systems. Three simulation strategies can be deployed with DT software configurations: i) *simulation with software in the loop*: used in application design, layout validation, parameter tuning; ii) *robust deployment*: provides reality-aware resource control, automatic diagnosis and predictive maintenance; iii) *embedded simulation faster than real-time*: the MES control may have certain intentions (adjusting parameters, rescheduling operations on resources) in a configuration *ii* actually deployed. Then, an aggregate DT virtualizing the collective intention solver will use a meta-level control that initiates a configuration *i* executing much faster than real time (at computing time) and subscribes to the appropriate publishing services of resource twins and performers within configuration *i*; this allows the meta-level to observe problems before they occur and decide upon the best global resource allocation.

- *Industrial Internet of Things*: with the Industrial Internet of Things (IIoT) production entities (resources, unfinished products and product execution orders) and the environment will be pervasively instrumented with embedded computing and interconnected using standard technologies (e.g., broadband Internet, wireless). IIoT architectures use *edge computing* technologies (edge gateway, aggregation node, next unit of computing) to perform the computation on shop floor data downstream cloud services and upstream IoT services at the edge of the physical network, and *fog computing* decentralized infrastructure in which data, computation results, storage and applications enable short-term analytics and ad-hoc decision making [18].

- *Big Data and Analytics*: the integration of increasingly smart devices, products and control software caused an explosion in data points available at shop floor level. In the Industry 4.0 context, smart sensors will pervasively instrument resources, products, processes and orders as 'plug-and-produce' modules [19], while IoT hardware aggregation nodes and middleware align data streams in normalized time intervals and transfer data in the cloud-based MES for analysis – this will become standard to support real-time decision making.

- *Cloud*: the IT model of Cloud Computing (CC) is extended in IoF perspective to services that orchestrate factory resources and their automatic controls through operational technologies. This dual Cloud Control and Computing (CCoC) model is the real time partition of the CMfg enterprise model mapped on its technical layer, and which: i) transposes pools of factory resources (robots, machines, controllers) into *on-demand* product making services; ii) may involve *on-demand* network access to a shared pool of configurable high performance computing resources - servers, storage, applications that can be rapidly provisioned and released as services to global MES workloads with minimal management [9, 20]. Global MES workloads virtualized in the cloud call for real time shop floor (big) data streaming. The new collaborative business models enabled by Industry 4.0 require the presence of enterprises in the cloud, which allows for infrastructure and knowledge sharing in the open and universal manufacturing space.

- *Cybersecurity*: in the context of multi-tenant shop floors where multiple and sometimes competing organisations share the same manufacturing facility, security requirements at plant level and throughout the MES and ERP components become very important, as physical security is no longer enough. Problems like unauthorised access to information, theft of proprietary information, denial-of-service (DoS) and impersonation need clear and formalised solutions to be applied in practice. Likewise, with the decentralization of control, increased connectivity between enterprise layers and use of standard communications protocols specific to Industry 4.0 there is an increased need to protect manufacturing systems from cybersecurity threats: secure, reliable communications, identity and access management of resources and users, authentication and encryption of intelligent, travelling products.
- *Horizontal and vertical system integration*: with Industry 4.0, dynamic business and engineering processes will enable last-minute changes to production and deliver the ability to respond flexibly to disruptions and failures. IoF businesses will establish global networks that incorporate their production, logistics and warehousing systems. Enterprise integration means that manufacturing systems are vertically networked with business processes within factories (for agility and responsiveness to the increasing knowledge intensity and complexity of negotiations), and horizontally connected to dispersed value networks (supply, design and engineering, production, logistics for delivery, after-sales services) managed in real time [21].

The actors in the manufacturing value chain (suppliers, producers, vendors, service providers) must work with governments to adapt education to create a skilled workforce as they embrace these advanced technologies of Industry 4.0 integrated in implementing frameworks (Fig. 1). This need can be achieved by adapting school curricula, training, and university programs and strengthening entrepreneurial approaches to increase the digital and IC^2T-related skills and innovation abilities of the workforce.

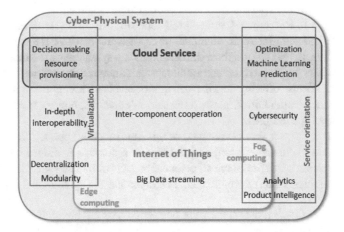

Fig. 1. Industry 4.0 framework integrating the technology advances

The implementing framework for Industry 4.0 is represented by Cyber-Physical Systems (CPS) that comprise smart embedded resources, products, warehousing systems and production facilities that have been developed digitally by virtualizing the physical plant components, interact and collaborate to accomplish global goals, and feature end-to-end IC^2T-based integration on all enterprise layers. Industry 4.0 involves the use of the Internet of Things and Services to create networks incorporating the manufacturing process. The IIoT integrates shop floor devices equipped with sensing, identification, processing, communication, and networking capabilities [18]; it uses industrial wireless networks and the Internet.

The IIoT relies on the convergence of IT (computing hardware, software, communication) and OT (controllers); while IT implements data access, big data streaming and local using a top-down approach, OT uses ground-up solutions such as edge computing from end-point devices, embeds intelligence and connectivity into virtualized plant entities, analyses data and extracts knowledge in fog computing nodes for higher level decision support in the CPS cloud.

Overall, the manufacturing CPS framework in Industry 4.0 perspective is characterized by pervasive instrumenting of plant devices, automation with distributed intelligence embedded on strongly coupled information counterparts of generic or instantiated entities – resources, products, orders – that are virtualized with various techniques: multi-agent systems, holarchies, digital twins for individual or collective, local (shop floor) or global (batch, customer order, value chain) workloads. CPS benefit from cloud services in the dual control-computing model for process optimization and reality-aware control, prediction of unexpected events and detection of resource anomaly based on AI and ML techniques. Modelling of control in industrial CPS uses the holonic manufacturing paradigm, service orientation, and in-depth interoperability of intelligent beings that create a software platform on which intelligent decision makers (agents) are the applications. The control in industrial CPS is implemented in semi-heterarchical information system topology with: i) MAS on the decentralized MES layer where individual workloads directly related to agentified plant devices are virtualized; ii) Cloud services where collective and MES workloads (e.g., system scheduler, optimization engine) related to global intentions (e.g., supervision, reconfiguring) are virtualized.

The RAMI 4.0, Reference Architecture Model Industry 4.0 was developed by the German Electrical and Electronic Manufacturers' Association (ZVEI) to support Industry 4.0 initiatives. RAMI 4.0 is a three dimensional map synthetizing the most important aspects of Industry 4.0 about reference architectures, structured deployment frameworks, standards and norms, offering to all participants involved a common understanding of technologies, functionalities, networking and interactions, and perspectives of I4.0 framework development [22].

RAMI 4.0 defines a service-oriented architecture where application components provide services to the other components through a communication protocol over a network; it represents in a 3D frame all essential aspects of Industry 4.0 [23]. RAMI 4.0 combines all elements and IT components in a layer and life cycle model (Fig. 2):

- The *Hierarchy Levels* axis: on the right horizontal axis are hierarchy levels from the international standards series IEC 62264, based upon ANSI/ISA95 for *Enterprise information and control systems*. These hierarchy levels represent the locations of

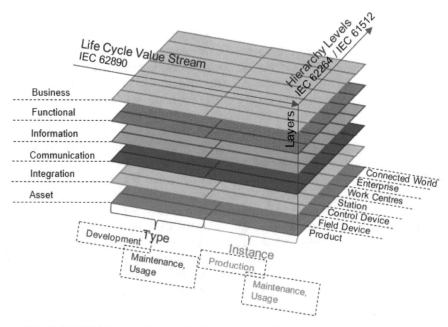

Fig. 2. RAMI 4.0, the Reference Architecture Model Industry 4.0 (adapted from [23])

functionalities within factories. To represent the I4.0 environment, the ISA95 functionalities are expanded to include manufacturing outcome labelled "Product," and the connection to the Internet of Things and Services labelled "Connected World".

- The *Life Cycle Value Stream* axis: represents the life cycle of facilities (factory) and products, based on IEC 62890, *Life-cycle management for systems and products* used in industrial-process measurement, control, and automation. Furthermore, a distinction is made between "types" and "instances." A "type" becomes an "instance" when design and prototyping have been completed and the actual product is being manufactured, or when the execution of an ordered product has been instantiated (i.e., operations scheduled and allocated on machines, parameter set up).
- The *Layers* axis: the six layers on the vertical axis define the structure of the information-based (virtual) representation of an *I4.0 component's* (e.g., CNC machine, robot, control) *properties structured layer by layer*. The layers symbolise various perspectives: data maps, functional descriptions, communication modes, vertical integration, hardware assets or business processes. This corresponds to IT and system theory where complex projects are split up into clusters of controllable parts.

The special characteristics of the reference architecture model are its combination of life cycle and value stream with a hierarchically structured approach for the definition of **I4.0 components**. Also, the model highlights the interaction between four I4.0 aspects: (A1) horizontal integration through value networks; (A2) vertical integration within a factory or shop floor; (A3) life cycle management, and end-to-end engineering; (A4) human beings orchestrating the value stream. The I4.0 component provides a flexible framework in which data and functions enabling these aspects are made available.

The I4.0 component in the reference model is located within the layers of RAMI4.0. It can adopt various positions in the life cycle and value stream and occupy various hierarchical levels. An I4.0 component can be a production system, an individual machine or workstation, or a processing board of a controller; it can be located in the area of the office floor or shop floor and is part of/interacts with significant factory systems: the PLM (Product Lifecycle Management), ERP and Industrial Control and Logistics. The CPS is a network of I4.0 components that, together with their contents, are structured such that connections between any end points are possible.

From logical point of view, an I4.0 component comprises one or more objects (physical entities) and an *administration shell* which contains the data for virtual representation and the functions of the technical functionality. For deployment, a software part of the virtual representation details the necessary administrative aspects on the I4.0 component, while a resource manager allows IT services to access the data and functions of the administration shell and make them available to the outside. I4.0 components play a vital role in ensuring a consistent, uniform information exchange and workflow coordination in CPS and Industry 4.0 value streams with their value adding processes.

3   Key Skills and Formation Programs for Industry 4.0

The 21st century skills involve competencies related to learning and innovation, digital literacy as well as career and life. Examining key skills for Industry 4.0, experts consider two main approaches that educational systems may take towards I4.0 [24, 25]:

- Educating *followers*: followers have the right skill set for reacting to changes in their working environment, adapting their performance, and learning to cope with the technological development.
- Educating *change makers*: in addition to people who are highly knowledgeable of I4.0 technologies, specialists are needed who can make informed decisions related to, for example, usability, reality awareness, safety, sustainability, and ethics.

Operating cyber-physical systems stresses the importance of human interactions. As digitalisation, robotics, artificial intelligence, networking and the industrial internet are freeing people from routine labour, human abilities for idea sharing and critical thinking are called for. Industry 4.0 is still evolving, which limits the understanding of the skills needed to work in I4.0 manufacturing. Yet, forerunning companies are already adopting modern technologies to realize the potential embedded in Industry 4.0; they are looking to upskill their current workforce and to recruit new employees with the right skill sets. Higher education institutions must work closely with industry, governmental agencies and student organizations to stay updated on the emerging competence needs.

The shift in manufacturing models accounts for social and market changes combined with technological advancements (integration of virtual and physical world; creating value by knowledge extracted from data; automation; human-machine collaboration) that bring added value in products, business models, processes, and human work.

The EU project 'Universities of the Future' grounds the reference framework for Industry 4.0 skills on *working life skills* (WLS) that are defined as the knowledge, skills or attitudes that are needed to perform a job successfully [26]. WLS refer to both I4.0 (discipline)-specific competencies and knowledge as well as transferable skills (may be applied in many different professional contexts). Table 1 shows the *discipline-specific competencies and knowledge* related to Industry 4.0, as a result of literature analysis; in addition, *transferable skills* include critical thinking skills, personal (soft) skills, systems thinking, commercial knowledge, and technological literacy.

Table 1. Taxonomy of discipline-specific competencies and knowledge for Industry 4.0

Discipline-specific competencies and knowledge for Industry 4.0	
Technical and Engineering competencies	Data science and big data analytics; Human-machine interface; Digital-to-physical transfer technologies (3D printing, shape reconstruction); Simulation and virtual plant modelling; Data communication and networks; System automation; AI; Robotics; Programming; Product and process quality control; Real-time Logistics and supply chain optimization
Business and Management competencies	Technology awareness; Change management and strategies; Organizational structures and knowledge; Managers' roles; Tech-enabled processes: forecasting and planning metrics
Design and innovation competencies	Understanding the impact of technology; Human-robot interaction and user interfaces; Tech-enabled product and service design; Tech-enabled ergonomic solutions and user experience

Preparing for Industry 4.0 requires developing digital skills and establishing technology education programs. As organisations and people become more interconnected, communication and networking skills become essential. The accent of work changes from task-based to value-based. This challenging environment requires abilities for *continuous learning.*

With Industry 4.0, it has become imperative for educational organizations to move towards a new revolution: Education 4.0 (E4.0) - a vision of the future of education that exploits the potential of digital technologies, personalized data, and the opportunities offered by this connected world to foster lifelong learning. *Education 4.0* is about integrating the technological advances of Industry 4.0 for educational purpose [27]. In addition to considering the requirements of Education 4.0, the *University 4.0* model has been defined to provide autonomous management of learning processes based on the integration of the physical and digital worlds to improve and adapt learning [28]. The concept of the fourth university revolution aims at applying the Industry 4.0 paradigm in universities to foster the automation, adaptation, and personalization of learning processes, contributing thus to the transition to Education 4.0. Several higher education programs in the Industry 4.0 area, running in E4.0 conditions are further described.

Industry 4.0-related courses are developed in collaboration with Wiley (global leader in publishing, education and research), achieved through applied, experiential learning with case analyses, projects and internships [29]. Programs offering these courses are integrated with Wiley Industry 4.0 certifications and Harvard Business School certificates of readiness, e.g., Data Engineering and Informatics in program Algorithms and Data Structures from University of California San Diego, US; Cybersecurity and Digi-tal Forensics, program in 'Cybersecurity' from Rochester Institute of Technology, US.

Some of the MIT, Cambridge MA, US learning programs and courses with Industry 4.0 subjects are listed in Table 2 [30]:

Table 2. MIT learning programs and courses for Industry 4.0

Learning program	Course	Duration
Computer Science (CS) Data Modeling & Analytics (DMA) Design & Manufacturing (DM)	Advanced Data Analytics for IIOT and Smart Manufacturing	8 weeks
DM, DMA	Professional Certificate Program in I4.0	9 months
	AI for Computation Design & Manufacturing	10 weeks
	Smart Manufacturing: Production in I4.0	10 weeks
CS, DMA, Digital Transform. (DT)	Machine Learning: Decision Making Technol	8 weeks
DT	Professional Certificate Program in DT	8 weeks
CS, DT, Legal Tech (LT)	Cloud & DevOps: Continuous Transform	8 weeks
	DT: From AI & IOT to Cloud & Cybersecurity	8 weeks
Innovation (In)	Systematic Innovation	8 weeks
In, Leadership & Communic. (LC)	Leadership and Innovation	8 weeks
CS, DMA, DT, In, LC	Profess. Certificate: Chief Technology Officer	12 month
	Applied Data Science Program	12 weeks
DM, In	Solving Complex Problems: Structured Thinking, Design Principles, an AI	5 days
	Additive Manufacturing: From 3D Printing to the Factory Floor	5 days
DMA	Data & Models: Regression Analytics	5 days
	Predicting Market Demand	5 days
	Machine Learning: from Data to Decision	8 weeks
CS	Deep Learning for AI & Computer Vision	5 days
CS, DM	Industrial Internet of Things: From Theory to Applications	9 weeks
Energy & Sustainability (ES)	Sustainability: Strategies for Industry	8 weeks
	Sustainable Planning and Operations Syst.	8 weeks
DM, Systems Engineering (SE)	Management of Technology: Strategy and Portfolio Analysis	9 weeks
	Design Product Family: Strategy, Implement.	8 weeks

The Northeast WI Technical College in Wisconsin, US offers degrees, classes, and workforce training in key Industry 4.0 technologies including advanced manufacturing, smart factory, IOT Data analysis and visualization, 3D printing, autonomous systems, SCADA, machine learning, Cloud computing, augmented and virtual reality [31]. The Master degree of Engineering Management and Leadership (MEML) at Rice university, Houston Texas US is designed to help engineers become Industry 4.0 leaders [32].

The Technical University of Vienna implemented a *scenario-based* Industry 4.0 Learning Factory concept in Austria's first Industry 4.0 Pilot Factory, I40PF, as human-centred cyber-physical assembly system composed of interconnected mobile assembly stations [33]. There is a digital assistance system based on AR technology.

The research reported in [34] applies the RAMI 4.0 reference framework to enable information dissemination and allow the *organizational learning* process at producers of manufacturing equipment (gaining experience and using this experience to create knowledge for enhanced equipment). The study modelled the Smart Production Lab. at Aalborg University that includes a small production line using RAMI 4.0 as artifact. The results of this study confirm that the RAMI 4.0 reference model can be applied to structure and support information dissemination and organizational learning.

The impulse paper [35] formulates recommendations for action based on a 3-dim. Reference framework model consisting of corporate and leadership culture, organisation and structure, and personal responsibility. These three fields contribute to developing a sustainable learning culture for Industry 4.0 in companies. In view of the need for skilled workers for Industry 4.0, the 'Work of Tomorrow Act' (May 2020) encourages German companies to qualify their employees in the time freed up by a decline in work.

Industry 4.0, as global digital manufacturing initiative, is seen as a technology driven strategic advance for economy which, at the moment, brings some challenges: the need to integrate legacy systems, the high cost of business cases, and the shortage of qualified personnel. These problems are perceived by SME manufacturers as difficulties making digitalisation inaccessible. A low cost, tactical approach at country level is represented by the 3-year research project 'Digital Manufacturing on a Shoestring' funded in the UK by the Engineering and Physical Science Research Council [36]. The scope of this initiative is twofold: developing individual digital solutions for which the total cost of deployment is kept low and applying the approach in a large number of companies and labs, i.e., outreach, engagement and dissemination towards Industry 4.0. The Shoestring program has five stages: 1) *Digital requirement assessment*: identifying specific digital needs of SMEs to identify solution priorities; 2) *Solutions development*: defining stan-dardised 'building blocks' that can be reused and plugged together to form a solution; 3) *Prototyping/Pilot testing* of the developed technologies and methods in partner SMEs; 4) *Incremental integration*: integrating solutions incrementally with existing legacy man-ufacturing facilities; 5) *Engagement and dissemination*: outreach of low cost digital solutions through informal development settings, application in production companies.

Such initiatives raise the awareness and skills' level for the participation of companies and personnel in Industry 4.0 developments.

4 The SELFI4.0 Learning Factory for Industry 4.0 Training and Research

This section presents the SELFI4.0 Learning Factory created in the Dept. of Automation and Industrial Informatics of the University Politehnica of Bucharest as a reality-conform manufacturing environment for education of change makers, hands-on training and research in Industry 4.0 key technologies, smart control solutions and Cyber-Physical Production Systems. The infrastructure is a pilot manufacturing cell integrated with a private cloud platform set up in partnership with industry - producers and integrators of high tech automation, robotics, vision, software, information technologies and equipment: East Electric (main partner, https://www.eastelectric.ro/), Cognex, Omron Adept Technology, ABB, Rockwell Automation and IBM Romania.

This complex Industry 4.0 learning environment is based on the authentical replica of an industrial production system with manufacturing value chain components, and on the integration of newest industry trends with academic content, physical infrastructure and engineering practices. The SELFI4.0 learning factory is defined by:

- Industrial *activities and processes* that are automatically performed on multiple workstations, use material components to manufacture *products* that travel between storages and machines, are handled, packed and palletized for distribution by robots.
- A six hundred m^2 location with *shop floor layout* and reconfigurable *resource team* that can be reengineered using modularity and pluggability, supporting thus the plug & produce metaphor and hardware and software control reconfigurability in the *value chain* from both technical and organizational viewpoint.
- A set of types of *physical products* that are manufactured differently, from open shop to fixed precedencies mode (e.g., pneumatic cylinders, carburettors, tool boxes, etc.)
- A *didactic model* including a 42-week systems engineering-oriented program (master in Robotics & Automation, RA) with 8h formal teaching, 8h hands-on training in the pilot cell, and 12h case study and application development (team work for skills growth) per week; 12h-week individual research project for innovation in IoT, CPS.

Figure 3 left shows the layout and structure of the industrial manufacturing cell with four robotized workstations: two part processing stations with (3 + 1)-axis CNC Concept Mill 105 milling machines from EMCO tended by 6-d.o.f. vertical articulated robots Viper s650 from Omron Adept, and two part assembly stations with 4-d.o.f. SCARA robots Cobra s600 from Omron Adept interconnected by a closed-loop transport system with access branches - a twin-track conveyor belt from Bosch-Rexroth controlled by four IndraLogic PLCs with 512I/256E lines, Ethernet, Profibus, and RS 232 communication. The workstations have local storages for material components; in case of storage depletion, a 4-d.o.f. SCARA robot Cobra s850 retrieves missing parts from a dual-spare part feeding ASRS in a strategy established by the information extracted from images collected by two 2D smart video cameras. These parts travel on pallets on the conveyor to the depleted storage where the station robot handles them. The operations on products are executed at the assigned workstations; products are placed on pallet carriers by robots and move on the conveyor from one station to the next. There is a sixth empty pallet supply/retrieval of pallet with finalised product workstation with pallet buffer cabinet tended by a 4-d.o.f. 400 × 800 × 1200 mm Cartezian Python robot from Omron Adept. The quality control of shape geometry, surface state and component alignment is done by machine vision measurements from 2D and 3D video cameras. Each industrial robot is equipped with stationary, down looking and mobile, arm-mounted cameras.

Groups of manufacturing resources have similar capabilities (e.g., CNC machines, SCARA and anthropomorphic robots), so certain operations on products can be allocated to any of them; this feature is exploited for global cost optimization of batch orders.

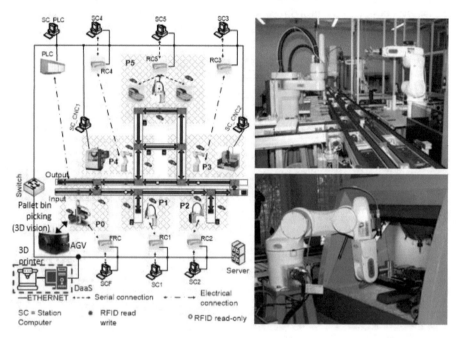

Fig. 3. Pilot manufacturing cell (the factory): layout (left); global view, machine tending (right)

SCARA robots in stations P1, P2 share a common working space; soft- and hard cooperation techniques have been made available for dual-arm task synchronization, part-driven operating, virtual coupling and dynamic interaction of robots. The material/product flow can connect any two workstations via the closed-loop conveyor, for both job shop and flow shop production processes. Empty pallets are brought in bins by an Omron LD-type AGV with laser and sonar; 3D vision Cognex is used for bin picking.

Intelligence is embedded on products during their execution lifecycle through Intelligent Embedded Devices (IED) - OMAP3503 Overo computer-on-module, processor ARM Cortex A8 - placed on product carriers (pallet carriers). IEDs host Work in Process (WIP) agents that maintain the dynamic image of product execution orders optimized in the cloud and transferred to the cell's product routing PLC. The communication between shop floor entities uses *Ethernet* (cell workstations host resource images, cell PLC dispatches optimized product execution orders), Bluetooth and 802.11b/g *Wi-Fi* (IEDs track product execution) LANs. Product locating: RFID tags on carriers read by belt sensors.

The private cloud computing and service infrastructure is based on IBM CloudBurst 2.1 platform with 460 preconfigured virtual machines, 4 GB RAM, 60 GB of HDD; server management node IBM x3550 M3 dual-socket Intel Xeon 5620 2.4 GHz 4-core processors; 13 cloud compute nodes IBM BladeCenter H, 29 TB System Storage; blade operating system VMware vSphere 4.1 enterprise ed., IBM system director, network control, active energy and security manager. Customers have access via cloud to product DaaS.

A *cyber-physical system* has been developed to transform the pilot manufacturing cell in a *smart factory*; products and resources are virtualized, networked and communicate in a smart control system with distributed, semi-heterarchical architecture (Fig. 4).

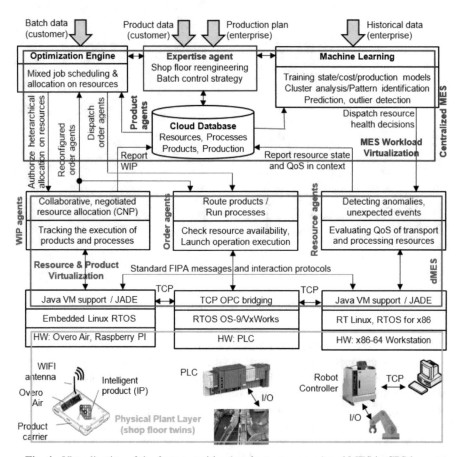

Fig. 4. Virtualization of the factory entities (products, resources) and MES in CPS layers

The CPS for smart production control operates with distributed intelligence and collaborative decisions of strongly coupled shop floor; the modelling approach is the holonic paradigm with agent-based implementing framework. This approach is based on the virtualization of a set of abstract entities: *products* (types in RAMI4.0 lifecycle value stream), *resources* (technology, humans), and *orders* (instances in RAMI4.0 lifecycle

Table 3. SELFI4.0 practical training and mapping with I4.0 technologies and courses

I4.0 hands-on training chapters	RA support courses	I4.0 technology	RAMI4.0 layers
Product lifecycle: 1) **Design**: digital models, group tech., DaaS, customer input, virtualization; 2) **Making**: product intelligence, WIP MAS, product-driven automation 3) **Usage**: smartness, servitization, after-sales services	- Mechatronics engineering, DDM - Human-machine interaction	Cloud, Simulation (S) H&V system int. (HVI) HVI	Business (B) Communication (C) B, Functional (F)
Value chain: 1) **Activities**: layout design, resource team configuring, production plan., activity scheduling & resource alloc., supply & inventory mngmt., manufacturing coordination & supervision, logistics for product distribution 2) **Processes**: DDM (additive manuf., 3D digital shape reconstruct., adaptive machining), robot services, 2D/3D vision (motion guid., quality ctrl., bin picking), transport (AGV, conveyor) 3) **Asset**: integration (robot, PLC, 3D print, CNC), instrumenting, health monitoring, maintenance, tooling	- Intelligent manuf. systems and value chain integration (IMS) - Robot motion planning and control - Guidance vision for robots and AGVs - IMS	HVI, S Additive manufacturing (AM), Autonomous robots (AuR) AM, AuR	Asset (A), F Integration (It) A, It
IC²T for CPS: 1) **Data collecting**: edge computing, IoT gateways, aggregation nodes, big data streaming, data aggregation 2) **Data transmission**: I/O and web-service communic. protocols, comm. middleware, TCP/OPC, Wi-Fi, fog 3) **Data processing**: analytics, optimization, anomaly detection, decisions 4) **AI and AR**: prediction, classification, clustering, pattern rec., model DT	- Embedded systems and IIoT design - IIoT - Big Data streaming and Analytics - Machine learning techniques & applic.	Industrial IoT (IIoT), Big Data Analytics (BDA) IIoT, Cybersecurity (Cy), BDA Cloud S, AR, Cloud	A, It C, If F, B F, If
CPS: 1) **Models**: holarchies, data-driven DT, semi-heterarchical control, MAS, software in-the-loop, SOA, CPS 2) **Layers**: shop-floor control, MES, SC 3) **Cloud**: workload virtualization, VM, containers, cybersecurity, Manuf. Service Bus, Software Defined Networks	- Multi-agent system and programming - CPS models, design - Holonic MES ctrl. - Cloud models and HA services	S, HVI, IIoT Cloud, HVI Cloud, Cy	C, F F, It, If C, B

value stream) that are modelled by autonomous holons collaborating by means of their information counterparts - *intelligent agents* organized in dynamic clusters to reach optimally and with reality awareness a common, production-related goal [37].

The agents that are virtualizing the generic classes of factory entities instantiate them at production run for two types of *local workloads* - (a) *individual* resource state, behaviour and performance monitoring, respectively tracking product execution; (b) *collaborative* activity (re)scheduling and allocating on resources for products that add up to the factory's capacity of simultaneous execution. *Global* MES workloads are virtualized in the cloud: *optimization* at batch horizon based on insights from aggregated data streams originating from shop floor resources, processes and environment and ML-based prediction of resource usage cost and resource health management for predictive maintenance.

The cyber-physical system available for the pilot manufacturing cell hosts 20 h per week of hands-on training, case studies, proof-of-concept and demos on I4.0 functionalities, design and implementing solutions. Table 3 lists these practical activities and their mapping with I4.0 technologies, RAMI 4.0 layers and RA teaching program.

5   Conclusions

Considered as a part of the fourth industrial revolution, Industry 4.0 is a trend towards automation, digital transformation of processes and controls, integration and interaction of products and factory devices throughout their lifecycles, and new business models with enterprise presence in the cloud. New IC^2T and production technologies will be used for these changes: Cyber-Physical Systems, Industrial Internet of Things, Cloud services, Artificial Intelligence and human-machine cooperation.

The Learning Factory is an efficient concept for the formation of professionals having Industry 4.0-related competencies, capable to operate cyber-physical production systems and pushing forward the development of key enabling technologies (production-, AI-, cyber-technologies) and innovative solutions for the industry of the future.

The paper describes a model of learning factory for Industry 4.0 based on an industrial real manufacturing cell controlled by a CPS that virtualizes individual shop floor devices and global MES workloads in the cloud for cost optimization, reality awareness, agility to product and market changes; this factory facilitates practical training that is driven by the learning content of courses scheduled in the 42-week master program 'Robotics and Automation'. This integrated learning factory benefits from the Education 4.0 concept and methods inspired by I4.0 transformations: personalization of learning with flexible, dynamic and adaptive learning pathways, digital teaching means, knowledge validation by real life examples, research opportunities driven by industrial partners.

Future research consists in extending the learning factory model with an incremental training program on I4.0 technologies for SME personnel combined with the creation of a low-cost solution portfolio for automation and information management in production companies.

References

1. Heidel, R., Hoffmeister, M., Hankel, M., Döbrich. U.: Industrie 4.0. The Reference Architecture Model RAMI 4.0 and the Industry 4.0 component, DIN Deutsches Institut für Normung e.V., Beuth Verlag GmbH, Berlin (2019)

2. sCorPius. Future trends and Research Priorities for Cyber-Physical Systems in Manufacturing, EuroCPS Project (2017). https://www.eurocps.org/wp-content/uploads/2017/01/sCorPiuS_Final-roadmap_whitepaper_v1.0.pdf
3. Kagermann, H., Lukas, W., Wahlster, W.: Industry 4.0: Mit dem Internet der Dinge auf dem Weg zur 4. Industriellen Revolution, VDI Nachrichten **13**, 1090–1100 (2011)
4. Trentesaux, D., Borangiu, T., Thomas, A.: Emerging ICT concepts for smart, safe and sustainable industrial systems. J. Comput. Ind. **81**, 1–10 (2016). https://doi.org/10.1016/j.compind.2016.05.001
5. Vaidya, S., Ambad, P, Bhosle, S.: Industry 4.0 - A Glimpse. Procedia Manuf. **20**, 233–238 (2018). https://doi.org/10.1016/j.promfg.2018.02.034
6. Waldner, J.B.: Principles of Computer-Integrated Manufacturing. Wiley, London, pp. 128–132 (1992). ISBN 0-471-93450-X
7. Zhong, R.Y., Xu, X., Klotz, E., Newman, S.T.: Intelligent manufacturing in the context of industry 4.0: a review. Engineering **3**(5), 616–630 (2017). https://doi.org/10.1016/J.ENG.2017.05.015
8. Liu, Y., Xu, X., Zhang, L., Wang, L., Zhong, Y.: Workload-based multi-task scheduling in cloud manufacturing. Robot. Comput.-Integr. Manuf. **45**, 3–20 (2017). https://doi.org/10.1016/j.rcim.2016.09.008,Elsevier
9. Borangiu, T., Răileanu, S., Morariu, O.: Virtualizing resources, products and the information system. In: Cardin, O., Derigent, W., Trentesaux, D. (eds.) book Digitalization and Control of Industrial Cyber-Physical Systems, Chapter 5, ISTE Ltd, London and Wiley, New York (2022). ISBN: 9-781-78945-085-9
10. Kagermann, H., Wahlster, W., Helbig, J.: Recommendations for implementing the strategic initiative INDUSTRIE 4.0. Final report of the Industrie 4.0 Working Group, Acatech - National Academy of Science and Engineering (2013). https://en.acatech.de/wp-content/uploads/sites/6/2018/03/Final_report__Industrie_4.0_accessible.pdf
11. 2030 Vision for Industrie 4.0 – Shaping digital ecosystems globally. Position Paper Plattform Industrie 4.0 (2022). www.plattform-I40.de. Accessed Mar. 2022
12. Oztemel, E., Gursev, S.: Literature review of Industry 4.0 and related technologies. J. Intell. Manuf. **31**(1), 127–182 (2018). https://doi.org/10.1007/s10845-018-1433-8
13. Babiceanu, R.F., Seker, R.: Cloud-Enabled Product Design Selection and Manufacturing as a Service. In: Borangiu, T., Trentesaux, D., Leitão, P., Giret Boggino, A., Botti, V. (eds.) Service Oriented, Holonic and Multi-agent Manufacturing Systems for Industry of the Future. SOHOMA 2019. Studies in Computational Intelligence, vol. 853, pp. 210–219 (2020). Springer, Cham. https://doi.org/10.1007/978-3-030-27477-1_16
14. Kostakis, V., Latoufis, K., Liarokapis, M., Bauwens, M.: The convergence of digital commons with local manufacturing from a degrowth perspective: Two illustrative cases. J. Clean. Prod. **197**, 1684–1693 (2018). https://doi.org/10.1016/j.jclepro.2016.09.077
15. German Standardization Roadmap Industrie 4.0. Version 4, Standardization Council Industrie 4.0, DIN e. V., DKE Deutsche Kommission Elektrotechnik (2020). https://www.din.de/resource/blob/65354/1bed7e8d800cd4712d7d1786584a7a3a/roadmap-i4-0-e-data.pdf
16. Kusiak, A.: From digital to universal manufacturing. Int. J. Prod. Res. **60**(1), 349–360 (2022). https://doi.org/10.1080/00207543.2021.1948137
17. Borangiu, T., Trentesaux, D., Thomas, A., Leitão, P., Barata, J.: Digital transformation of manufacturing through cloud services and resource virtualization. Comput. Ind. **108**, 150–162 (2019). https://doi.org/10.1016/j.compind.2019.01.006
18. Industrial Internet Consortium. The Industrial Internet of Things, Volume G1: Reference Architecture, IIRA Version 1.9 (2019). https://www.iiconsortium.org/pdf/IIRA-v1.9.pdf

19. Madiwalar, B., Schneider, B., Profanter, S.: Plug and produce for industry 4.0 using software-defined networking and OPC UA. In: 24th IEEE International Conference on Emerging Technologies and Factory Automation, pp. 126–133 (2019). https://doi.org/10.1109/ETFA.2019. 8869525

20. Coullon, H., Noyé, J.: Reconsidering the relationship between cloud computing and cloud manufacturing. In: Borangiu, T., Trentesaux, D., Thomas, A., Cardin, O. (eds.) Service Orientation in Holonic and Multi-Agent Manufacturing. SCI, vol. 762, pp. 217–228. Springer, Cham (2018). https://doi.org/10.1007/978-3-319-73751-5_16

21. Kuhn, T., Schnicke, F., Oliveira Antonino, P.: Service-based architectures in production systems: challenges, solutions and experiences, 2020 ITU Kaleidoscope: Industry-Driven Digital Transformation, pp. 1–7 (2020). https://doi.org/10.23919/ITUK50268.2020.9303207

22. Lydon, B.: RAMI 4.0 Reference architectural model for industrie 4.0, international society of automation (ISA) (2022). https://www.isa.org/intech-home/2019/march-april/features/rami-4-0-reference-architectural-model-for-industr

23. VDI/VDE, ZVEI. Status Report. Reference Architecture Model I4.0 (RAMI4.0) (2015). https://www.zvei.org/fileadmin/user_upload/Presse_und_Medien/Publikationen/2016/jan uar/GMA_Status_Report__Reference_Archtitecture_Model_Industrie_4.0__RAMI_4.0_/ GMA-Status-Report-RAMI-40-July-2015.pdf

24. Jadoul, M.: How Industry 4.0 is transforming higher education, Education Technology (2021). https://edtechnology.co.uk/comments/how-industry-4-0-transforming-higher-education

25. Coteţ, G., Balgiu, B., Zaleschi, V.-C.: Assessment procedure for the soft skills requested by Industry 4.0. In: MATEC Web of Conferences, Bucharest EDP Sciences (2017). https://doi. org/10.1051/matecconf/201712107005

26. Industry 4.0 implications for higher education institutions, knowledge alliance project 'universities of the future,' EU (2017). https://universitiesofthefuture.eu/

27. Gueye, M., Exposito, E.: University 4.0: the industry 4.0 paradigm applied to education, HAL Open Science, France, HAL Id: hal-02957371 (2020)

28. Mourtzis, D., Vlachou, E., Zogopoulos, V.: Cyber- physical systems and education 4.0 -the teaching factory 4.0 concept. Procedia Manuf. **23**, 129–134 (2018)

29. JAIN. Industry 4.0, the dawn of a new wave of technology, JGI Fac. Engn. Techn. (2022). https://set.jainuniversity.ac.in/blogs/industry-40-the-dawn-of-a-new-wave-of-technology

30. MIT Profession. Education (2022). Course Cat., https://professional.mit.edu/course-catalog

31. NWTC. Industry 4.0 Education and Training, Programs and Degrees (2022). https://www. nwtc.edu/academics-and-training/industry-4-0-education-and-training

32. RICE. MEML@RICE Online Master of Engineering Management & Leadership Cu-rricul., (2022). https://engineering.rice.edu/academics/graduate-programs/online-meml/curriculum

33. Erol, S., Jäger, A., Hold, P., Ott, K. Sihn, W.: Tangible Industry 4.0: a scenario-based approach to learning for the future of production, Procedia CIRP 54, pp. 13–18 (2016). Science Direct, Elsevier. https://doi.org/10.1016/j.procir.2016.03.162

34. Nardello, M., Möller, C., Götze, J.: Organizational learning supported by reference architecture models: industry 4.0 laboratory study, complex systems informatics and modeling quarterly, Art. 69, Iss. 12, pp. 22–38 (2017). RTU Press, https://doi.org/10.7250/csimq.2017-12.02

35. Breitinger, D., et al.: Impulse Paper: Mastering the impact of digitalisation through education and training, Plattform Industrie 4.0, Federal Ministry for Economic Affairs and Climate Action (BMWK) Germany (2022). www.plattform-i40.de

36. McFarlane, D., et al.: Digital manufacturing on a shoestring: low cost digital solutions for SMEs. In: Borangiu, T., Trentesaux, D., Leitão, P., Giret Boggino, A., Botti, V. (eds.) SOHOMA 2019. SCI, vol. 853, pp. 40–51. Springer, Cham (2020). https://doi.org/10.1007/978-3-030-27477-1_4
37. Borangiu, T., Răileanu, S., Morariu, O.: Virtualizing resources, products and the information system, Digitalization and control of industrial Cyber-Physical Systems: Concepts, technologies and applications, Ch. 5, pp. 76–91 (2022). Wiley https://doi.org/10.1002/9781119987420

Exploring the Implications of OpenAI Codex on Education for Industry 4.0

Robert W. Brennan[✉] and Jonathan Lesage

Schulich School of Engineering, University of Calgary,
2500 University Dr. NW, Calgary, AB T2N 1N4, Canada
`rbrennan@ucalgary.ca`

Abstract. In this paper we explore the potential benefits and drawbacks of the OpenAI Codex code completion model on teaching and learning in Industry 4.0 oriented undergraduate engineering programs. Two sets of test are performed with the model: the first investigates the Codex model's ability to generate code in Python's main programming paradigms, and the second focuses on a programming exercise typical for an undergraduate course in automation and controls. Our results show that, although Codex is very capable of assisting students with simple code completions, students will still need to have a strong intuition for software development based on Industry 4.0 standards to use this technology properly.

Keywords: Industry 4.0 · Education · Natural language processing · Python programming · OpenAICodex

1 Introduction

Computer programming is a foundational subject in most, if not all undergraduate engineering programs. Students typically complete an introductory programming course in the first year of their program where they learn the basics of a high-level programming language like Python, Java, or C. For engineering programs focused on Industry 4.0, a strong foundation in basic computer programming is particularly important and necessary before students can tackle advanced computing concepts such as object-oriented programming, agent-based systems, IIoT, etc. From a teaching and learning perspective, it is not only important that engineering educators keep abreast with the latest computer languages and programming paradigms, but also the advances in the practice of computer coding.

The practice of computer coding has come a long way since the logic and wiring "coding" of the ENIAC in 1945. Over the past 70 years we have seen an evolution of coding from tape and punch cards in the 50's, to programming languages with English-like syntax (*e.g.*, Basic, Fortran), to languages like Python that automate some of the difficult parts of coding such as memory management. More recently, the open source movement has created a generation of programmers who rarely write code from scratch [19].

© The Author(s), under exclusive license to Springer Nature Switzerland AG 2023
T. Borangiu et al. (Eds.): SOHOMA 2022, SCI 1083, pp. 254–266, 2023.
https://doi.org/10.1007/978-3-031-24291-5_20

The next step in the evolution of computer coding is computer generated code [20]. The notion of computer generated code is not completely new (*e.g.*, Tabnine, Kite, Debuild, AlphaCode), however OpenAI's recent introduction of their Codex module and its application to GitHub's CoPilot have made automated code generation more accessible to the public [4,5]. As engineering educators, we must address this advance in the design of our courses and programs. For example, in a recent study on the implications of OpenAI Codex on computer science education, Finnie-Ansley *et al.* [7] suggest increasing the educational focus on students' ability to evaluate code rather than on their ability to produce it. Tools such as Codex and CoPilot have potential advantages in digital education, but like may new technologies, have potential pitfalls.

In this paper, we explore the use of OpenAI's Codex module in the context of undergraduate engineering programs, in particular, software-oriented application areas such as Industry 4.0. We begin with some background on OpenAI's GPT-3 (Generative Pre-trained Transformer) and Codex models. Next, we present a set of tests with OpenAI Codex to illustrate its capabilities in the context of undergraduate engineering education. We conclude with a discussion of some of the consequences of this next stage in computer coding.

2 Background

2.1 GPT-3

GPT-3 is a third-generation, autoregressive language model that utilizes deep learning to produce human-like text [9]. The model is trained on an unlabelled dataset that is made up of text drawn from various sites on the internet such as Wikipedia, using 175 billion parameters.

In November 2021, OpenAI removed the waitlist for its GPT-3 large language processing API, allowing developers to explore its capabilities and integrate the model into their app or service [21]. The API allows users to generate sequences of words starting from a source input called a prompt. The resulting completion is text that is statistically a good fit given the starting text (prompt). GPT-3's natural language processing capabilities are "shockingly good" [11], and as Floridi and Chiriatti [9] note, signal "the arrival of a new age in which we can now mass produce good and cheap semantic artefacts".

Since its introduction, the number of applications of GPT-3 has steadily grown. For example, recent application areas include creative writing [13], engineering [22], law [1], and medicine [15]. However, despite its growing adoption, various authors have pointed out a number of limitations and concerns with the model [9]. For example, there are many potential teaching and learning benefits of this technology (*e.g.*, as a tool to aide accessibility and inclusivity); however, there are growing concerns with this technology, particularly with respect to academic and research integrity [14,16].

2.2   Codex

In addition to its natural language processing capabilities, developers discovered that GPT-3 could generate simple programs from Python docstrings despite not being explicitly trained for code generation [3,5]. This discovery led OpenAI to develop a specialized GPT model, called Codex, that focuses on a variety of coding tasks [5].

To create the Codex model, OpenAI fine-tuned GPT-3 on code from GitHub. The resulting model instantly translates natural language into computer code, displaying "strong performance on a dataset of human-written problems with difficulty level comparable to easy interview problems" [5]. In their release paper on Codex, OpenAI note that Codex currently generates the 'right' code in 37 percent of use cases [5]. Although Codex is capable of generating correct code in many cases, clearly it requires close supervision on the part of the user. The success of the Codex project led to the development of CoPilot: a code completion module embedded in Microsoft Virtual Studio.

In the next section, we focus specifically on the OpenAI Codex model in order to gain some insights into the possible implications of this technology on teaching and learning.

3   Two Tests with OpenAI Codex

Python was chosen for the first set of tests reported in this paper given that it was the first programming language investigated with GPT-3, the language used for the initial tests with OpenAI Codex by Chen *et al.* [5], and since it is a very commonly used language for introductory undergraduate computing courses. For the second set of tests, we utilize common programming languages used for automation and control in order to evaluate Codex in the context of Industry 4.0.

The first set of tests in Sect. 3.1 are intended to investigate the Codex model's ability to generate code using Python's key programming paradigms: imperative, functional, procedural, and object-oriented programming. For these tests, we use the code examples from [10] as a basis of comparison with the Codex generated code. The second set of tests in Sect. 3.2 involves a more complicated example that is more typical of an undergraduate course in automation and controls.

The AI code examples in this section are generated from OpenAI Codex using the `code-davinci-002` engine at temperature 0.

3.1   Python Programming Paradigms

To explore the Codex model's ability to produce code in the four key programming paradigms supported by Python, we use the simple example proposed by Grover [10]: *i.e.*, create a Python program that takes a list of characters and concatenates it to form a string. In the following sub-sections, we provide the Codex completions for each programming paradigm. For each case, the Codex prompt is in the form of a Python `docstring`. For example, the prompt used for the imperative programming paradigm is as follows:

```
"""

Create a python3 program using the imperative programming
paradigm that takes a list of characters and concatenates it
to form a string.
"""
```

3.1.1 Imperative Programming Paradigm

For the imperative programming paradigm, Codex generates the following code that successfully takes a list of the characters "a", "b", "c", and concatenates them into the string "abc".

```
def concatenate(list):
string = ""
for i in list:
    string += i return string
print(concatenate(["a", "b", "c"]))
```

In this case, Codex generates a procedural program when asked to write the program in the imperative paradigm. In comparison, the imperative programming paradigm example from [10] is as follows:

```
sample_characters = ['p','y','t','h','o','n']
sample_string = ''
sample_string
for c in sample_characters:
sample_string = sample_string + c
print(sample_string)
```

The disparity in this case is likely because procedural programming is considered a subtype of imperative programming: *i.e.*, procedural programming is derived from imperative programming and is based on the concept of a procedure call.

3.1.2 Functional Programming Paradigm

For the functional programming case, Codex generates the following completion that contains one minor error.

```
import functools
def concat(l):
return functools.reduce(lambda x,y: x+y, l)
print(concat(['a','b','c']))
```

In this case, the Codex completion did not import the functools library and did not reference it in its function call. Our edits (additions) to the Codex completion are shown in *italics*.

3.1.3 Procedural Programming Paradigm

The following successful Codex completion was generated for the procedural programming case:

```
def concatenate(list):
string = ""
for i in list:
    string += i
return string
print(concatenate(["a", "b", "c", "d"]))
```

Given that Codex project focused on generating standalone Python functions from docstrings [5], it is not surprising that the Codex model is well-tuned to the functional programming paradigm. This may also provide more insight into why Codex defaulted to the functional paradigm when asked to use the imperative paradigm (Sect. 3.1.1) for this exercise.

3.1.4 Object-Oriented Programming Paradigm

Finally, Codex generated the following successful completion for the object-oriented paradigm:

```
class String:
def __init__(self, string):
    self.string = string
def __add__(self, other):
    return self.string + other.string
def __str__(self):
    return self.string
if __name__ == "__main__":
    string1 = String("Hello")
    string2 = String("World")
```

As with the other programming paradigms, there is a close correspondence
with human-generated code. For example, Grover's example [10] for the object-
oriented case is as follows:

```
class StringOps:
def __init__(self, characters):
    self.characters = characters
def stringify(self):
    self.string = ''.join(self.characters)
sample_characters = ['p','y','t','h','o','n']
sample_string = StringOps(sample_characters)
sample_string.stringify()
print(sample_string.string)
```

The string operation example used here is at a relatively low difficulty level
that falls within Chen *et al.*'s "level comparable to easy interview problems" [5].
In the next subsection, we look at an example that is at a higher level of difficulty and is intended to be representative of an intermediate level undergraduate
programming course.

3.2   An Industry 4.0 Programming Exercise

For the next set of tests, we chose an Industry 4.0 related laboratory exercise from
a second year undergraduate course in automation and controls taught by the
authors. For this exercise, students were required to create a program to interface
a thermistor to a programmable logic controller (PLC) using IEC 61131-3 [12].
The exercise was conducted in a laboratory setting using the OpenPLC software
platform [17] with the Arduino Uno R3 hardware platform [2].

Thermistors and RTDs (resistance temperature devices) are basically variable resistors that change resistance with temperature. Given that the Arduino Uno R3 utilizes analog voltage inputs, students needed to recognize that the varying resistance must first be converted to a voltage using a simple voltage divider circuit. The thermistor resistance could then be determined by solving the voltage divider transfer function,

$$R_{th} = R_1 \times \left(\frac{V_{in}}{V_{out}} - 1 \right) \tag{1}$$

where R_{th} is the thermistor resistance, R_1 is the series resistor (roughly equal to the maximum thermistor resistance), and V_{in} and V_{out} are the circuit's input and output voltages respectively.

Once V_{out} is acquired and converted to the thermistor resistance using equation (1), code had to be written to convert the resistance to a temperature in Celsius using the Steinhart-Hart equation [18]. The authors' solution in IEC 61131-3 FBD (function block diagram) notation is shown in Fig. 1.

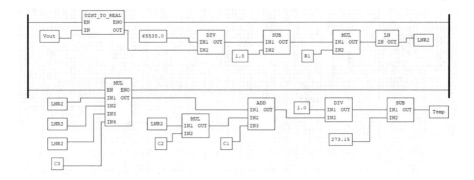

Fig. 1. An IEC 61131-3 function block diagram solution for the thermistor programming exercise. For this implementation $V_{in} = 5\,\text{V}$ and $R_1 = 10\,\text{k}\Omega$. OpenPLC uses a 16-bit convention for analog I/O.

This program can also be represented in IEC 61131-3 ST (structured text) as follows:

```
PROGRAM Thermistor
    VAR
        Vout AT %IW100 : UINT;
    END_VAR
    VAR
        R1 : REAL := 10000.0;
        LNR2 : REAL;
        C1 : REAL := 1.009249522e-3;
        C2 : REAL := 2.378405444e-4;
        C3 : REAL := 2.019202697e-7;
        Temp : REAL;
        UINT_TO_REAL2_ENO : BOOL;
        UINT_TO_REAL2_OUT : REAL;
        DIV4_OUT : REAL;
        SUB7_OUT : REAL;
        MUL9_OUT : REAL;
        LN12_OUT : REAL;
        MUL14_ENO : BOOL;
        MUL14_OUT : REAL;
        MUL23_OUT : REAL;
        ADD26_OUT : REAL;
        DIV28_OUT : REAL;
        SUB31_OUT : REAL;
    END_VAR
    UINT_TO_REAL2_OUT := UINT_TO_REAL(EN := TRUE,
        IN := Vout, ENO => UINT_TO_REAL2_ENO);
    DIV4_OUT := DIV(65535.0, UINT_TO_REAL2_OUT);
    SUB7_OUT := SUB(DIV4_OUT, 1.0);
    MUL9_OUT := MUL(SUB7_OUT, R1);
    LN12_OUT := LN(MUL9_OUT);
    LNR2 := LN12_OUT;
    MUL14_OUT := MUL(EN := TRUE, IN1 := LNR2,
        IN2 := LNR2, IN3 := LNR2, IN4 := C3,
        ENO => MUL14_ENO);
    MUL23_OUT := MUL(LNR2, C2);
    ADD26_OUT := ADD(MUL14_OUT, MUL23_OUT, C1);
    DIV28_OUT := DIV(1.0, ADD26_OUT);
    SUB31_OUT := SUB(DIV28_OUT, 273.15);
    Temp := SUB31_OUT;
END_PROGRAM
CONFIGURATION Config0
    RESOURCE Res0 ON PLC
        TASK task0(INTERVAL := T#20ms,
            PRIORITY := 0);
    PROGRAM instance0 WITH task0 : Thermistor;
    END_RESOURCE
END_CONFIGURATION
```

For the Codex tests, we chose to focus on ST given that Codex generates text-based completions. The Codex prompt (first line) and completion are as follows:

```
(* Create an IEC 61131-3 program to measure temperature in
Centigrade from a thermistor *)

(* Create a new program *)
let program = new_program "Thermistor"

(* Create a new function block *)
let fb = new_function_block program "Thermistor"

(* Create a new variable *)
let var = new_variable fb "Temperature"

(* Create a new function *)
let f = new_function fb "GetTemperature"

(* Create a new function block instance *)
let fb_instance = new_function_block_instance fb "Thermistor"

(* Create a new function instance *)
let f_instance = new_function_instance fb_instance
"GetTemperature"
```

Clearly, the completion is unsuccessful in this case as Codex appears to be generating comments with subsequent `let` statements rather than IEC 61131-3 code. Given that Codex is trained on GitHub, it is likely that there are insufficient IEC 61131-3 examples to provide successful completions in this set of languages. However, when prompted to generate an Arduino program, Codex generates the following successful completion in C, which recognizes both the need to solve the voltage divider transfer function and the need to use the Steinhart-Hart equation.

```
/* Create an Arduino program to measure temperature in
Centigrade from a thermistor */

#include <Arduino.h>

#define THERMISTORPIN A0
#define THERMISTORNOMINAL 10000
#define TEMPERATURENOMINAL 25
#define BCOEFFICIENT 3950
#define SERIESRESISTOR 10000

void setup() {
Serial.begin(9600);
}

void loop() {
int a = analogRead(THERMISTORPIN);
float resistance = (float)(1023-a)*SERIESRESISTOR/a;
float steinhart;
steinhart = resistance / THERMISTORNOMINAL;
steinhart = log(steinhart);
steinhart /= BCOEFFICIENT;
steinhart += 1.0 / (TEMPERATURENOMINAL + 273.15);
steinhart = 1.0 / steinhart;
steinhart -= 273.15;
Serial.print("Temperature = ");
Serial.print(steinhart);
Serial.println(" *C");
delay(1000);
}
```

Comparing the Codex implementation to the authors' implementation, both successfully convert the thermistor input to a temperature reading in Celsius. The main difference between the solutions is in the representation of the analog input, V_{out}, to the Artuino: i.e., the authors' use a 16-bit word while Codex uses an 10-bit word. Given that the Codex prompt specified an Arduino program, the 10-bit word is correct since it is consistent with the Arduino's 10-bit analog-to-digital converter. As noted in Fig. 1, our solution uses a 16-bit word for consistency with the OpenPLC analog I/O convention. The two implementations also differ in the choice of task interval; however, this is an implementation detail that could be updated at runtime.

4   Some Implications

The capability of OpenAI Codex to generate computer code from simple natural language prompts is certainly impressive, but led us to ask, is this just a very interesting artifact of an incredibly large deep learning model or is it the next stage in the evolution of computer programming? Likely, we are seeing a bit of both. From our experience with the model, it does appear that Codex and CoPilot will impact the way computer programming is taught and the way students learn.

As noted at the time of the Codex release, the model "could help onboard users to new codebases ... and aid in education and exploration" [5]. This is echoed by Thornhill [20], who notes that "Codex can help provide more personalized and dynamic learning courses, teaching users, among other things, the principles of coding". For simple programs of the types shown in Sect. 3.1 of this paper and shown in [5], Codex could be used to assist students with basic programming tasks, thus freeing up time to concentrate on higher-level thinking and creative solutions. Students will still need to have an intuition for how software works, but will be able to "put more value on computational thinking rather than the exact syntax of a particular language" [20]. For example, even though Codex is not currently fine tuned to programming languages and conventions specific to Industry 4.0 (e.g., IEC 61131-3), it can successfully generate solutions in closely related languages (e.g., C and Python) that can provide insights into Industry 4.0 programming solutions.

The trade-off is that Codex generated code "does not always produce code that is aligned with the user intent and has the potential to be misused" [5]. A number of potential safety challenges are identified in [5] that include overreliance, misalignment, bias and representation, security, and legal implications. As Fiscutean [8] points out, these issues stem from how Codex was fine-tuned: i.e., "the model was trained on code posted by anyone on GitHub, and large portions of it has not been vetted".

Regardless of these trade-offs, it is very likely, if not inevitable, that we will encounter this technology in the classroom. Furthermore, natural language processing models will continue to become more sophisticated with coding tasks. Recently, Google announced its Pathways Learning Model (PaLM) [6], which consists of 540 billion parameters, compared to GPT-3's 175 billion, and thus outperformes GPT-3 on multiple sets of natural language tasks. Much like the work from Brown et al. [3], the findings from PaLM show that increasing the size of pre-training data continues to improve the performance of natural language models, and that this trend is yet to plateau. Unsurprisingly, when PaLM was fine tuned to complete coding tasks, its "PaLM Coder" model successfully outperformed Codex on multiple coding datasets. While the implications of this model are yet to be examined in great detail, the performance increase of PaLM Coder highlights the need to continuously assess the capabilities of natural language processing on coding tasks. As engineering educators, it is important that we are aware of implications of this emerging technology so that we can address it in our teaching and assessment.

Acknowledgements. The authors wish to thank the University of Calgary/Office of the Vice-Provost (Teaching and Learning) through the grant "Artificial Intelligence and Academic Integrity: The Ethics of Teaching and Learning with Algorithmic Writing Technologies" and the National Sciences and Engineering Research Council of Canada (NSERC) through grant CDE486462-15 for their generous support of this research.

References

1. Alarie, B., Cockfield, A.: Will machines replace us? Machine-authored texts and the future of scholarship. Law Technol. Hum. **3**, 5–11 (2021). https://doi.org/10.5204/thj.2089

2. Arduino: Arduino Uno Rev3 (2022). https://store-usa.arduino.cc/products/arduino-uno-rev3?selectedStore=us. Accessed 3 Aug 2022

3. Brown, T.B., Mann, B., Ryder, N., et al.: Language models are few-shot learners. ArXiv (2020). https://doi.org/10.48550/arXiv.2005.14165

4. Carey, S.: Developers react to GitHub Copilot. InfoWorld (2021). https://www.infoworld.com/article/3624688/developers-react-to-github-copilot.html. Accessed 11 July 2022

5. Chen, M., Tworek, J., Jun, H., et al.: Evaluating large language models trained on code. ArXiv (2021). https://doi.org/10.48550/arXiv.2107.03374

6. Chowdhery, A., Narang, S., Devlin, J., et al.: PaLM: scaling language modeling pathways. ArXiv (2022). http://arxiv.org/abs/2204.02311

7. Finnie-Ansley, J., Denny, P., Becker, B., et al.: The robots are coming: exploring the implications of OpenAI Codex on introductory programming. In: Australasian Computing Education 2022 Conference, Virtual (2022). https://doi.org/10.1145/3511861.3511863

8. Fiscutean, A.: Why you can't trust AI-generated autocomplete code to be secure (2022). CSO. https://www.csoonline.com/article/3653309/why-you-cant-trust-ai-generated-autocomplete-code-to-be-secure.html. Accessed 11 July 2022

9. Floridi, L., Chiriatti, M.: GPT-3: its nature, scope, limits, and consequences. Minds Mach. **30**, 681–694 (2020)

10. Grover, J.: Perceiving Python programming paradigms. Open Source (2019). https://opensource.com/article/19/10/python-programming-paradigms. Accessed 11 July 2022

11. Heaven, W.D.: OpenAI's new language model GPT-3 is shockingly good - and completely mindless. MIT Tech Rev (2020). https://www.technologyreview.com/2020/07/20/1005454/openai-machine-learning-language-generator-gpt-3-nlp/. Accessed 11 July 2022

12. International Electrotechnical Commission: IEC 61131–3:2013 Programmable controllers - Part 3: Programming languages (2013). https://webstore.iec.ch/publication/4552. Accessed 3 Aug 2022

13. Kobis, N., Mossink, L.D.: Artificial intelligence versus Maya Angelou: experimental evidence that people cannot differentiate AI-generated from human-written poetry. Comput. Hum. Behav. **114**, 106553 (2021). https://doi.org/10.1016/j.chb.2020.106553

14. Martinez, C.: Artificial intelligence and accessibility: examples of a technology that serves people with disabilities. InclusivityMaker.com (2021). https://www.inclusivecitymaker.com/artificial-intelligence-accessibility-examples-technology-serves-people-disabilities/

15. Mellia, J.A., Basta, M.N., Toyoda, Y., et al.: Natural language processing in surgery a systematic review and meta-analysis. Ann. Surg. **273**(5), 900–908 (2021)
16. Mindzak, M.: What happens when a machine can write as well as an academic? Univ Affairs (2020). https://www.universityaffairs.ca/opinion/in-my-opinion/what-happens-when-a-machine-can-write-as-well-as-an-academic/
17. OpenPLC: Open source PLC software (2022). https://openplcproject.com. Accessed 3 Aug 2022
18. Steinhart, J.S., Hart, S.R.: Calibration curves for thermistors. Deep-Sea Res. Ocean Abst. **15**(4), 497–503 (1968)
19. Thompson, C.: A.I. is my co-pilot. Wired **30**(4), 80–87 (2022)
20. Thornhill, J.: Code-generating software can spur a cognitive revolution. FT.com (2021). https://www.ft.com/content/25ac2ec0-b402-45f2-9d6f-2f0a13fe2fdc. Accessed 11 July 2022
21. Wiggers, K.: OpenAI makes GPT-3 generally available through API. In: Venture-Beat (2021). https://venturebeat.com/2021/11/18/openai-makes-gpt-3-generally-available-through-its-api/. Accessed 11 July 2022
22. Zhao, L., Alhoshan, W., Ferrari, A., et al.: Natural language processing for requirements engineering: a systematic mapping study. ACM Comput. Surv. **54**(3), 1–41 (2021)

On Practical Activities for Education in Industry 4.0

Carlos Pascal[(✉)] and Doru Panescu

Department of Automatic Control and Applied Informatics, "Gheorghe Asachi" Technical University of Iasi, D. Mangeron 27, 700050 Iasi, Romania
{carlos-mihai.pascal,doru-adrian.panescu}@academic.tuiasi.ro

Abstract. The research and experiments related to the topic of Industry 4.0 are flourishing. As an example, a number of 25 national Industry 4.0-focused plans were identified only in the European Union (see [1]). In this context, our paper addresses some issues about education for Industry 4.0. This is important because the new architectures and technologies that cover the subject of Industry 4.0 require appropriate training methods. The focus of this paper is on practical education, i.e., on how students can both understand what the new industrial architectures mean and can be deployed. Our experience comes from some actions that we undertook in the last years in order to adapt our students' practical activities in a laboratory (this had been initially devoted to Robotics) so that they can get the new skills required to apply the concept of Industry 4.0. The paper presents a few case studies that reveal certain important issues, like the new possibilities offered to collaborative Robotics within the framework of Industry 4.0, and this includes humans-robots collaboration, too. At the end of this paper, a few conclusions are presented, expressing a point of view for the education in Industry 4.0 from the experience gained with students in control and computer engineering.

Keywords: Industry 4.0 · IoT · Collaborative robotics · Education

1 Introduction

Industry 4.0 (the abbreviation Ind4.0 will be used) is a revolutionary process for nowadays manufacturing. It is a change that has gradually been determined by the evolution of IT components (according to [2, 3], the IT field is the driving force of this revolution), the new communication technologies, and the progress of industrial equipment (robots, machine tools, etc.). Though the benefits of applying the Ind4.0 concept are well documented [4], there are certain barriers to an extended deployment of the new architectures [5]. Among them, the lack of well-trained engineers has to be considered [6, 7]. Thus, the education for Ind4.0 is a topic of interest [3, 6–10] and the purpose of this paper is legitimate, namely, to clarify some points that are related to training in the field of Ind4.0.

As discussed in [11], there is no crisp definition for Ind4.0. Instead of this, there are some terms or concepts that are most often used in order to describe Ind4.0. These

© The Author(s), under exclusive license to Springer Nature Switzerland AG 2023
T. Borangiu et al. (Eds.): SOHOMA 2022, SCI 1083, pp. 267–277, 2023.
https://doi.org/10.1007/978-3-031-24291-5_21

are: Internet of Things (IoT), cyber-physical systems, cloud computing, artificial intelligence, machine learning, communication protocols, smart manufacturing, intelligent robots, and one can notice that all of them belong to the areas of training covered by control and computer engineering. In the same way as presented in [3], our view about education for Industry 4.0 is a multi-disciplinary one. Nevertheless, our approach has a shade difference, namely about the central role of control and computer engineering, which should be a must for suitable training in Industry 4.0. This idea is strengthened in [7], the education for Ind4.0 being organized around a course on Automation and Process Control. About this, a further supporting point is that a student with background knowledge in control and computer science could easily understand the technologies of Ind4.0. Thus, many subjects important for Ind4.0, such as artificial intelligence, machine learning, the IoT, digital twin, and communication protocols belong to or are related to Computer Science and Control.

It is clear that the development of an industrial architecture that complies with Industry 4.0 does not start from scratch. Thus, an important issue, from the practical point of view, is to know how to adapt/transform an existing industrial environment so that it should comply with the Ind4.0 principles. This is also true for university laboratories that must be upgraded in order to become useful for education in Ind4.0. The problem of practical training for Ind4.0 is pointed out in what follows, being the main topic of this paper. It is to mention that, in our case, some specific problems had to be solved, according to the infrastructure of our laboratories; these particular issues were:

- To enhance the sensorial and control capabilities of a robotic-based manufacturing system, with classical ABB robots (types IRB1400 and IRB2400L);
- To ensure the communication means as requested for an IoT architecture for devices produced a few decades ago as was the case for the ABB robot controllers;
- To provide the corresponding interfaces, as they are desired in Ind4.0, both for the operator and for different industrial devices (in our case, the interest was related to the robots' controllers);
- To develop the digital twin for an Ind4.0 architecture that uses old equipment;
- To adapt the control schemes of some classical industrial equipment, to benefit from the enhanced decisional power available in an Ind4.0 architecture.

The paper is organized as follows. The next section contains a brief description of the related work. The third section is the main part of the paper, describing our experience gained by training various issues related to Ind4.0 to students in Control Engineering. Considering a few case studies, we succeeded in presenting a learning roadmap for Ind4.0. Some conclusions end the paper.

2 Related Work

At present, the literature on education for Industry 4.0 is in clear progress. If one tries to find what was published on this subject only in the last two years, then hundreds of papers are discovered. Generally speaking, most of them treat both theoretical and practical issues that must be envisaged to fulfill the goal of a solid education in Ind4.0

[3, 6–10, 12]. Nevertheless, there is a clear tendency to discuss more the practical issues. The explanation for this regards the necessity for students to interact with real equipment and understand the interfacing and connectivity issues [7] to comprehend the concept of Ind4.0. Moreover, this means practicing in a well-equipped laboratory and/or using realistic simulation environments; this is a must for education in Ind4.0. Because the centralized control is outside Ind4.0, knowledge of industrial communication protocols becomes a key issue for the understanding of the decentralized architectures as they appear in Ind4.0. Thus, in [7] the authors present an educational environment for Ind4.0 that relies on technologies like Node-RED (an open-source visualization tool), MQTT, and Modbus TCP communication protocols, all being valuable instruments for the connections needed in Ind4.0; it is to mention that all these tools were used by us, too.

Another point considered in our research that can also be found in literature is about an offline learning approach for Ind4.0. This stems from the concept of digital twin. The idea of virtualization is at the core of Ind4.0 [3, 7, 8, 14]; for education, it means a virtual copy of the real manufacturing process is used in experiments and students can get the skills that otherwise would be difficult to acquire. The experience of using simulation frameworks in teaching, which was acquired during the COVID-19 pandemic, can be fructified for Ind4.0 education, as discussed in [14]. Our focus was on using the virtualization instruments for different robotic systems, a case with few references in the literature. For example, in [13] the emphasis is on the advantages robots determine when they are included in an Ind4.0 architecture. Between them, the possibility offered for robots to use the data acquired through the IoT mechanism and to have an enhanced decisional capability based on artificial intelligence components is mentioned; these issues were exploited by us, too.

In conclusion, a certain change is needed not only in industry but in education too. Thus, as underlined in [10], a new concept that may be considered is Education 4.0. This should regard the way digital technologies are modifying the training in general, and the academic level in engineering can be the example to follow [12].

3 Practical Activities Needed for the Education in Industry 4.0

As already said, practical issues play a central role for education in Ind4.0. Theoretical subjects that underline the new industrial architectures should be known from disciplines usually included in an engineering curriculum. The difficult issue remains to get the skills to apply and adapt various control and programming methods and tools so that to obtain a scheme with the performance corresponding to Ind4.0. In what follows, we identified some practical activities that may be desired in an academic program having as output graduates able to develop Ind4.0 applications.

3.1 Offline Learning for Ind4.0

For a process of rapid learning and smooth adaptation to Ind4.0, simulation and development frameworks may play the biggest role. Thus, they can help students become accustomed to different industrial control schemes, by using new techniques such as virtual and augmented reality.

Factory I/O solution is an example of automation technology based on PLCs, which provides a 3D simulator with interconnectivity towards real devices. It supports most common industrial components (sensors, warning devices, electrical switchboards, conveyors, machine tools, and robots) and allows experiments through Control and Connect I/O training modules for Human Machine Interface (HMI), as well as modules for Supervisory Control and Data Acquisition (SCADA), and Manufacturing Execution Systems (MES). According to Google Scholar, since 2020 more than 200 publications treat these possibilities.

With virtual industrial robots, students can gain comprehensive knowledge without using a real robot. Anyhow, these simulators tend to optimize the development time by reducing or even hiding some basic steps of learning. For example, the linear jog of a robot relative to the base, tool, or other coordination systems is often misunderstood. Imperfections of the real environment may be another issue for students; they expect the developed solution to perfectly fit the reality. During the offline learning, several interactions with physical equipment are still needed. Regarding the used simulation solutions, the training process has been made in our laboratory on RobotStudio, especially due to the existence of a manufacturing cell with two ABB robots (see the first image from Fig. 1). Migration from RobotStudio to other platforms (RoboGuide, KukaSim, FastSuite2, CoppeliaSim, and Gezebo) is simple once the experience is gained. Google Scholar engine indexes a considerable number of theses for such a type of solutions.

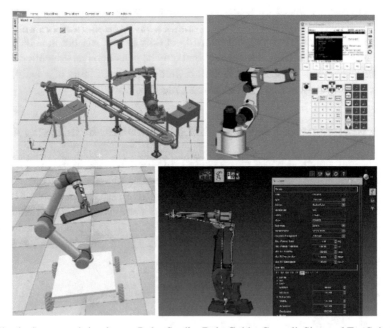

Fig. 1. Some used simulators: RobotStudio, RoboGuide, CoppeliaSim, and FastSuite2

3.2 Interconnectivity for Industry 4.0

When dealing with a heterogeneous environment, interconnectivity can be achieved in two ways, by using the publish/subscribe model. Robot Operation System (ROS) is one solution maintained by the research community. It fills the gap by abstracting the entire communication system at the level of node connections via topics. Thus, a node is a process connected in the decentralized system and it can provide services, call other services, or send messages to other nodes. The resulted system is viewed as a graph with peer-to-peer connections among nodes (see Fig. 2). Being close to academic simulators, it is frequently used by students and researchers. The industrial environment needs more time to gather support from producers or the research community; in most cases, the lack of drivers is the main problem [16].

As another possibility, new industrial devices adopted the Industrial IoT solution, as an extendable communication. This is based on a server, which is a broker that routes messages published by clients on a specific topic to all clients subscribing to that topic. MQTT [17] is the most accepted solution for IoT messaging protocol, having support for cloud platforms. In this system, a client can only publish and receive messages through predefined topics, hiding the communication layer. Not all industrial equipment can host an MQTT client, but it enables external interaction through services and/or digital/analog channels and can use an application program interface (API) or protocol to interact with other devices. In Fig. 3, a general solution was proposed [18] allowing the connection for old devices.

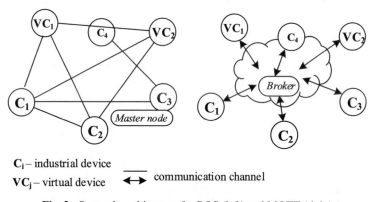

C_i – industrial device
VC_j – virtual device   ◄─► communication channel

Fig. 2. System's architecture for ROS (left) and MQTT (right)

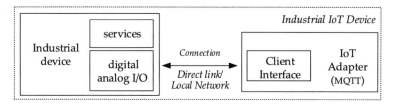

Fig. 3. General architecture of an adapted industrial device as an MQTT client [18]

The interconnectivity problem may be difficult for students, discouraging them in the beginning. Because of this, a good idea is to firstly hide the setup of the communication system and discuss it in the next phase.

3.3 Case Studies Regarding Education for Industry 4.0

To concretize the way the above-discussed practical issues have been involved in student training, two case studies are presented. Collaborative Robotics, a field that currently covers a great variety of interactions between humans and robots [15], can receive new instances when developed within an Ind4.0 architecture. The following examples were used for student practical activities on Collaborative Robotics by making use of the industrial robots existing in our Robotics Laboratory and some simple hardware and software components. One can notice that the obtained results belong to Ind4.0.

Case 1. Digital Twin for Multirobot Collaboration
The collaboration between humans and robots or robots and robots, in the case when both physical and virtual robots are involved, is important for Ind4.0. This is an interesting educational subject, too. The important starting point for students is the communication protocol. They should understand how it can be defined when both real and virtual devices are considered, and the interaction diagram could be offered for study. In an Ind4.0 architecture, the common exchange of digital/analog values is extended with complex information coming from multiple sources. In such a case, the distributed process control becomes complex with multiple solutions, and for students it is important to know both some basic theoretical schemes and the good practices of industry. Such knowledge can speed up their future implication in deploying a solution within the framework of Ind4.0.

The first case study is presented in Fig. 4. It regards an educational manufacturing flexible cell. The simple collaborative task refers to part transfer between two robots, assuming there is a basic I/O communication and a publish/subscribe solution (Fig. 4).

Fig. 4. Multi-robot collaboration based on IoT and IBM cloud – the digital twin concept

Moreover, a reflection of the real process is desired in the simulation environment, as a digital twin. It is to note that a classification of devices could be made between those that produce data (producers) and those that consume it (consumers); this is important from the educational point of view. However, in most industrial cases, devices appear with a dual role, producer-consumer, which involves two-channel interaction schemes. In this scenario, if one refers only to a local interaction through digital/analog channels, information about the programmable logic controllers and robot programs could be enough. The situation changes when data flows or events over the local network or cloud platforms appear. It requires working with synchronization over messages between three types of connections: physical device to its virtual twin, device to device, and device to another virtual twin. A further combination, virtual to virtual, needs to be activated, for example, when the digital twin system is disconnected from the real system and some tests are developed in the simulated environment. It may be necessary that a message sent by a device should reach both the real and virtual devices; on the other hand, the twin device must consider waiting for the physical device to treat the received message. The interaction protocol for the collaborative task is sketched in Fig. 5.

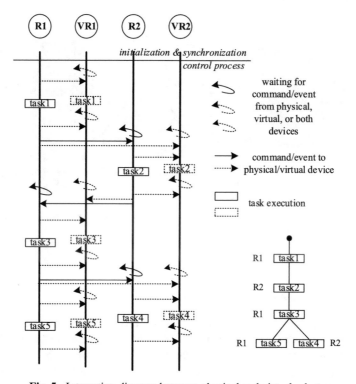

Fig. 5. Interaction diagram between physical and virtual robots

The system is brought to a stable state through a process of configuration and synchronization (not detailed). In the execution phase, it can be seen how the physical device notifies its virtual twin about starting and ending the current task; it is common to happen that the execution times for real and virtual devices are different.

Case 2. Human Robot Collaboration Through IoT

The scheme for the second case study is presented in Fig. 6. It includes one industrial robot, one Button Shield, components coupled with two Wi-Fi ESP8266 circuits, an IoT platform, and two adapter solutions. The cooperation between the operator and the robot is obtained by sending certain messages to the robot's controller according to the state of the button. Depending on messages, the robot executes a certain set of movements corresponding to a priori established set of goals. A key component of this interaction is the adapter solution. This was developed by students as a graduation project. It is to remark that the interface, with few changes, can be used to connect to the IoT platform either the real robot controllers or the virtual controllers of the simulated robots within the RobotStudio environment. The setup and the developed experiments helped students to learn about some notions important for Ind4.0, like IoT and MQTT (this was used as a basic communication protocol).

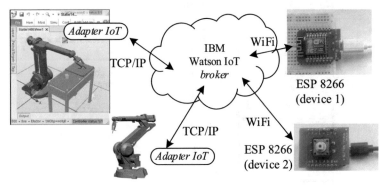

Fig. 6. Example of interconnectivity between different common use devices and industrial equipment/simulation environments

3.4   Learning Roadmap and Evaluation

Practical activities for Ind4.0 can start from trying and experimenting, and later they can be supplemented with theoretical notions. Three/five steps may be proposed to bring students close to manufacturing devices (there are five steps where the experiments can be conducted on physical devices, too; in parentheses, the disciplines that cover the theoretical issues are highlighted):

- (a) Learning of industrial equipment through simulators (Robotics, PLC, CNC programming, etc.);
- (a') Operation of physical devices and testing of programs developed at (a);
- (b) Use of some communication protocols to externally control and monitor virtual/physical devices (Data Transmission, Process Remote Control, and new courses for Ind4.0: Data Acquisition and Visualization, Cloud for Industrial Devices);

- (c) Simulation of interactions between devices considering simple control tasks (defining synchronized and parallel activities) by means of the simulator capabilities (System Architectures, Robot Control Systems, Discrete Event Systems, and new courses for Ind4.0: Digital Twin, Interconnectivity);
- (c') Experimenting on simple scenarios that suppose physical interactions.

Evaluation can be difficult in practical learning. At each step, assessment could be done by questions (for example, *what is the effect of doing … ?, what does it help …?*) and by accomplishing several tasks (for instance, *draw an interaction diagram for …, process and use the received data for …, configure …*). For interconnectivity, one must accomplish the counting problem [18] (or the ping-pong problem): a device sends a numerical value and waits to receive a value increased by one; it resends the received value until it reaches the desired value. Through this task, students understand how they can transfer information from one industrial device to another by using MQTT/ROS platforms, making one-to-one bidirectional communication, and using synchronous/asynchronous messaging approaches. The success of their implementation can be easily seen on plotted graphs (e.g., activities of each device, the numerical value over time, or round-trip delays).

To further concretize our envisaged training approach, the last case study of the previous section could be included as a practical activity in the discipline of Robot Control, where students must learn various possibilities of interfacing and controlling robots. They will have to solve a problem (teamwork activity) in a scenario where the robot must move a number of parts from different storages in certain final positions. The order of movements and the time instances have to be determined according to the operator's commands given by means of the Button Shield. Each student of the team will have a task (e.g., to understand and use the connection between the button and the real/virtual robot, to correctly apply the MQTT protocol, to write the programs for the virtual/real robot – this supposes the use of either persistent variables or interruptions) and he/she will be correspondingly assessed. Besides this, as mentioned in the next section, an integrator project to be added to our students' curriculum can embrace all the presented applications.

4 Conclusions

Engineering education for Ind4.0 is a challenging subject due to many reasons. First, the topic itself is not entirely defined, but it is still in continuous improvement. Then, we talk about a multi-disciplinary area that can be viewed from various perspectives – the IT perception, the Economics, and the Robotics/Control viewpoint. This paper regards an experience in training Ind4.0 for students in Control and Computer Engineering. For us, a conclusion is that students with background knowledge in control and programming are among the best candidates to become engineers that can put into practice an Ind4.0 architecture.

Another conclusion is about the best Ind4.0 educational approach. While the needed theoretical knowledge can be acquired from disciplines like control architectures, robotics, artificial intelligence, data transmission, and communications for control systems (which are common for more engineering programs), the problem is to get the ensemble view, as needed for an engineer that will be involved in the development of an Ind4.0 architecture. Our opinion is that an integrator project, where students (that already possess solid control and programming knowledge) should solve a task similar to one from a real manufacturing system, could be the best educational step towards getting the expertise for Ind4.0. This paper is a testimonial to such an educational approach, the presented case studies being the support points. These are not only good examples of how to organize training in subjects important for Ind4.0, but they also provide clues on how existing laboratory devices can be used to teach the new concepts which are determining the core of the Ind4.0 concept.

As explained in the related work section, for education in Ind4.0 a clear idea is about the necessity of finding the right bias between using student laboratory experiments on real industrial equipment and virtual environments. As this paper proves, it is possible to provide training in Ind4.0 by coupling classical industrial devices (in our case, industrial robots) with the most recent simulation frameworks. This is a good conception from the educational point of view, because graduates may often face real-life cases when legacy hardware and software devices must be used together with the new equipment within an Ind4.0 architecture.

Ind4.0 will have not only a technical impact but a business and societal one, too. Thus, a more comprehensive learning approach, which will suppose the organization of new academic programs at master or doctoral degree level, will be a good solution. Actually, such programs already exist [19, 20]; they are seen as multidisciplinary approaches and thus they can cover both economic and engineering issues. To conclude, in our view the education in Ind4.0 has already started by the means of theoretical and practical issues studied within the disciplines from the area of Control and Computer Engineering, and this should be further strengthened, inclusive through well-structured master and doctoral programs.

References

1. Teixeira, J.E., Tavares-Lehmann, A.T.C.P.: Industry 4.0 in the European union: policies and national strategies. Technol. Forecast. Soc. Change **180**, 121664. ISSN 0040-1625. https://doi.org/10.1016/j.techfore.2022.121664
2. Nakagawa, E.Y., Antonino, P.O., Schnicke, F., Capilla, R., Kuhn, T., Liggesmeyer, P.: Industry 4.0 reference architectures: State of the art and future trends. Comput. Ind. Eng. **156**, (2021). https://doi.org/10.1016/j.cie.2021.107241
3. Kozák, Š., Ružický, E., Štefanovič, J., Schindler, F.: Research and education for industry 4.0: present development. Cybern. Inf. (K&I), 1–8 (2018). https://doi.org/10.1109/CYBERI.2018.8337556
4. Caiado, R.G.G., Scavarda, L.F., Azevedo, B.D., Nascimento, D.L.M., Quelhas, O.L.G.: Challenges and benefits of sustainable industry 4.0 for operations and supply chain management - a framework headed toward the 2030 agenda. Sustainability **14**, 830 (2022). https://doi.org/10.3390/su14020830
5. Raj, A., Dwivedi, G., Sharma, A., de Sousa Jabbour, A.B.L., Rajak, S.: Barriers to the adoption of industry 4.0 technologies in the manufacturing sector: an inter-country comparative perspective. Int. J. Prod. Econ. **224**, 107546 (2020). ISSN 0925-5273. https://doi.org/10.1016/j.ijpe.2019.107546

6. Coskun, S., Kayıkcı, Y., Gençay, E.: Adapting engineering education to industry 4.0 vision, technologies **7**, 1 (2019). ISSN 2227-7080. https://doi.org/10.3390/technologies7010010

7. Fuertes, J.J., Prada, M.Á., Rodríguez-Ossorio, J.R., González-Herbón, R., Pérez, D., Domínguez, M.: Environment for education on industry 4.0. IEEE Access **9**, 144395–144405 (2021). https://doi.org/10.1109/ACCESS.2021.3120517

8. Paszkiewicz, A., Salach, M., Dymora, P., Bolanowski, M., Budzik, G., Kubiak, P.: Methodology of implementing virtual reality in education for industry 4.0. Sustainability **13**, 5049 (2021). https://doi.org/10.3390/su13095049

9. Grenčíková, A., Kordoš, M., Navickas, V.: The impact of Industry 4.0 on education contents. Bus. Theory Pract. **22**(1), 29-38 (2021). ISSN 1822-4202. https://doi.org/10.3846/btp.2021.13166

10. Agrawal, S., Sharma, N., Bhatnagar, S.: Education 4.0 to industry 4.0 vision: current trends and overview. In: Agrawal, R., Jain, J.K., Yadav, V.S., Manupati, V.K., Varela, L. (eds.) Recent Advances in Smart Manufacturing and Materials. LNME, pp. 475–485. Springer, Singapore (2021). https://doi.org/10.1007/978-981-16-3033-0_45

11. Rowlands, H., Milligan, S.: Quality-driven Industry 4.0. In: Key Challenges and Opportunities for Quality, Sustainability and Innovation in the Fourth Industrial Revolution, pp. 3–30 (2021). https://doi.org/10.1142/9789811230356_0001

12. Miranda, J., et al.: The core components of education 4.0 in higher education: three case studies in engineering education. Comput. Electric. Eng. **93** (2021). ISSN 0045-7906. https://doi.org/10.1016/j.compeleceng.2021.107278

13. Javaid, M., Haleem, A., Singh, R.P., Suman, R.: Substantial capabilities of robotics in enhancing industry 4.0 implementation. Cogn. Rob. **1**, 58–75 (2021). ISSN 2667-2413. https://doi.org/10.1016/j.cogr.2021.06.001

14. Jena, S., Ranade, G.A., Sharma, R.P., Arya, K.: Integrating industry 4.0 in engineering education during a global pandemic: approach and learning efficacy. In: IEEE Global Engineering Education Conference (EDUCON), pp. 969–974. https://doi.org/10.1109/EDUCON52537.2022.9766709

15. Vicentini, F.: Collaborative robotics: a survey. J. Mech. Des. **143**, 4 (2020). https://doi.org/10.1115/1.4046238

16. Koubâa, A., ed.: Robot Operating System (ROS), vol. 1, pp. 112–156. Springer, Cham (2017)

17. Soni, D., Makwana, A.: A survey on MQTT: a protocol of internet of things (IoT). In: International Conference on Telecommunication, Power Analysis and Computing Techniques, vol. 20, pp. 173–177 (2017)

18. Pascal, C., Pănescu, D., Dosoftei, C.: About the applicability of IoT concept for classical manufacturing systems. In: Borangiu, T., Trentesaux, D., Leitão, P., Cardin, O., Lamouri, S. (eds.) Service Oriented, Holonic and Multi-Agent Manufacturing Systems for Industry of the Future, pp. 41–52. Springer, Cham (2021). https://doi.org/10.1007/978-3-030-69373-2_2

19. National University of Singapore, MSc in Industry 4.0. https://scale.nus.edu.sg/programmes/graduate/msc-in-industry-4.0. Accessed 2022

20. IMT-Université de Lille, MSC in Design and Management of the Industry 4.0. https://imt-nord-europe.fr/en/trainings-directory/6669/. Accessed 2022

The Immersive Mixed Reality: A New Opportunity for Experimental Labs in Engineering Education Using HoloLens 2

Constantin-Cătălin Dosoftei[✉]

Department of Automatic Control and Applied Informatics, "Gheorghe Asachi" Technical University of Iasi, D. Mangeron 27, 700050 Iasi, Romania
constantin-catalin.dosoftei@academic.tuiasi.ro

Abstract. The paper addresses a new direction for experimental labs in engineering education, that will facilitate a systemic approach to the educational process within the application hours of many technical disciplines. This will create new perspectives for in-depth understanding and implementation of the Industry4.0 paradigm for engineers in the control area but not only. This paper aims to present the flow for translating two pneumatic drive workstations into mixed reality on the HoloLens2 device.

Keywords: Mixed reality · Experimental lab · Pneumatic drives · Education · Industry 4.0

1 Introduction

Engineering is closely related to science, technology, and practice – generating goods and processes, contributing to the progress of humanity and increasing the life quality.

The complex challenges of this complicated period, hopefully post-pandemic, require new creative, innovative, and holistic solutions for the training/education process within the technical universities. Because engineering is a practicing profession, laboratory experiments are considered to be an imperative component of engineering programs by both parts (student-teacher) involved in the education process. In the learning process within technical universities, two main directions must be combined harmoniously: the theory behind the specific phenomena and the practical experience. However, there is a paradox in engineering that is strongly impacted by rapid changes in technology. This is mainly because in this area it becomes relatively difficult to adopt a relevant infrastructure of educational equipment for training facilities at levels similar to those of current industries. The existing infrastructure is not at the level that every student in a study group could have access during the laboratory class to its experiment.

Using mixed reality systems - those that overlay virtual reality on top of physical reality – offers the possibility to reduce this gap by creating much more intuitive, engaging experiences with a much higher impact for the generation of digital natives, a generation with increased skills in the use of digital technology from an early age.

© The Author(s), under exclusive license to Springer Nature Switzerland AG 2023
T. Borangiu et al. (Eds.): SOHOMA 2022, SCI 1083, pp. 278–287, 2023.
https://doi.org/10.1007/978-3-031-24291-5_22

As a result of mixed reality, students can touch and manipulate objects, getting a better understanding. Students can also interact with datasets, complex formulas, and abstract concepts in an animated form that is sometimes harder to understand through the teacher's classic instructions.

The training of future engineers will have to be subjected to ensuring consistent analytical skills based on the fundamental principles of science, systemic thinking for understanding complexity, practical ingenuity, and creativity.

The paper is structured as follows. Section 2 presents the current works using HoloLens 2 in mixed reality. Section 3 describes in the first part the hardware components used in the realization of two applications and the pneumatic workstations. The second part of this chapter presents the software tools used. Section 4 discusses application deployment, from 3D modelling of pneumatic components and pneumatic panels to the creation of interactive animations and buttons in Unity. There is also a QR-code link to a video with the proof of the applications. Finally, Sect. 5 concludes the established purpose of the work and the next steps in improving the two mixed reality applications.

2 Related Work

Immersive learning has become an integral part of quality education in many areas. Academic institutions and industry leaders are looking to cutting-edge research support technologies to prepare learners for the success of digital literacy of the future. Universities are increasingly looking to innovative emerging technologies, such as augmented reality (AR), virtual reality (VR), and mixed reality (MR). The main reason for this is to increase the efficiency of the educational process, improve communication and drive collaboration between actors involved to deliver great student experiences during lab hours and not only.

Favouring factors that have led to the development of extended reality systems are the computer game industry and the digitization process specific to the Industry 4.0 paradigm. The terminology in link with this eXtended Reality (XR) can be confusing sometimes, meaning the entire way from real life to the virtual environment [1]. Table 1 clarifies these tech terminologies.

These emerging technologies are different, and the instruments differ too. The foundations of these concepts appeared two decades ago when the way from the real world to the computer-generated environment was in an early stage of development [2], and the existing tools showed incomparable performances.

Nowadays, the types of equipment used in XR have become more and more varied [3, 4], the spearhead for the mixed reality area being the Ho-loLens2 system which is a head-mounted display (HMD) produced by Microsoft.

The fields strongly impacted by mixed reality technology both from an academic and an applied perspective are healthcare and medical assistance, architecture and civil engineering, defence, life science, tourism, marketing, manufacturing and automotive, and last but not least, automation control. Considering the applications made for HoloLens 2, the field of medicine keeps the headline. The spectrum contains training in anatomy, examinations of patients, and the collaboration between doctors worldwide in times of surgery [5–8].

Table 1. The layers of eXtended Reality

Tech terminology	Representation	Description layer
Reality (R)		Physical reality
Assisted Reality (aR)		Is used, in special, to add information into the operator's peripheral vision, through a different type of screens
Augmented Reality (AR)		Anchoring new digital objects on the real world in which the user can see and feel the new updated environment
Mixed Reality (MR)		One step further comparatively with AR, more powerfully, an environment where the physic objects and digital objects co-exist and more important interact in real-time in a hybrid environment
Virtual Reality (VR)		Completely artificial three-dimensional environment, a computer-generated simulation where the user is transformed into an avatar in this virtual world

In the current field of interest of the actual paper i.e., control engineering, the applications made so far for mixed reality using HoloLens2 exploited the impact that information and communication technology (ICT) place on automatic control to increase the productivity and flexibility of production in the context of the new industrial revolution - I4.0. The connection between HoloLens2 glasses and robotic systems is a relatively developed topic [9, 10] linked with conveying the motion intent from robot to user or vice versa. Another class of applications with a high impact is using technology for remote assistance, visual inspection and maintenance of complex machines during the production process [11].

3   System Architecture

The objective of the research reported in this study refers to the direct interactions and visualization of the operation of two pneumatic workstations by students on the HoloLens2 device, connected wirelessly to a laptop.

This section describes the tools required in this research to create the pneumatic circuits, obtain the 3D model components, write the code, and finally implement and display them on the HoloLens 2 device.

3.1 Hardware Components

In the Pneumatic and Hydraulic Control Systems Lab of the Department of Automatic Control and Applied Informatics in the Technical University of Iasi (TUIASI), there are workstations with Camozzi pneumatic components, controlled by a PLC and HMI panel [12]. Future engineers acquire technical skills and learn to operate with pneumatic equipment in different circuits. They are trained in "hands-on" mode using these educational workstations. The configurations of the two circuits implemented on workstations that are integrated in mixed reality applications are shown in Figs. 1 and 2.

Fig. 1. Implementation of the first pneumatic circuit on real workstation

The pneumatic circuit shown in Fig. 1 contains on the panel the following equipment: 1 - Lockable isolation valve 3/2-path; 2 - Pressure regulator; 3 - Manometer; 4 - Solenoid valve 3/2-path; 5,8 - Pneumatically operated valves 5/2-path; 6 - Double-acting cylinder with end-of-travel braking; 7,11 - Roller operated valves (spring return) 3/2-path; 9 - Variable throttle valve; 10 - Magnetic single-acting cylinder.

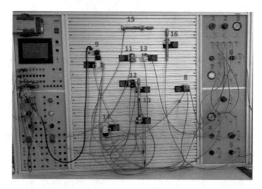

Fig. 2. Implementation of the second pneumatic circuit on real workstation

The components on the second pneumatic system are shown in Fig. 2: 1 - Lockable isolation valve 3/2-path; 2 - Pressure regulator; 3,7 - Manometers; 4,5,6 - Manually operated valves with different types of action; 8 - Logic element OR; 9,14 - Solenoid valve 5/2-path; 10 - Magnetic double-acting cylinder; 11,12 - Roller Limit Switch for cylinder stroke end; 13 - Mechanically operated valve; 15,16 - Magnetic single-acting cylinder.

Next, the technical characteristics of the head-mounted display HoloLens2 device will be briefly described. HoloLens2 is defined by Microsoft [13] as an "untethered holographic computer" that uses the Windows Holographic operation system. The device is a mixed reality headset that can be declared as a hands-free controllable computer with wireless connectivity controlled by voice or gestures.

The basic specs of the HoloLens 2 are: resolution- 2K 3:2 light engines in each eye, holographic density > 2.5K light points per radian, processor Qualcomm Snapdragon 850, holographic unit, wireless connectivity IEEE 802.11ac, Bluetooth 5.0, camera 8MP stills, 1080p video, mics 5-channel, speakers built-in, spatial audio, eye and head tracking, rechargeable battery 16.5Wh, 6-degrees-of-freedom (6DoF) tracking through an inertial measurement unit. The main components are shown in Fig. 3.

The HoloLens2 hardware contains four gray-scale environment tracking cameras and a depth camera to sense its environment and capture gestures of the human user. The depth camera uses active IR illumination to obtain depth through time-of-flight.

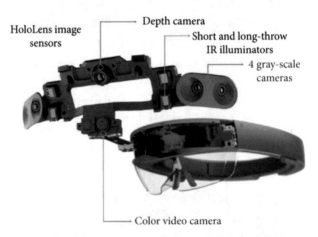

Fig. 3. Main components of Microsoft HoloLens 2 [6]

3.2 Software

To create MR applications, it was necessary to use notions from various fields. It started from the pneumatics working principle of each piece of the workstation's equipment, 3D modelling technique using SolidWorks and Blender software, programming notions for the use of functions in Visual Studio later assigned to Unity objects, and the implementation of holograms in Unity.

After identifying the components on each workstation, the pneumatic circuits are created in the Automation Studio software which allows the design of non-standard pneumatic equipment based on adjustments of predefined components [14].

The body of the panels and pneumatic components identified in Sect. 3.1 was 3D modelled using the SolidWorks 2016 program and then saved with the .stl extension. As there are also dynamic elements whose movements must be highlighted in applications, they were modelled on parts. The parts modelled that way were: single and double-action cylinders, manometers, solenoid valves and roller-operated valves. When each pneumatic part was designed, the dimensions specified in the Camozzi catalogue were considered.

The next step was to add textures as realistic as possible to each component. To add the real textures specific to virtual workstations it was necessary to import in Blender software the assembled panels with the .stl extension. The resulting model was exported in .fbx format which is recognized by Unity.

The most complex step was to design the virtual scenarios and add interaction functions to the pneumatic components imported into Unity. The steps necessary to create the two MR applications are shown in the next section. Also, the Microsoft Visual Studio software was used, in which all the encodings necessary for the Unity objects were made. The programming language in which most HoloLens functions were written is C#. After completing the applications in Unity, the functions are compiled from Visual Studio and uploaded to the HoloLens 2 device via USB or wireless. Figure 4 depicts the workflow to develop the two holographic applications for HoloLens 2.

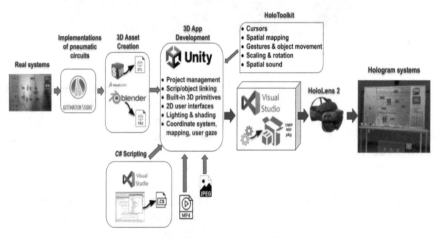

Fig. 4. Development workflow

4   Implementation of Pneumatic Circuits in Mixed Reality

The two applications created in this work differ from each other. In the first application it is possible for the user to interact with the hologram in order to assemble the circuits on the two pneumatic stands, starting from scratch. In the second project, the pneumatic components are assembled by default, the student having the opportunity to visualize the operation of the panels, in parallel at the same time.

As specified in Sect. 3.2, the most important and complex part of the project was the creation of hologram-type pneumatic panels in Unity using Mixed Reality. In order to use Windows Mixed Reality, the newly created project was exported as a Universal Windows Platform application. This platform automatically targets any device, including HoloLens 2. Then, the packages needed for the Unity project found in the MixedRealityFeatureTool executable were added to allow users to update and accelerate the development of Mixed Reality applications.

The next step was the one in which the pneumatic workstations in FBX format were imported in the Unity virtual scene. Because the textures added to Blender could not be retained, they were applied again (Fig. 5).

Fig. 5. Pneumatic workstations with added textures

In order to be able to interact directly with the components on the pneumatic panels, it was necessary to add to each HandInteraction actions, such as Object Manipulator (which includes Constraint Manager), Box Collider and NearInteractionGrabbable. In this way, the user is able to move, scale or rotate objects by hand, both in Unity and on the HoloLens 2 device. Also, the Rigidbody component has been added, which applies the gravitational force to each object. Therefore, when a piece is taken by hand from the table of the panel and is then released, the pneumatic component has a free fall.

To help the user become more familiar with the circuits, labels have been created that contain the name of the pneumatic parts. Animation gif files have been added to illustrate the operation principle of the components in the longitudinal section.

A useful object is the tool for displaying the catalogue with pneumatic components, that can also be found on panels. It will only be displayed when the wrist is turned, and the user's gaze turns to the catalogue while wearing the HoloLens 2 device (see Fig. 6).

The next step was to create an interactive menu that is able to indicate the locations of each pneumatic part in the circuit or manipulate the components placed on the panel. The following will explain the buttons on the menu: Hints - indicates the place corresponding to each pneumatic part in the assembly, Explode - will move the fixed components following a set command, Reset - will place the pneumatic piece in the initial place from where it was taken. The menu is represented in Fig. 7.

For extra creativity, animations such as translation and rotation movements were made on the cylinders, manometers, roller operated valves and the flash action of the

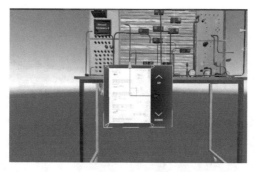

Fig. 6. Immersion of specialized literature (catalogue) in virtual environment

solenoid valve led. The bottleneck of the work was the creation of the rotational movement around a fixed point. Because it could not be done directly by changing the angle of rotation on the z-axis of the roller, it was necessary to create a special script called "Gizmo", which attaches a transparent sphere to the roller around which it can rotate.

Fig. 7. The interactive menu

A plus of originality of the project is the possibility to do animations on manometers, cylinders, valves and solenoid valves when attaching the air tubes to them. Each air tube that connects to these parts has a "Drag&Drop" script specifying location of the tube and starts the animation of the pneumatic part while fixing it. Finally, the interactive buttons Pause and Resume were created to stop and resume animations, see Fig. 8.

Fig. 8. Interactive buttons

Fig. 9. Testing applications by students and QR code to access the video experiment

To be able to display the applications on HoloLens 2, it was necessary to build projects in Unity. Following this process, a.sln file is generated and opened with Visual Studio. The application was loaded on the device using the computer's IP address connected to the HoloLens2 device. Figure 9 shows students working with Mixed Reality applications near the pneumatic workstations. During the testing of the created scenarios, everything that the students viewed on HoloLens 2 was displayed on the laptop in the Windows Device Portal by accessing the Live Preview option.

A detailed demonstration of the two holographic pneumatic workstations presented can be accessed at the link https://youtu.be/im3G4kHGgFc or by scanning the QR-code below.

5   Conclusions and Future Work

This paper aims to contribute to facilitating the learning process of future engineers by creating the virtual technical scenes, and at the same time familiarizing, capturing their interest and training them in MR tech that will storm the entire production of companies in the near future. Integrating digital twin capabilities with MR tech started being used across some industries. The capability to visualize the digital model in the proximity of a real process is an incredible opportunity given by Industry 4.0.

Educational MR tools in a "hands-on" lab can improve students' performance by a significant percentage, having an opportunity of increasing the experiment's difficulties without compromising the understanding level. The collaborative mixed reality can be used to transmit procedural knowledge and could eventually replace in some scenarios face-to-face training.

The next step is to introduce voice commands for ease and more user control over the components. It is also desired to change the structure of the 3D air tubes to give them malleability similar to those in the real environment.

Acknowledgement. The authors would like to express their gratitude to Tech-Con Industry for their continuous support for developing Hydraulic and Pneumatic Control Systems Lab in the Department of Automatic Control and Applied Informatics from TUIASI.

References

1. https://www.amaxperteye.com/. Accessed 29 May 2022
2. Milgram, P., Takemura, H., Utsumi, A., Kishino, F.: Augmented reality: a class of displays on the reality-virtuality continuum. Proc. SPIE **2351**, 2351–34 (1994). Telemanipulator and Telepresence Technologies
3. Parida, K., Bark, H., Lee, P.S.: Emerging thermal technology enabled augmented reality. Adv. Funct. Mater. **2021**(31), 2007952 (2021)
4. Liu, Y., Dong, H., Zhang, L., Saddik, A.E.: Technical evaluation of hololens for multimedia: a first look. IEEE MultiMedia **25**(4), 8–18 (2018)
5. Park, S., Bokijonov, S., Choi, Y.: Review of microsoft HoloLens applications over the past five years. Appl. Sci. **2021**(11), 7259 (2021). https://doi.org/10.3390/app11167259
6. Galati, R., Simone, M., Barile, G., De Luca, R., Cartanese, C., Grassi, G.: Experimental setup employed in the operating room based on virtual and mixed reality: analysis of pros and cons in open abdomen surgery. J. Healthcare Eng. **2020** (2020). Article ID 8851964. https://doi.org/10.1155/2020/8851964
7. https://www.gigxr.com/blog/roi-in-xr-a-look-at-standardized-patients. Accessed 5 May 2022
8. Hanna, M.G., Ahmed, I., Nine, J., Prajapati, S., Pantanowitz, L.: Augmented reality technology using Microsoft HoloLens in anatomic pathology. Archives Pathol. Lab. Med. **142**(5), 638–644 (2018). https://doi.org/10.5858/arpa.2017-0189-OA
9. Hietanen, A., Pieters, R., Lanz, M., Latokartano, J., Kämäräinen, J.K.: AR-based interaction for human-robot collaborative manufacturing. Rob. Comput.-Integr. Manuf. **63**, 101891 (2020). https://doi.org/10.1016/j.rcim.2019.101891
10. Gruenefeld, U., Prädel, L., Illing, J., Stratmann, T., Drolshagen, S., Pfingsthorn, M.: Mind the ARm: Realtime visualization of robot motion intent in head-mounted augmented reality. In: Proceedings of the Conference on Mensch und Computer, Germany, pp. 259–266 (2020). https://doi.org/10.1145/3404983.3405509
11. Vorraber, W., Gasser, J., Webb, H., Neubacher, D., Url, P.: Assessing augmented reality in production: remote-assisted maintenance with HoloLens. Procedia CIRP 2020 **88**, 139–144 (2020). https://doi.org/10.1016/j.procir.2020.05.025
12. Dosoftei, C.C., Cojocaru, A.E.: Implementation of a virtual control lab to support teaching in engineering control. In: 2020 International Conference and Exposition on Electrical and Power Engineering, Romania, pp. 699–703 (2020). https://doi.org/10.1109/EPE50722.2020.9305528
13. https://docs.microsoft.com/en-us/HoloLens/HoloLens2-hardware. Accessed 9 Apr 2022
14. https://www.famictech.com/portals/0/PDF/brochure/automation-studio-educational. Accessed 1 July 2022

Performance, Ethics and Operations Management in Internal Logistics 4.0

Challenges of Material Handling System Design in the Context of Industry 4.0

Zakarya Soufi, Pierre David$^{(\boxtimes)}$, and Zakaria Yahouni

University Grenoble Alpes, CNRS, Grenoble Institute of Engineering, G-SCOP,
38000 Grenoble, France
`Pierre.David@grenoble-inp.fr`

Abstract. Mastering Material Handling Systems is a crucial issue for numerous companies since the costs of material handling activities are far from being negligible. The demand for a sustainable industry and the introduction of Industry 4.0 technologies for material handling are renewing the Material Handling System design concerns. Many questions arise on whether it is suitable or not to bring technologies such as Autonomous Mobile Robots or Real-Time Location Systems to the shop floors. In the presented study, various companies are questioned on their utilization of new technologies and on their processes to manage and modify their Material Handling System. It appears that practices in Material Handling System design and management are diversified and that no consensus exists on how to efficiently design these systems. A cross-analysis is performed to identify the differences and the common patterns between the literature and the field study. This work discusses the need to better understand the relation between the different aspects of Material Handling System design and presents key challenges to be addressed in the context of Industry 4.0. To address these challenges, research directions are proposed. They are composed of four main challenges area: Material Handling specifications, Material Handling Equipment selection, Material Handling Equipment deployment, and Material Handling System analysis.

Keywords: Material Handling · Material Handling System design · Material Handling Equipment · Level of Automation · Industry 4.0

1 Introduction

Material Handling System (MHS) addresses storing, packaging, and moving products. It plays a key role in the performance of the entire manufacturing system [1, 2]. In this article, MHS is considered to be a collection of Material Handling Equipment (MHE) and operators performing the internal logistics task of a factory. An efficient MHS leads to effective production management, improvement of on-time delivery, and enhancement of production quality [3]. But the MHS could be a source of excessive expenditure if it is not efficiently designed [4]. In this article the design of MHS is seen as the process used to specify and select the equipment needed to perform the Material Handling operation along with the definition of their operating policies and infrastructure (e.g., network

© The Author(s), under exclusive license to Springer Nature Switzerland AG 2023
T. Borangiu et al. (Eds.): SOHOMA 2022, SCI 1083, pp. 291–303, 2023.
https://doi.org/10.1007/978-3-031-24291-5_23

and communication protocols). The design of MHS is renewed by the high amount of newly available technologies coming from Industry 4.0 as well as the potential use of large scale data treatment technologies. To cite a few, Autonomous Mobile Robots are proposed along with Real-Time Location Systems (RTLS) or cobots. These technologies are seen by many as a solution to support the production systems in reaching their goals of operational excellence and agility. Nevertheless, considering the literature MHS design is missing a largely admitted design process that handles the challenges of Industry 4.0. The main approaches focus on some MHS sub-problems, such as the optimization of material flow [5], the definition of the appropriate Level of Automation (LoA) [6], or the Material Handling Equipment (MHE) Selection Problem [7]. These problems are separately addressed in the literature, which does not allow to solve the MHS design problem with a systemic view. Automation is an important principle in classic MHS design approaches [8, 9]. In these approaches, the automation of Material Handling activities is presented as a mandatory path without precisely analyzing the company context. However, many articles support the idea that LoA must be carefully defined to maintain the effectiveness of systems [10]. There is currently no evidence that the MHS design problem is mastered in the industry.

The first contribution of this paper is to analyse some of the existing practices of MHS design in five companies and compare them to the literature. The objective is to identify the potential improvements in the processes, methods, and tools to make the right decision in MHS design projects. The second contribution is to propose research directions organizing the needed improvements in MHS design processes within the context of Industry 4.0. The results of the field studies are not to be generalized, but to identify some of the practices that can be found in today's industries. The remainder of this article is organized as follows: It starts with a literature review on Material Handling. Afterward, the field studies are presented through their cross analysis to identify the currently encountered practices for MHS design and management. Finally, based on the obtained results, research challenges for MHS design are discussed. The conclusions are summarized in the last section.

2   Material Handling System Design

The design of MHS has been studied by many researchers, but no consensus exists on defining an appropriate design process. Namely, differences may be found whether the study addresses a modification of an existing system or the design of a new system. MHS design may be of various complexity: it can address single point-to-point transportation or a whole plant transportation network. Nevertheless, different approaches may be found in the literature. The approaches are sometimes design principles with associated analysis tools and sometimes optimization-oriented approaches. For instance, in [11] the authors propose an approach that integrates the design of facilities and the MHS to minimize the Material Handling cost. This approach mainly focuses on the layout of the MHS by defining the location of the pick-up/delivery stations and the paths and their associated direction.

In [8, 9] the authors cite ten principles that should be taken into consideration during the design of MHS. These principles are developed by the MHI association (www.mhi.org). The principles to be applied to MHS design are summarized in 10 concepts: Planning, Standardization, Work, Ergonomic, Unit load, Space utilization, System, Automation, Environmental, and Life cycle cost. The concepts are well defined but not articulated to form a design approach. They just highlight the important points to consider in the design process.

In addition to these ten concepts, [8] proposes to structure the approach to find the right Material Handling solution with five standard questions (Why, What, Where & When, How, Who). It might be noticed that the questions are addressing a single point-to-point movement when a global approach of the whole MHS design should be made.

In [12] the authors propose a design approach that is based on the examination of materials and routes. The design approach is composed of four steps: Assemble flow analysis output, Class selection of route, Calculation of requirements, and Selection.

In this kind of approach, it is assumed that the plant layout is fixed and known. The workstation placement does not change and cannot be changed. Aso, production volume is assumed to be constant. It can be noticed that neither the technology nor the communication and control systems are addressed by this proposal. Important aspects such as the internal logistics and automation strategies of the company should also be analysed [13]. Research works highlight the importance of company strategy analysis to correctly address the MHS design. As stated in [9, 12], it is essential to associate the competitive advantage or the company's strategy with its MHS design. MHS design should fit and support the firm's manufacturing strategy [12] and should also reflect the strategic objectives and the needs of the organization [9, 13].

The main global design approaches found in the literature give high-level outlines of the process but remain vague on their realization. The proposed principles are to be used but the approaches' lack of precision and do not explicitly tackle the new challenges of Industry 4.0. The more precise approaches based on optimization algorithms concentrate on sub-problems of MHS design.

To complement the vision on MHS design problems, a field study at various companies is conducted to extend the observations made from the literature. The second objective is to analyse whether literature results spread in companies. The next section presents the field studies conducted within five industrial companies to share their experience of MHS design, management, and usage.

3 Field Studies

3.1 Field Study Design

Five interviews are conducted using a grid analysis. Then three steps are adopted to exploit the data: data reduction, data display, and cross-analysis. This procedure is inspired by the work of Säfsten et al. [14]. Table 1 presents general information about the company typology. Five companies have been chosen to observe different practices. At company D, a metal part workshop (D_1) and a plastic part workshop (D_2) were analyzed.

The interviews were semi-directed, with additional questions to deepen the interviewees' answers. In most cases, the interviews were complemented by a plant visit.

Table 1. Cases general information

Case	Field	Global no. of employees	Product size, volume, and complexity	Production strategy	Production organization
Case A	Construction machinery manufacturer	102.300	Large, high, complex	Make to order, make to stock, engineering to order	Assembly line
Case B	Semiconductor manufacturer	45.500	Small items, large batches, very high, complex	Make to order	Flexible job shop
Case C	Stainless steel fixings	40	Small, very high, medium	Make to order	Flexible flow shop
Case D	Fastening & assembly	7.250	Small, very high, medium batches	Make to stock	Flexible flow shop
Case E	Metal wire parts manufacturer	130	Small, small to medium, simple	Make to order	Job shop

To analyse the companies' practices, questions were asked to understand the MHS Design process and the MHS solutions implemented on sites. To distinguish the various typologies of MHS solutions, the different cases are analysed in terms of LoA. For that purpose, the LoA analysis concept proposed by Parasuraman et al. [15] is adopted. The concept allows visualizing the LoA of different systems through the cognitive and physical aspects. The visualization was originally made through four classes of information functions (information acquisition, information analysis, decision and action selection), and action implementation. In this study, the LoA in different cases is visualized for two types of activities: productive and material handling activities (Fig. 1).

For each type of activity, the physical and cognitive tasks are analyzed in an aggregated view. Productive and Material Handling activities are analyzed to identify whether the automation strategies are different in the company. Figure 1 summarizes the observation made on the visited plant LoA. Due to space limitations in this article, we do not present the whole cases description. But a cross-analysis revealing the important facts of analysed cases is provided in the next section. The purpose is to report on the observed practices and to raise a promising subject for future research on MHS.

3.2   Cross Analysis

To identify the main challenges of MHS in the context of the fourth industrial revolution, this cross-analysis is based on four axes; Material Handling Automation, Material Handling Control System (MHCS), MHE Selection, and MHS Design. The analysis of each axis starts with a review of practices of the different cases. Then, common and

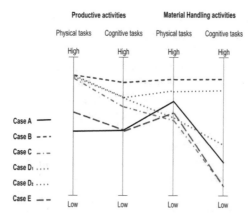

Fig. 1. Advancement in terms of Automation

different patterns are highlighted. Besides, when needed, the literature is mentioned to compare the theoretical and the empirical materials. A summary of the cross-analysis is presented at the end of this section opening new research questions.

Material Handling and Automation

As shown in Fig. 1, the gap between the automation of physical and cognitive tasks in productive activities is not as significant as in material handling activities. The automation of material handling is more restrained compared to the automation of productive activities. Sometimes this restriction is driven by the corporate culture of the company. On one hand, small and medium-sized companies (such as cases C and E) tend to automate the productive activities more than the material handling ones. On the other hand, large-sized companies (such as cases B and D) aim to adopt high LoA for both productive and material handling activities. Besides, in terms of material handling activities, the physical tasks are more automated than cognitive tasks.

In four cases (A, C, D_1, E), decisions related to the cognitive tasks were left to the human, e.g., parts positioning, storing, navigation, and so on. Although the LoA of cognitive tasks of material handling activities is low compared to others, several companies tried to test new technologies that offer higher LoA for cognitive tasks. Both cases B and D tried to deploy AIV (Autonomous Intelligent Vehicles) in their plants, which offers a high LoA for both physical (e.g., moving) and cognitive tasks (e.g., navigating through the plant). The use of this equipment was accepted by case D, and rejected by case B due to many factors (behaviour unpredictability, collision risks, etc.). The choice of the appropriate LoA seems to be difficult to make. Automating the cognitive Material Handling actions seems to be linked to the intensity of the product flow (such as in cases B and D2 where the flow must be kept at its maximum to satisfy the demand).

While the automation of physical material handling actions appears to be related to the characteristics of the handled materials (as in case A where the handled materials are heavy and require high physical automation). However, a thorough study has to be done to identify other parameters that might influence decisions related to LoA.

Additionally, methods can be found in the literature for the readjustment of LoA such as DYNAMO++ [6]. The procedure of DYNAMO++ can be applied for existing MHS only. Yet no methods can be found in the literature to assist the LoA decision for a new MHS. Based on the results shown above, it is shown that, although automation is considered a crucial element of the MHS design, the LoA definition remains unclear. Thorough studies have to be conducted to identify parameters and propose procedures to support the choice of LoA for the design of MHS.

Material Handling Control System
The field studies show that the use of technologies related to the MHCS is heterogeneous. In some cases, the investment in such technologies is not needed since the used MHE do not have a significant cognitive LoA (e.g., pallet trucks). But for other cases it is important to deploy this kind of technology. In case B, the use of such technologies is mandatory due to the high LoA adopted in the plant; they use a workshop supervision software automatically making the affectation of transportation to autonomous carts. In case D, recent tests were done to include this type of technology. The MHCS in this case allows the operators to control, program, and supervise the AIVs through a screen. The investments in different cases are mainly driven by two factors: the need to remove Non-Value Added (NVA) activities and to have a ROI in a short period. To present convincing proofs, simulation has to be done to display the possible benefits that can be derived from the use of such technologies.

In the literature [19], several structures and technologies related to the MHCS can be found while in the field studies it can be seen that MHCS technologies are slowly emerging for large-sized companies. On the other hand, the MHCS is still not considered by small and medium-sized companies. Consequently, it can be seen that MHCS is not widely used, and the implementation of such technologies remains unclear in the literature. More research has to be done to develop this subject and propose methodologies for the deployment of MHCS.

Material Handling Equipment Selection
Throughout the field studies, it is noticeable that investment for developed MHE is not simply related to the size of the company nor the complexity and the size of the production. Even small-sized companies such as case E, with a small and medium production volume, are using equipment such as cobots. The selection of this equipment is mainly done through benchmark with a restrained list of suppliers; this practice is sometimes successful (e.g., the selection of the tugger train in case A). However, it does not always ensure positive results. For example, in case C the selection of an automated transfer cart was based on copying the practices of another plant, but appeared not to be suitable. The loading/unloading operations were too difficult to be performed by the operators which negatively impacted the performance of the plant.

The criteria used for MHE selection in small and medium-sized companies are limited and mainly focus on economics. Furthermore, the large-sized companies include diversified criteria during their MHE selection (see Table 2). It may be observed that no clear MHE selection method is shared between cases. The decision is often not formalized and is made with a partial view of the impacts on the production system.

Table 2. Identified criteria for the MHE selection

Categories	Criteria	Cases
Economical	Return on investment (ROI)	A, B, C, D, E
	Total cost of ownership (TCO)	D
	Cost of the implementation	E
	Cost of the equipment	E
Technical	Capacity of the equipment	A
	Speed movement	A
Operational	Safety	B
	Precision	B
	Operating skills	A
	Ease of implementation	E
Supplier	Supplier's reputation	D, E
	Supplier's location	D, E

Material Handling System Design

The cases show that companies do not always use a clear method to design their MHS. Large-sized companies (case A, B, and D) use approaches based on benchmarks and on copy the best practices. In addition, case A uses corporate guidelines common for all production sites. These guidelines are principles to apply (e.g., the reduction of the number of part touch) but do not constitute a design process. It is noticeable that in these cases, the MHS design is associated with the competitive priorities in a way to fit and support the manufacturing strategy followed by the company (e.g., in case A, the MHS is designed in a way to reduce the number of part touch which positively impacts the On Time Delivery). Key Performance Indicators (KPIs) are used to constantly evaluate the system (see Table 3).

Table 3. Key performance indicators for the evaluation of MHS

Indicators	Cases
Overall Equipment Effectiveness (OEE)	A, B
Non-Value Added (NVA) activities	C, E
Inventory on location	A
On-time deliveries	A
Number of touches	A

(*continued*)

Table 3. (*continued*)

Indicators	Cases
Line balance	A
Number of transports	B
Duration of transport	B
Throughput	B
Availability of operators	C
Intermediate storage time	D
Cost of work-in-progress (WIP)	D
Cycle time	D

A failure in one of these KPIs triggers the modification of some elements of the design; e.g., in case A MHE was dropped off because its speed movement and reactivity caused a decrease in the OEE of the line.

The small and medium-sized companies (case C and E) do not use any approach for the design process of MHS. Material handling is not considered as important as productive activities. However, the interviewees were aware of the benefits that could be generated by relevant MHS. In case E, the use of cobots for simple pick and place tasks resulted in removing the NVA activities performed by operators. In cases A and D, the MHS is designed as a part of the production, while for cases B, C, and E the MHS is designed after designing the production system. In this context, the MHS design has been adapted to the workshop configuration. In case A, the organization is an assembly line for which moving and gathering parts is an important part of the activity. In case D, flow management was seen as an important priority by the top management. It can be noticed that both companies operate in the automotive market and then share a comparable approach for flow management. With this example, it is seen that the field of production can have an impact on the approach followed for the MHS design. For case B, the management assumes that MHS should adapt to production since growth management and technical factors already constrain production organization. For cases C and E the MHS has not been seen as a high priority.

The MHS design process followed in the different cases is mainly done through MHE Selection. Nevertheless, the literature shows other potential factors (e.g., the LoA definition) that have to be included during the design of MHS. These factors do not seem to be a major concern for the interviewed industrial cases, especially for the small and mid-sized enterprises. These results support the assumptions that the design of MHS still requires a holistic approach that: a) Considers the challenges of Industry 4.0, and b) Adapts to different industrial typologies.

4   MHS Design Challenges in the Context of Industry 4.0

The theoretical and empirical materials show that the design of the MHS is lacking a global approach which includes the challenges of Industry 4.0. In the literature, MHS problems are addressed as independent problems and are solved separately due to their complexity (e.g., MHE selection, LoA definition, communication structures proposal for the MHS), while the analysis of the literature and the results of the field studies show that MHS aspects are interdependent and should be treated based on a holistic approach. On one hand, theoretical materials unveil several challenges that can be faced during the MHS design. Therefore, it is important to have a global approach addressing all MHS problems and challenges. On the other hand, the empirical materials display various practices and insights about the different MHS design errors that were encountered (e.g., investments on inappropriate MHE, lack of a design approach, etc.).

As a guideline to list research challenges on MHS, we propose changing the paradigm of MHS design by adopting a Systems Engineering approach. MHS design is a complex system design and therefore deserves a well-established process to be carried out. IEC 15288 [16] proposes efficient guidelines to address such system development. A MHS design process should take inspiration from such design approaches. It will encourage high-quality specification, stakeholder integration, rationalized design and selection choices, and a well-prepared deployment phase. In the remaining of this section, we will discuss some important research directions linked to the various systems engineering phases.

4.1 Material Handling Needs Specification

The field studies illustrate the magnitude of the decisions related to the MHS; they can be opportune (e.g., the use of cobots in case E) and they can also be deceiving (e.g., the use of the automated transfer cart in case C). This type of issue can be caused by a lack of problem specification including material handling needs definition and constraints elicitation. The study shows that MHS design decisions are multi-facet and influenced by multiple aspects (costs, staff skills, production requirements, to cite a few). An efficient MHS design process should carefully assist designers to precisely specify the MHS design problem [17].

In the literature, techniques for material handling specifications can be found. They consist of charts and diagrams [9] which can be time-consuming and less efficient with the requirements of Industry 4.0. New methods are needed to specify the material handling activities in a non-technological way, to specify the expected overall production flow, the companies' strategies in terms of internal logistics and admissible LoA, the available deployment and support systems. An exact definition of the company's strategies is important as well; Granlund, Säfsten et al. [13, 14] show the importance of the analysis of companies' strategies on automation and internal logistics for MHS automation. In their work, only automation is highlighted but the authors give relevant advice on the decomposition of strategies.

4.2 MHS Architecture Definition

Subsequently to the clarification of needs comes the design phase of the MHS. In a Systems Engineering approach, the purpose is to generate several possible architectures to advance concurrent ideas. The MHS architecture is a composition of interacting MHE and operators, where each element is allocated for a specific set of Material Handling activities. In the field studies, different cases showed a lack of a formalized approach; many cases used benchmarks or testing technologies individually.

In the literature, general principles are given for the proposal of MHS architectures without sequencing the analysis nor giving procedures to generate sufficient architectural possibilities. However, the literature addresses some MHS sub-problems that should be combined to define a procedure for MHS architecture definition (e.g., the layout definition, selection of MHE, MHCS definition, operational strategy selection, equipment allocation to task, dispatching rules…). The literature shows that, despite the increasing use of AI and computational models to solve MHS sub-problems, the expert's knowledge still remains necessary for such decisions. Current MHE fleet sizing algorithms are always underestimating the number of equipment to use since they omit many constraints as traffic or product availability uncertainty. The empirical and theoretical materials emphasize the need to propose techniques for MHS architecture definition, while including several aspects such as:

- Considering diverse MHE technologies
- Handling task allocation and dispatching rules
- Ensuring a wide search of solution (handling the combinatory)
- Considering the human/machine interactions
- Considering the needed communication information system
- Considering the experts' knowledge
- Handling criteria related to the company strategies

4.3 MHS Performance Analysis

An efficient MHS architecture is dependent on the capacity to analyse foreseen system performances. The analysis of MHS is facing new challenges due to its complexity that follows Industry 4.0 trends. Two types of complexity must be handled for a relevant performance analysis:

1. The complexity of the MHS makes it difficult to predict every emergent behaviour coming up from the equipment interaction. Therefore, there is an emerging need to develop performance analysis techniques. These techniques should include the various MHE types and their interactions through communication systems and with onsite operators. The influence of the MHS on the production site behaviour including degraded situation is also to be included. This imposes using heavy performance analysis techniques as for Case B and opens new research trend in finding techniques to automate evaluation the creation of models.
2. The second complexity to handle is the evolving definition of performance for MHS. With current megatrends on sustainability or ethical concerns, the performance cannot be seen through the reduced prism of operational excellence as encountered in most of field studies. New indicators must be thought to embrace more long-term perspectives such as MHS flexibility or sociological aspects such as the influence of the MHS on the work group organization.

4.4 MHS Selection

The selection of the appropriate MHS for a given Material Handling need is not straight-forward. The process must be formalized and must involve many dissimilar criteria. The field studies analysis has shown that the selection process is poorly defined for most of the studied companies. Only a few criteria are used objectively or unconsciously by companies and their selection remains unclear. It is not obvious to understand why certain criteria are retained and others ignored. An analysis of literature on MHE [18] revealed that more than 200 criteria were identified for MHE selection processes, while in the field studies of this work 12 distinct criteria were identified, and the companies used at most 6 criteria to decide. In [18], a methodology with a list of potential criteria is presented for two analysis levels: MHE category selection and MHE model selection. To advance the field it would be relevant to propose techniques for the selection of relevant criteria, based on the company's characteristics and its industrial activity and strategy.

4.5 MHS Deployment

The integration of MHS into a plant appears to be challenging due to the high complexity of today's industrial systems. It can be confronted with important problems as the integration to the information system, MHE/Operator interaction or the management of technologically advanced equipment maintenance. Few works can be found on the integration of MHE to the information system. Wang et al. [19] propose an architecture of Cyber-Physical System (CPS) for MHS. The paper shows the potential of such technologies to improve shop-floor production fluency and efficiency. This topic still requires more effort to define other potential technologies that can be used in the context of MH. The interviewed companies expressed their current fear for MHE integration in an MHCS.

The deployment of ERP (Enterprise resource planning) or MES (Manufacturing Execution System) in small and medium-sized companies is considered to be a period of intensive work. Therefore, the integration of MHE into a communication network is often feared. The interaction between MHE and the operators should be carefully considered for a successful deployment. As encountered in case E, the company staff may be reluctant to share the workspace with automated systems. In this context, new questions and opportunities about human-machine cooperation arise. The extent of the interventions that should be made by the operators is to be analyzed. Human-Machine cooperation principles in a manufacturing system context are proposed in Pacaux-Lemoine et al. [20]. These principles should be adapted to the context of MHS.

5 Conclusion

The objective of this paper is to analyse the existing practices of MHS design in literature and five companies, to identify the main challenges that are faced in the context of Industry 4.0. These challenges may allow researchers to extend their MHS design approaches while including the latest technologies of Industry 4.0. The literature review shows that fractions of MHS design problems are solved separately. Meanwhile, in the field studies, it has been shown that there is a lack of a formal MHS design approach. However, some existing practices are used and highlighted in this work, as benchmarking. These practices do not have constant outcomes; it has been shown that sometimes, these practices led to a significant financial loss.

The comparisons and analysis stressed the need of having a properly detailed methodology for MHS design that tackles diverse sub-problems. Such a global approach should enhance the use and efficiency of MHS. Moreover, the need to enhance the MHS design methodology is increasing with the arrival of Industry 4.0 demands and technologies. It has been shown that the deployment of communicating and automated MHE is difficult for many companies. To address this issue, both theoretical and empirical materials were analyzed based on five Material Handling aspects; Material Handling activities, Material Handling automation, MHCS, MHE selection, and MHS. As a result, MHS design research directions are proposed.

For future development, a thorough analysis of each aspect of the MHS design challenge has to be done. For this purpose, techniques for data collection and specification have must be either identified from the literature or developed. The Model Driven approach could be adapted to propose a generic framework regrouping complementing views on MHS. Such framework should rely on a MHS domain metamodel that could enhance data viewing, comparison, and translation to adapted analysis models such as discrete event simulation to enhance deep MHS analysis.

Acknowledgments. The authors thank the French National Research Agency (ANR) that funded this research under the LADTOP project (grant ANR-19-CE10-0010-01). The authors would also like to thank the industrial companies for their participation in the field studies.

References

1. Esmaeilian, B., Behdad, S., Wang, B.: The evolution and future of manufacturing: a review. J. Manuf. Syst. **39**, 79–100 (2016)
2. Beamon, B.M.: Performance, reliability, and performability of material handling systems. Int. J. Prod. Res. **36**, 377–393 (1998)
3. Montoya-Torres, J.R.: A literature survey on the design approaches and operational issues of automated wafer-transport systems for wafer fabs. Prod. Plan. Control. **17**, 648–663 (2006)
4. Bouh, M.A., Riopel, D.: Material handling equipment selection: new classifications of equipments and attributes. In: International Conference on Industrial Engineering and Systems Management, IEEE IESM, pp. 461–468 (2016)
5. MacGregor Smith, J.: Queueing network models of material handling and transportation systems. In: Smith, J., Tan, B. (eds.) Handbook of Stochastic Models and Analysis of Manufacturing System Operations. International Series in Operations Research & Management Science, vol. 192, pp. 249–285. Springer, New York. https://doi.org/10.1007/978-1-4614-677 7-9_8
6. Choe, P., Tew, J.D., Tong, S.: Effect of cognitive automation in a material handling system on manufacturing flexibility. Int. J. Prod. Econ. **170**, 891–899 (2015). https://doi.org/10.1016/j.ijpe.2015.01.018
7. Saputro, T.E., Masudin, I., Rouyendegh, B.D.: A literature review on MHE selection problem: levels, contexts, and approaches. Int. J. Prod. Res. **53**, 5139–5152 (2015)

8. Heragu, S.S.: Material handling. In: Facility Design Fourth Edition. CRC Press (2016)
9. Stephens, M.P., Meyers, F.E.: Material handling. In: Manufacturing Facilities Design and Material Handling. Purdue University Press (2013)
10. Granell, V., Frohm, J., Winroth, M.: Controlling levels of automation - a model for identifying manufacturing parameters. IFAC Proc. **9**(2006), 65–70 (2006)
11. Aiello, G., Enea, M., Galante, G.: An integrated approach to the facilities and material handling system design. Int. J. Prod. Res. **40**, 4007–4017 (2002)
12. Shell, R., Hall, E., Wrennall, W., Tuttle, H.: Material handling and storage systems. In: Handbook of Industrial Automation (2000)
13. Granlund, A.: Facilitating automation development in internal logistics systems. Mälardalen University Press Dissertations, no. 150 (2014). https://www.diva-portal.org/smash/get/diva2:680303/FULLTEXT02.pdf
14. Säfsten, K., Winroth, M., Stahre, J.: The content and process of automation strategies. Int. J. Prod. Econ. **110**(1–2), 25–38 (2007)
15. Parasuraman, R., Sheridan, T.B., Wickens, C.D.: A model for types and levels of human interaction with automation. IEEE Trans. Syst. Man, Cybern. Part A Syst. Hum. **30**, 286–297 (2000)
16. International Electrotechnical Commission. International Organization for Standardization (2009)
17. Noble, J.S., Tanchoco, J.M.A.: A framework for material handling system design justification. Int. J. Prod. Res. **31**, 81–106 (1993)
18. Soufi, Z., David, P., Yahouni, Z.: A methodology for the selection of material handling equipment in manufacturing systems. IFAC-PapersOnLine **54**(2021), 122–127 (2021)
19. Wang, W., Zhang, Y., Zhong, R.Y.: A proactive material handling method for CPS enabled shop-floor. Robot. Comput. Integr. Manuf. **61**(2020), 101849 (2020)
20. Pacaux-Lemoine, M.P., Trentesaux, D., Zambrano Rey, G., Millot, P.: Designing intelligent manufacturing systems through human-machine cooperation principles: a human-centered approach. Comput. Ind. Eng. **111**(2017), 581–595 (2017)

Integration of Ethical Issues in the 4.0 Transition of Internal Logistics Operations

Cindy Toro Salamanca[1], Lamia Berrah[2], Pierre David[1(✉)], and Damien Trentesaux[3]

[1] Univ. Grenoble Alpes, CNRS, Institute of Engineering, G-SCOP, 38000 Grenoble, France
pierre.david@grenoble-inp.fr
[2] University Savoie Mont Blanc, LISTIC Laboratory, 74940 Annecy, France
lamia.berrah@univ-smb.fr
[3] Univ. Polytechnique Hauts-de-France, LAMIH, UMR 8201, 59313 Valenciennes, France
damien.trentesaux@uphf.fr

Abstract. Today, with the advancement of technology and the advent of the fourth industrial revolution, organizations are increasingly implementing automated digital systems in internal logistics. This is giving rise to a new generation of production models with I4.0 technology. This deployment, in turn, raises questions of awareness of the ethical impact produced by the interaction between cyber-physical systems and humans. The purpose of this research is to suggest ways to analyse and evaluate the adoption of Industry 4.0, particularly in internal logistics with a focus on material handling systems (MHS), and to identify ethical risks that need to be addressed for companies to operate in an appropriate and sustainable manner. This study focuses on the importance of the ethical dimension in the implementation of I4.0 technologies. Typical ethical dilemmas that were addressed are relevant to data abuse, stress, social interaction and human surveillance. In this paper, it is proposed to build a methodological guide model to assess ethical risks in the adoption of 4.0 technologies for MHS. The guide exploits concepts from literature on ergonomics, sustainability and ethical risks. This guide has been tested on an order picking system called "Voice picking" to identify potential ethical risks during its use. It was concluded that the methodological guide model could effectively assess the existence of ethical dilemmas in the use of these systems. The results of the study can serve as a starting point for researchers and companies interested in integrating the ethical dimension into internal logistics activities.

Keywords: Ethics · Industry 4.0 · Material handling · Ergonomics · Sustainability · Internal logistics

1 Introduction

Industry 4.0 (I4.0) is the further digitization and integration of information technologies, which includes artificial intelligence (AI), the internet of things (IoT), the cloud, cobots and cyber-physical systems, among other things. These systems enable interaction between humans, machines, and products through virtual and physical means, which may even contribute to increased sustainability [10].

© The Author(s), under exclusive license to Springer Nature Switzerland AG 2023
T. Borangiu et al. (Eds.): SOHOMA 2022, SCI 1083, pp. 304–316, 2023.
https://doi.org/10.1007/978-3-031-24291-5_24

I4.0 brings a great development at industrial and technological level, but at the same time brings with it many questions around the ethical dimension. Ethics can be defined from a system or societal point of view as behaviour that conveys wellbeing, fairness and morality while being aligned with the culture and values of the entity [2]. In this work, we initially considered ethics as "the striving for a good life, with oneself and with others, under just institutions" [11]. With the application of I4.0, ethical problems may arise due to the complexification of the systems and to the intensification of interaction between human and artificial systems as stated in [26], that exposes a large set of ethical-related stakes in I4.0.

Throughout this study, the focus is on the implementation of I4.0 technologies in material handling systems (MHS) as it is the sector of production that involve the most interactions between human and cyber-physical systems. Moreover, material handling activities are known to be painful for humans, but are one type of tasks that are the less analysed in I4.0 literature. MHS deal with the storage, packaging, and transportation of goods.; it is crucial to the overall performance of the manufacturing system [14]. In the last decades, material handling technologies have evolved, considering the growing interest in Industry 4.0. MHS design can incorporate new technologies such as robotics, automation, Cloud, Big Data analysis, augmented reality, AI and Digital twins.

Handling materials may necessitate extensive distances through the warehouse; collecting items from storage places and placing them onto a cart or bin may necessitate stretching and bending of the body while carrying the load [13]. If these actions are performed at high cadence, it can induce muscular fatigue and a high peak and/or cumulative load on the worker's spine, resulting in a high risk of the operator suffering musculoskeletal diseases. At the same time, past research has suggested that these excessive actions may have mental and emotional consequences, such as producing visual discomfort, stress or monotony [13]. These situations that can occur in the industry raise both ergonomic and ethical questions about the precautions to be taken before implementing a state-of-the-art system. In that context, I4.0 technologies might be solutions for a more convenient MHS, but might also bring new issues.

Even though many studies have shown that devices like augmented reality, light and computer aids, and voice headsets can significantly improve performance and reduce pick errors when compared to paper pick lists, while there are evident short-term benefits, managers should keep in mind that the design of some "gadgets" might compromise human elements and affect the long implications (performance) [13].

Hence, the objective of this research is to find tools to analyse the implementation of I4.0, specifically in the activities of internal logistics, to identify the ethical risks that must be considered for the proper functioning and sustainability of the industry. It is because logistics is an area that exposes employees both physically and cognitively that we chose this angle. As a result, a primary research question is proposed, namely "**How can we identify ethical risks during MHS design?**".

It is expected that answering the proposed research question can provide an overview of the ethical issues that exist with the implementation of this new technology, and at the same time provide a tool to analyse the ethical issues related to the MHS during the design phase.

In the first part of the paper, a bibliographic study focuses on the definition of concepts, context, and possible methods to be applied. In the second part, the construction of a methodological guide capable of identifying ethical risks is given; its application is carried out on an example of an Industry 4.0 system. Finally, the paper is concluded with a review of the main conclusions, as well as some possible future directions.

2 Ethics and Industry 4.0

There is a vast range of studies on Industry 4.0 and what it entails in the literature today. As a result, certain key lines of study will be addressed in this part to offer a thorough grasp of the context. The link between I4.0, sustainability, and an approach that is particularly relevant to the ethical dimension will be discussed in the first half. The I4.0 tools and their ethical approach will be discussed in greater detail in the second half.

2.1 Industry 4.0, Ethical Dimension and Sustainability

Today, the application of I4.0 in organizations aims to increase production and profits. But at the same time, the need arises that with the adoption of these new technologies an ethical dimension is taken into consideration. If applied correctly in the industry, it would bring great sustainability to both the organization as an entity and its workers.

Interoperability, virtual applications, decentralized systems, real-time capabilities, service orientation, and modular production are among the principles of Industry 4.0 [4]. These ideas, in combination with empowering technologies, could improve corporate operations by focusing on process, product, and business models [4].

Despite the great benefits that I4.0 brings, at the same time it brings with it issues about the ethics applied to the industry within this 4.0 context. For this reason, Margherita and Braccini [8] make a coupling between ethics and business development, considering a mix between the two concepts and defining business ethics. Thus, ethics is initially considered as "the striving for a good life, with oneself and with others, under just institutions" [11]. This definition has been particularized in the context of I4.0 in [26] precising that the extended enterprises are the *institutions* while *oneself and the others* are all the humans involved throughout the lifecycle of autonomous and intelligent systems used in I4.0 production systems. A detailed comparison of ethics definitions is provided in [26] for the interested reader.

It is important to consider that, according to Neumann et al. [10], an organization promotes safe employment, job enrichment, worker satisfaction, career advancement and well-being by acting fairly towards occupational well-being. Therefore, ethical business development and fair business growth are a complement defined as "seeking to balance the welfare benefits of capital and labour without harming either, encouraging capital benefits while maintaining stable, pleasant, and sustainable working conditions" [8].

Considering the above, it is possible to realize that sustainability becomes an important position, being the result of a well-applied ethical dimension, in a context of 4.0 technologies implementation.

According to Luthra and Mangla in [7], Industry 4.0 and sustainability are two newer technology and organizational trends that are driven or impacted by increasing productivity and ensuring long-term sustainability. At the same time, according to Müller, Kiel and Voigt in [9] in terms of social sustainability, smart and autonomous production systems can assist employees' health and safety by automating monotonous and repetitive operations, resulting in increased employee satisfaction and motivation. Industry 4.0 technologies, on the other hand, address society with numerous obstacles and constraints. Reduced employment, information security concerns, data complexity, electronic waste, and poor quality are just a few examples [1].

Industry 4.0 has a huge number of benefits in performance and production efficiency for organizations that implement these technologies. But at the same time, as its use is becoming more frequent, the ethical issues are also becoming more frequent. This directly links ethics and the transformation of technology, safeguarding both the well-being of employees and the performance of organizations, and of course benefiting sustainability at the same time.

2.2   Tools of Industry 4.0 and Its Ethical Approach

The adoption of Industry 4.0 tools brings additional benefits, such as increased worker productivity, improved working conditions, efficiency, new sources of income and the creation of new jobs. At the same time, due to its impact on human rights, privacy, and employment, among other things, the development of this fourth industrial revolution poses several dangers and ethical challenges [3]. As a result, they must be managed to maximize the advantages while avoiding or reducing the risks that come with their usage. In 2019, the European Commission drafted ethical guidelines on AI specifically, identifying three fundamental components of trustworthy AI: *legality*, *ethics*, and *robustness* [3]. Within this same research, Bonson et al. in [3] considered 200 industries from European Union countries, strategically chosen for factors such as the use of 4.0 technologies like AI, which had to follow the set of Ethical Guidelines mentioned above. Thus, it was revealed that, although several entities reported compliance with ethical principles, only one of them mentioned full compliance with the EU Ethical Guidelines [3]. So, we can see that currently there is a lack of ethics regulation on applied 4.0 tools.

Despite having so many benefits, according to Bonson et al. in [3], I4.0 tools are also associated with the risk of economic stagnation, inequality, employee destitution, and lack of basic human needs, among other things. The ethical impact of the application of this technology depends not only on the technology itself, but also on the context and the way it is used [3]. According to Berrah et al. in [2], redefining or even restricting the role of AI, for example, since it can be considered a substitute for humans at all levels, could lead to increased pressure and demotivation among workers in companies using this technology.

Also, according to the investigation of Berrah et al. in [2], today's performance model is linked to efficiency, effectiveness and relevance, so when implementing this new technology, if the ethical dimension is not considered as part of the performance it will lead to cost overruns that will be associated with unfair or immoral behaviour and will have an immediate impact on the profitability of a company.

In a conference in 2021 given by D. Trentesaux on digital transformation and ethical issues [15], different ethical issues were recognized with the application of 4.0 technologies, classified as classic (with AI - learning with non-representative data - and the storage of personal data, either for use or sale) or scientific (with the yet unproven behaviour of automated systems). These identified ethical risks lead to different ethical issues that the operator may have when using these tools, such as, for example, around:

- A change in behaviour (if he seems reluctant to use the new system).
- Statistics of supervision not only of the performance, but also of the control that can be over the operator (decreasing his privacy in the work environment).
- Encouragement to continue working despite exhaustion.

At the same time, it is understood that although an automated system can be identified as a safe and cost-effective means, it does not guarantee ethical behaviour with users, so it can clearly induce the risks mentioned above. In other words, it should be considered that relying on an intelligent system will always have an added cost [15].

Industry 4.0 tools are increasingly being employed in businesses' internal logistics. However, according to various studies, there is insufficient control of the ethical dimension in the deployment of these new technologies, which is a common concern. Due to a lack of fair and ethical behaviour, this can have an immediate impact on the profitability of enterprises.

3 Building a Guideline for Ethics Analysis of MHS in I4.0

3.1 Guideline Construction Based on the Literature

Elements of Industry 4.0 such as cyber-physical and human systems (CPHS), which are defined as the connection of computing elements, physical systems and people through networks, are secure systems but this does not mean that they behave ethically [16]. Glock. et al. indicated in [13] that the use of augmented reality (AR) in a warehouse environment could have an impact on worker safety due to gait adjustments or distracted behaviour. Therefore, a guideline has been proposed in [16], which can ensure that CPHS are ethical, contemplating their entire life cycle [16].

At the same time, it is also important to consider that the guideline must account for other different dimensions such as those that relate ethics to society, the human being and certainly, the environment in which the CPHS evolves and transforms [16]. In other words, the objective of the Guideline is to evaluate whether the implementation of a new I4.0 system (technical system 4.0) and the relationship between the operator and the function he exercises (whether general or specific) involves ethical risks in its use.

Therefore, to build a guideline, it is necessary to consider its scope. Trentesaux et al. suggest in [26] that a list of relevant "sieves" be made considering the dimensions mentioned above. We also consider that the definition and classification of sieves can be particular and depend on each case, function of the field of application for which the Guideline will be employed [16]. The next step is the creation of questions that will relate the operator, the system, the function that the operator performs when using this new technology, and the development framework. These questions will be defined in function of the previously selected sieve and must be answered by YES or NO. According to Trentesaux in [26], when creating the questions, it is suggested that two ethical paradigms are considered:

- The *deontological* one (the conclusion is based on ethical "must/mustn't" standards) [16].
- And then there is the *utilitarian* one (decided according to the review of the specification, design, and production of the CPHS) [16].

Once the questions have been created, the application of this methodological guide can begin by answering the question on the selected system to rise the ethical risks that may exist.

Once the Guideline has been applied, the next thing to do is to verify the answers obtained and filter all those that have a "YES" answer. This will indicate that there is an ethical risk that needs to be addressed. So, the analysis and mitigation of the ethical risk for the corresponding sieve is activated if at least one positive response is given.

After verifying the answers and generating the list of existing ethical issues, Trentesaux et al. propose in [26] to make a mitigation for each ethical risk identified, because ethical risk mitigation can be employed as an iterative process to improve the specification, design, and manufacturing of the CPHS [16]. Accordingly, the guideline proposed still needs to gain maturity and applicability. This is why we chose to focus on a specific type of CPS to assess the applicability of the guideline on a more specific problem.

From the steps to build the guideline, different models of the method can be created depending on the most important aspects to be evaluated, which makes it very versatile. It is proposed in this work to create a guideline dedicated to MHS analysis. Designing the guideline implies finding the relevant sieves to cover ethical risks of MHS and to define the correct set of questions to be as exhaustive as possible to track every possible risk. This methodological guideline, in addition to assessing the ethical risks of new technologies, can help organizations to design ethically safe material handling systems. The contribution of the paper is on operationalizing the proposal of [16] leading to a concrete analysis tools for MHS.

3.2 Creation and Operation of a New Guide Model

Ethical risks that specifically concern MHS are linked to ergonomics regarding operators activities, data treatment issues, sustainability of industrial activities including impact on society or environment. To create the new guideline model, at first the most important thing is to make a correct selection of sieves. Therefore, an analysis was made on the main aspects that should be evaluated in detail. The human dimension was considered to be fundamental in the selection of sieves, due to the main impact of the interaction between the human and the cyber-physical systems (risks include physical and cognitive ergonomy, respect of privacy and social interaction). After that, sieves are included to reflect the set of interactions existing within and outside the firm that may reveal ethical dimension. The selected sieves are defined to zoom off from operators perspective to the society stakes, by considering ethical risks at very different levels. All different sieves are given in Fig. 1. After human operator consideration, a focus is made on the production system to analyse the collective of workers, the machine network and the information system. The next level is to scan the impact on companies operation through business ethics analysis and finally impact on society must be considered regarding all dimensions of sustainability.

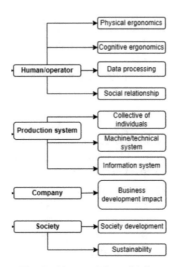

Fig. 1. Sieves of the guideline

Human/Operator Sieve

Since the impact on operators is an are important stake in internal logistics activities, a large ergonomic axis is included in the human/operator sieve.

According to the International Ergonomics Association (IEA), "ergonomics (also known as human factors) is a scientific discipline that focuses on the fundamental understanding of the interactions between humans and the components of a system, as well as the frameworks, data and methods that are key to improving the well-being of people and the overall performance of systems or industries" [17]. Ergonomics takes a fundamental role in material handling, evaluating, and detecting if the systems involved in internal logistics are the right ones in terms of operator welfare and industry profitability. Ergonomics is directly related to ethics, and they have common targets, which are the safety, health, and job satisfaction of the employees of a firm/industry and, as a product of good practices, the sustainability of the firm/industry [23].

Operator's occupational health and safety (OHS), as well as their sense of wellbeing, are critical components of industry's social sustainability [22]. Firms must treat the human aspect as a key and valuable component of their operations, improving working conditions and developing human-centred production processes. Work ergonomics is important in this situation [22].

Many behavioural traits, such as mental effort, attention, fatigue, monotony, situational awareness and others have proven to be pragmatically valuable conceptions, where ergonomics has been involved in the struggle to define and quantify them [18].

The ethical risks that these features raise are quite clear, and are reflected, for example, in the search for sustainability (economic, human, and environmental).

Within the branch of ergonomics, physical [21] and cognitive [19, 20] ergonomics have been considered to edit the questions of the sieve. Complementary topics of this sieve are on data treatment regarding operator privacy and on the automation bias [5] that may be induced by the system. Finally, questions on social interaction are given.

Production System Sieve

This sieve analyses if the system introduced may cause unsatisfying distortion in the collective work by inducing tasks re-affectation, new hierarchical links, new dependencies between workers, distortion in skills, redefinition of employees mission or a negative impact on social interactions in the plant.

Then, it addresses the machine network evaluating whether the newly introduced system may change functioning parameters of existing systems and if it has access to possibly sensitive data of other systems.

The last aspects concerns the company information system to analyse any unintended perturbation or access to the information flow used in the plant. It is about preserving data integrity and confidentiality.

Company and Society Sieves

The company and society sieves are controlling the impact on business ethics of the company and the impact of the new system on sustainable stakes. It specifically analyses impacts on employment, energy consumption, environmental threats and relations with the socio-economic environment of the company (e.g., supplier network, client network, cities…). These last sieves allow a larger view of the impact of the company's activities and the way new systems introduction may disturb it.

The use of the guideline begins with open questions to know the context, the internal logistics activities they currently perform and those they would like to apply for. The objective is to clarify the situations that should be compared. A reference situation (before the application of a 4.0 technology) is defined and a future situation (with the application of the 4.0 technology). The set of questions is also used to elicit the company's strategy and skills regarding the I4.0 technology.

Then, the methodological guide identifies the ethical risks existing in the system being evaluated; it is important to consider how the risks found can be listed and evaluated. Therefore, it is proposed that the companies include in the guide a classification of risks according to a criticality indicator that measures the frequency and severity of the revealed risks in a FMEA fashion. In this way, ethical risks can be prioritized or simply accepted with the trace of guideline utilization.

In the present research, the new model guide has a total of 60 questions, 13 of which are open questions and 47 are YES/NO questions. It is worth mentioning that the application of this methodological guide is aimed at firms engaged in the digitization and automation of their processes and the integration of one or more Industry 4.0 technologies (automated systems, artificial intelligence, augmented reality, cloud, robots, and cyber-physical systems, among other tools).

After identifying all the questions that have a YES answer, it is proposed to make a filter of all of them and produce a list showing all the risks and ethical problems that exist around the system. In a final stage, it is proposed that the firms supervising the application of the method carry out a mitigation of the ethical risks found. This means that after visualizing the list of ethical risks, they can be classified according to a criticality indicator (frequency and severity).

4   Example of Guideline Application

After creating the guidelines model, it was decided to generate a test of the second stage (YES/NO questions) to have a first test of the question set's relevance. For this purpose, a frequently used tool in order picking operations is selected and at the same time a reference situation and a future situation are defined. The example is set as follows:

- Technology 4.0 system to be evaluated: Voice Picking.
- Selected reference situation: Paper-based order picking.
- Selected future scenario: Voice picking order picking.

Voice picking is a system that assists operators during the picking operation by giving them voice instructions [24], which they receive through a device that is equipped with a headset, a microphone, and a control mechanism (hands-free system) [25]. At the same time, the voice picking system is linked to the warehouse management software [24].

This device works as follows [24]:

- As soon as the operator starts his workday and activates the voice picking device, the specific software transmits to him all the data that is related to order preparation.
- The operator receives the instructions in his headset, such as the reference, the location, and the quantity for sampling.
- After that, the operator dictates the control code and then chooses the product reference and the requested quantity.
- When the device receives the information from the operator, the warehouse management software compares the received data with the data at its disposal and makes sure that they match.
- Finally, at the time of the last picking, the operator prints the order labels and completes the operation. [24]

4.1   Guideline for Voice Picking

– **Sieve:** Human/operator

The questions and answers that are part of the guide applied to the Voice Picking system are included in Table 1.

Table 1. Questions and answers that are part of the guide applied to the Voice Picking system

1. Physical ergonomics	YES	NO
Are the tasks to be performed repetitive?	x	
Does the task require repetitive gestures for the head, trunk, upper limbs, or lower limbs?	x	
Does the operator's body position when lifting and lowering a load (standing, sitting, bending, and balancing) create a risk of MSD (musculoskeletal disorder)?	x	
Is there a risk of forgetting the PPE (personal protective equipment) during the execution of the task? (Or PSD - Personal safety device?)		x
Is the number of displacements for the realization of the various tasks that the operator must make during the working day important (ratio displacements - loads)?	x	
Does the use of the system induce an increase in the physical workload? (e.g.: more load to process, decrease of cycle time,…)	x	
Is the workload of lifting, unloading, and transporting of loads performed by the operator significant (ratio of quantity to weight of loads)?	x	

The guideline has been applied to the Voice Picking system by answering the questions, a sample of the sieve being given in Table 1. Finally we found out that 51% of question revealed an ethical risk that should be addressed. The rate of positive answer in this example should be considered as very high showing that the implementation of a voice picking system is very dangerous from an ethical perspective. This is corroborated by the high number of critics that have been emitted against this type of system and that led to huge controverse in France around the application of this technology in Lidl warehouses. This result show that the guideline is a good alert maker on possible difficulties in a system deployment.

After analyzing and identifying the ethical risks encountered, as the last stage of the guideline, it is proposed that the firms that apply this methodological guide end up with a risk mitigation. To give an idea of how to do this final stage, it is proposed to do the following:

- Make a list of all the ethical risks encountered.
- Evaluate with indicators of criticality (frequency and severity) these ethical risks.
- Generate solutions for mitigating the ethical risk identified.

Within this study, an example of mitigation is proposed, which can be applied by organizations (Fig. 2).

In this way, the final comparison of the scenarios will provide a better visualization of how the implementation of the new I4.0 system may impact the design of the MHS. At the same time this allows to directly target the specific issue, and thus proceed to provide solutions.

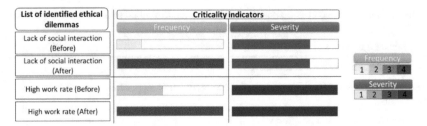

Fig. 2. Ethical issues analysis proposition according to criticality indicators (referential)

5   Conclusions and Perspectives

This paper addressed the need for further research on the impact and ethical risks of Industry 4.0 technologies implemented in the MHS of organizations.

First, a literature review was conducted, providing the context for the research. Through this, it was identified that the ethical dimension in Industry 4.0 is still a topic that has been little addressed, but is increasingly rising interest. At the same time, tools to assess the ethical dimension in I4.0 tools are scarce or practically non-existent.

Therefore, in a second part of the research, and based on the literature review, a guide capable of identifying ethical risks in MHS using I4.0 technologies is proposed. The guide extends the existing work of [16] by making it suitable and operational to study MHS. The proposal is an off the shelf tuning of the guideline directly applicable for any MH activity analysis. The updated guide was tested with an example of an I4.0 tool - Voice Picking, showing the possibility to quickly identify ethical risks.

During the research, some limitations have been identified. Among them, it is high-lighted that the sample size on which the guide was carried out is limited. It has been tested on the presented academic example and within one single company. More utilization and a deeper usage analysis are needed to assess its applicability and exhaustivity (namely concerning the question lists).

An important perspective of this work is to focus on the risk assessment stage of the methodology. The identified risks must be quantified to permit a credible risk management of ethical aspects within projects. Several ideas can be applied; firstly, standardized scale of evaluation of the likelihood and severity must be proposed for each risk category (e.g., the 47 YERS/NO questions). Secondly, methods should be defined to evaluate the score of a situation on the proposed scale. To do so, we made first a development using classical ergonomics analysis techniques, such as APACT grid) for the ergonomics-related risks revealed in the guide. We also developed some simulation models for risks related to production intensity and human interaction within the workshop which cannot be presented here. Thirdly, methods must be defined to design the company's policy on ethical risk acceptance. Namely, as in any risk management policy rules have to be formulated to define the acceptability frontier of the elicited risks.

So, it is recommended for future research that the methodological guide should be applied in conjunction with other tools such as simulation and calculation of indicators for rapid identification of ethical risks and their proposed mitigation.

Acknowledgments. The authors thank the French National Research Agency (ANR) that funded this research under the LADTOP project (grant ANR-19-CE10-0010-01). The authors would also like to thank the industrial companies for their participation in the field studies.

References

1. Bai, C., Dallasega, P., Orzes, G., Sarkis, J.: Industry 4.0 technologies assessment: a sustainability perspective. Int. J. Prod. Econ. **229**, 107776 (2020)
2. Berrah, L., Cliville, V., Trentesaux, D., Chapel, C.: Industrial performance: an evolution incorporating ethics in the context of industry 4.0. Sustain. **13**(16), 9209 (2021)
3. Bonsón, E., Lavorato, D., Lamboglia, R., Mancini, D.: Artificial intelligence activities and ethical approaches in leading listed companies in the European Union. Int. J. Account. Inf. Syst. **43**, 100535 (2021). https://doi.org/10.1016/j.accinf.2021.100535
4. Chauhan, C., Singh, A., Luthra, S.: Barriers to industry 4.0 adoption and its performance implications: an empirical investigation of emerging economy. J. Clean. Prod. **285**, 124809 (2021)
5. Goddard, K., Roudsari, A., Wyatt, J.C.: Automation bias: Empirical results assessing influencing factors. Int. J. Med. Inform. **83**(5), 368–375 (2014)
6. Kerin, M., Pham, D.T.: A review of emerging industry 4.0 technologies in remanufacturing, J. Clean. Prod. **237**, 117805 (2019)
7. Luthra, S., Mangla, S.K.: Evaluating challenges to Industry 4.0 initiatives for supply chain sustainability in emerging economies. Process Saf. Environ. Prot. **117**, 168–179 (2018)
8. Margherita, E.G., Braccini, A.M.: Managing industry 4.0 automation for fair ethical business development: a single case study, Technol. Forecast. Soc. Change **172**, 121048 (2021)
9. Müller, J.M., Kiel, D., Voigt, K.I.: What drives the implementation of Industry 4.0? The role of opportunities and challenges in the context of sustainability. Sustainability **10**, 1 (2018)
10. Neumann, W.P., Winkelhaus, S., Grosse, E.H., Glock, C.H.: Industry 4.0 and the human factor – a systems framework and analysis methodology for successful development, Int. J. Prod. Econ. **233**, 107922 (2021). https://doi.org/10.1016/j.ijpe.2020.107992
11. Ricoeur, P.: Soi-même commeun autre. Sciences humaines. Seuil (1990)
12. Rossini, M., Costa, F., Tortorella, G.L., Portioli-Staudacher, A.: The interrelation between Industry 4.0 and lean production: an empirical study on European manufacturers. Int. J. Adv. Manuf. Technol. **102**(9–12), 3963–3976 (2019). https://doi.org/10.1007/s00170-019-03441-7
13. Glock, C.H., Grosse, E.H., Neumann, W.P., Feldman, A.: Assistive devices for manual materials handling in warehouses : a systematic literature review literature review. Int. J. Prod. Res. **59**(11), 3446–3469 (2021)
14. Soufi, Z., David, P., Yahouni, Z.: A methodology for the selection of Material Handling Equipment in manufacturing systems. Proc. INCOM 2021 **54**(1), 122–127 (2021)
15. Trentesaux, D.: Transformation digitale en génie industriel : enjeux éthiques du monde de la recherche, CIGI - Qualita21: Conférence Internationale Génie Industriel, Grenoble (2021)
16. Trentesaux, D.: Ensuring Ethics of cyber-physical and human systems: a guideline. In: Trentesaux, D., Borangiu, T., Leitão, P., Jimenez, JF., Montoya-Torres, J.R. (eds.) Service Oriented, Holonic and Multi-Agent Manufacturing Systems for Industry of the Future. SOHOMA 2021. Studies in Computational Intelligence, vol. 987, pp. 223–233. Springer, Cham (2021). https://doi.org/10.1007/978-3-030-80906-5_15
17. *** (2022). Usabilis - Conseil UX et ergonomie digital. https://www.usabilis.com
18. Dekker, S.W.A., Hancock, P.A., Wilkin, P.: Ergonomics and sustainability: towards an embrace of complexity and emergence. Ergonomics **56**(3), 357–364 (2013)

19. Colver, M.: Cognitive ergonomics: supporting effective use of predictive analytics using ethical mental frameworks. New Dir. Inst. Res. **2019**(182), 51–61 (2019)
20. Bergman, M.W., Berlin, C., Chafi, M.B., Falck, A., Örtengren, R.: Cognitive ergonomics of assembly work from a job demands – resources perspective : three qualitative case studies (2021)
21. International Ergonomics Association. Definition and Applications of ergonomics (2022). https://iea.cc/what-is-ergonomics/
22. Gualtieri, L., Palomba, I., Merati, F.A., Rauch, E., Vidoni, R.: Design of human-centered collaborative assembly workstations for the improvement of operators' physical ergonomics and production efficiency: a case study. Sustainability **12**(9), 3606 (2020). https://doi.org/10.3390/su12093606
23. Corlett, E.N.: Ergonomics and ethics in a changing society. Appl. Ergon. **31**(6), 679–683 (2000)
24. Pixisoft. Logistique: qu'est-ce que le voice picking ? https://www.pixisoft.com/logistique-quest-ce-que-le-voice-picking%E2%80%89/#:~:text=Le%20voice%20picking%20est%20un,entrep%C3%B4t%20(ou%20logiciel%20WMS).https://www.pixisoft.com/logistique-quest-ce-que-le-voice-picking%E2%80%89/(2022) Accessed 2 Jun 2022
25. Industry Search. Honeywell - Vocollect Talkman A730 Solution (2022). https://www.industrysearch.com.au/honeywell-vocollect-talkman-a730-solution/p/74538. Accessed 20 Jun 2022
26. Trentesaux, D., Caillaud, E.: Ethical stakes of industry 4.0, IFAC world congress 2020. IFAC-PapersOnLine **53**(2), 17002–17007 (2020). https://doi.org/10.1016/j.ifacol.2020.12.1486

Ethical Internal Logistics 4.0: Observations and Suggestions from a Working Internal Logistics Case

Marc M. Anderson[1]([⊠]) and Karën Fort[2]

[1] LORIA, UMR 7503, Université de Lorraine, Campus Scientifique, 615 Rue du Jardin-Botanique, 54506 Vandœuvre-lès-Nancy, France
`marc.anderson@inria.fr`
[2] Sorbonne University, LORIA, UMR 7503, Université de Lorraine, Campus Scientifique, 615 Rue du Jardin-Botanique, 54506 Vandœuvre-lès-Nancy, France
`karen.fort@loria.fr`

Abstract. In this paper we present our experiences and insights from a Use Case in heavy industry, where OCR text recognition is combined with algorithms to correctly identify labels for additives to be introduced into a production process. Ethical issues are presented relative to the effects of the Use Case upon the shop floor operators using the new technology. We then discuss recommendations given and our success in getting them implemented. An argument follows, regarding what we view as the source of many of the ethical issues: the unreflective acceptance of Industry 4.0 and Internal Logistics 4.0 as a generalized and idealized 'plan' to which technological development and the human operator have to adapt. We contrast this to an approach where the needs of the human in the work context would drive and limit Internal Logistics 4.0 development as a set of gradual improvements tailored to the worker's situation.

Keywords: Ethics · Internal Logistics 4.0 · Industry 4.0 · Artificial Intelligence · Human-centered manufacturing · Material identification · Supporting technology

1 Introduction

Internal logistics will be a very important element of Industry 4.0, and the more so since many of the interactions of humans with autonomous systems being developed for Industry 4.0 are located precisely in the realm of logistics. Logistics involves the movement of materials or - in the Industry 4.0 context - information, between distinct parts of a process. Logistics creates linkages in areas where the regularity suitable to mere machine linkages is absent so that human participation will be needed for a long time to come.

In this article, we examine some particular issues surrounding a working case in internal logistics, a Use Case in an ongoing project in which we are involved as ethical advisors. Following a brief overview of ethical research related to Internal Logistics 4.0, we proceed in four sections. In Sect. 3 we describe the Use Case generally, in Sect. 4

© The Author(s), under exclusive license to Springer Nature Switzerland AG 2023
T. Borangiu et al. (Eds.): SOHOMA 2022, SCI 1083, pp. 317–328, 2023.
https://doi.org/10.1007/978-3-031-24291-5_25

we describe the various ethical issues and challenges encountered in the Use Case, in Sect. 5 we describe recommendations that we have made relative to those issues, and in Sect. 6 we generalize our insights from our efforts, to provide future research pathways for those engaged in the ethics of Internal Logistics 4.0.

2   Overview of Ethics of Internal Logistics 4.0

The ethics of Internal Logistics 4.0 specifically is a relatively unresearched area. It overlaps with ethical research into Industry 4.0, the latter being itself not well developed. Trentesaux, Caillaud and others have made various opening moves in the area of Industry 4.0 ethics, looking in particular at the issue from the perspective of the engineer [4]. But as they admit, there is "a critical lack of contributions in that field". There is an even greater void in the area of ethics of internal logistics and, to our knowledge, there is no research which specifically focuses on the latter. Authors of [5] outline a number of issues that are definitely ethics related, e.g., human supervisory roles, privacy rights, psychological effects on operators and job loss. But their list is only a subset of issues which require ethical engagement, and they go no further than an outline with no explicit reference to ethics. Another technical outline [14] notes the problem of job losses and the relieving of employees as benefits, but goes no further and makes no link to ethics. The best consideration we have found of the issues involved – although it addresses Industry 4.0 generally rather than Internal Logistics 4.0 specifically – is in [3], which definitely take a worker centred view which agrees in part with the view we argue for below. Authors of [1] address internal logistics indirectly in the context of practical suggestions toward ethical Industry 4.0.

3   An Internal Logistics Use Case

Our project, a partnership between research and industry, develops various types of AI systems to be used in heavy industry factory settings. The Use Case described is one of many initially selected for development. The specific logistics area involved in this Use Case would be categorized under the dimension of Material Identification and subdivision Method of Identification [16] or under Picking/Supporting Technologies/Information Flows Management [5].

Within a factory setting, bags of additives are combined with several main materials to create a final product. The additives change regularly depending upon production needs. The bags, each weighing on the order of a ton, are retrieved from a stockroom by mechanical means. They are selected by an operator and brought to a hopper zone, where they are placed upon lifts to be hoisted over specific hoppers through which the additives are released into the production process. Bag labels describing the additives are sometimes unclear and labelling is often inconsistent between suppliers. Currently, the operator carries the bag label to a control room several floors down to be verified by a control room operator before the bag contents begin to enter the process. The factory is a busy place and sometimes the operator is delayed. Occasionally this coincides with the wrong bag being selected, which leads to trouble in the production process.

Shortening the time for verification of the bag label and eliminating any potential error and delay in this part of the process is the goal, by combining OCR (Optical Character Recognition) with an algorithm that can decipher unclear labels. The text recognition is to be carried out first by means of a hand-held tablet of a type already used, and then more experimentally, by a voice activated assisted reality helmet mounted wearable. The OCRed text is sent to one of the tech development partners, processed by the algorithm, and results are sent back to the tablet or helmet to be confirmed by the operator and forwarded to a central computer handling the production process.

The human operator's original role in this logistics process combines manipulation with material identification. The operator is part of an operator team who's central 'gathering place' is the control room, in a nearby building, several floors below. Normally the operator walks down to the control room, interacting with colleagues. The environment is noisy, requiring special ear protection, and dangerous. Quick evacuation may be required under certain conditions of which the operator must remain aware. Different operators fill the role at different times. The new technology will change the operator's role and they will no longer have to walk down to the control room with the labels. They will also have to deal with any mistakes made by the algorithm, either correcting manually through the tablet or helmet HMIs (Human-Machine Interfaces), or rejecting the algorithm's suggestions.

4 Ethical Issues and Challenges Encountered

What counts as an ethical issue? In our experience related to this Use Case, the role of ethics shifts from one of normative theorizing to practical contextualization and application. Historical developments in mass manufacturing have inexorably removed the human from the mass aspect of production processes [15]. But since the automated aspects of production processes are imperfect, the human is still needed. Ethical efforts are thus best directed at the humans remaining in the production processes, whether shop floor operators, maintenance workers, or machine specialists. Therefore, our concern is with the initial context and role of the operator, as compared with the changes planned by the tech developer and the factory management for the context and role of that operator. We constantly ask: 'what is planned, what harm might it do to the operator, and how can we alleviate that harm?'.

4.1 AI for Public Relations

One of the issues we have uncovered is the tendency toward incorporating AI and other technologies merely for public relations reasons. The project is a research project, but expected to produce at least some results which are marketable. As observers from various fields have noted, e.g., [9, 13], AI is a much-hyped technology. Its inclusion is thus arguably an obvious choice to improve marketability for eventually marketable results. Given that the project is research based, there is an impetus to 'experiment' with new techniques and technologies. Yet, since those techniques and technologies are open to being adopted and marketed, the call to research could serve as an excuse for taking ethical liberties with regard to marketing.

In our Use Case there are several levels of devices proposed for HMIs. The first level devices are selected to solve the problem at hand and probably be actually used. The next level is selected so as to experiment with more complex technologies. An AI application integrated on a tablet-based HMI was developed for the first level, tested in working conditions, and expected to be adopted. Once this was well underway, the HMI focus shifted to potential wearable helmet devices, virtual or assisted reality. *The operator is thus expected to adapt to a new technology, and then potentially adapt once again to a still more complex technology, which is stressful.* It also *puts the focus on developing and displaying new technologies primarily, regardless of potential operator inconvenience.* It leads naturally to asking: "what is the primary purpose of adding the technology?" If it is marketing the abilities of the tech developer, there is an ethical danger of related human effects being rendered secondary.

4.2 Overlooking or Misusing Technical Specifications and Product Warnings

Product specifications raise other issues. Off the shelf technology products such as the HMIs mentioned have technical specifications and product warnings, which get bound up in ethical issues in at least two ways. Firstly, *they contain information and warnings that may be overlooked, regarding potential inconvenience and harm which the product can cause the user,* and secondly, *they can serve as an objective impediment to adapting to the context of the user.* We encountered both. We took time to read the product warnings for the proposed HMIs, which showed us that the virtual reality wearable in particular might cause a wide range of physical or psychological problems. On the other hand, we have seen the technology developers sometimes fall back upon technical specifications - and the inbuilt architecture of the device - to argue that certain changes cannot be made. The language of the operators became an issue here, because the software base on which the OCR/AI application was developed was only devised for a set number of languages, and not the primary language of the operators. The solution was using the device in English. The operators can speak English, but their accent creates difficulties. Authors of [12] have discussed both the problem and advanced potential solutions to it. As in many devices, English serves as one of the default languages for use, but here this fact comes to serve as an objective impediment for not being able/willing to adapt the device to the operator's native language.

4.3 Deferring to the Technology Developer

Whether the OCR/AI application will actually be used on the factory floor has been difficult to clarify explicitly. The managers seem to adopt a 'wait and see' approach, i.e. *if* they can get it to work sufficiently and *if* the operators accept it, then it will be used. This alternates with a 'we need this at any cost' approach. Thus, the industrial partner tends to defer to the technology developer in terms of particulars: 'if you can get it working right we need it,' which implies: 'how you get it done is more or less up to you.' These attitudes *put the operator's role at the mercy of the technological development plan and the specific technology.* They also raise *the danger of the redundancy of technologies,* i.e., the operator may have to use both the new and old systems at once, since 'we can't afford to make mistakes.' This doubles the operator's work, increasing the chances of confusion between the different systems.

4.4 Lowering the Reliability Bar

A further issue is the tendency to lower the bar regarding reliability. KPIs (Key Performance Indicators) are determined and prioritized in the project first in the initial project proposal (generally) and then more specifically through a dedicated project task. Performance of AI solutions is the first priority - assessed in KPIs related to production efficiency, product quality and resource consumption, among others - but the impact on workers in terms of user acceptance of AI solutions is also considered. Going beyond the latter, we attempted to get *worker side* KPIs instituted. Questioning how often the application would correctly read the bag labels, we asked early in the Use Case for a first reliability estimate, which was about 80%. As months passed however, and we requested updates and discussed potential errors, the estimated reliability became successively 85%, 90–95%, 98%, and finally 99.5%. Meanwhile, the issue of setting a reliability KPI, i.e., 'what is the minimum error rate for the application to be considered reliable enough to use?' tended to be pushed further away. To admit that the technology will make errors at all seems difficult for the developers to do. Thus, questioning reliability seems to invite lowering the bar, until the working assumption becomes: 'the technology will nearly always be right.' Later we learned that the industrial partner had provided very little data yet to make a reasonable reliability assessment. There are thus three interwoven issues here: *failure to adopt strong KPIs related to human operator inconvenience* (the default assumption is that the human will 'take up the slack' when the system makes an error); *a lack of enthusiasm on the industrial side in providing data to support setting and assessing such KPIs*; and finally, *a gradual lowering of the bar for reliability, abetted by the first two issues.*

4.5 Lack of Access to Operators and Workers

A number of ethical issues arise from not having free access to the operators. In a human centred approach, not being able to speak to the workers or observe their testing of the application on the HMIs means *not being able to directly assess their comfort with those technologies, either explicitly with regard to KPIs or implicitly through more subjective observations about worker satisfaction.* It also means *not being able to verify whether the operators are consulted about the developments* by the industrial partner or tech developers, before, during, or after. The delineation between business as a private concern, with workers subject to contractual obligations and management orders, as long as the obligations are legal, is a *de facto* state of affairs inimical to ethical assessment. Bluntly: we often cannot easily find out what we need to know in order to make ethical assessments. This issue is not particular to ethical assessment. In [7] it is noted that government often lacks information about both business and technology with which to make regulation effective. But ethics does not even have the types of leverage which government wields, which is worse.

4.6 Responsibility for Errors

Responsibility for errors remains an ongoing worry also. In case KPIs are adopted and faithfully measured, neither of which is certain, still *the ultimate responsibility and the question of what to do when an eventual inevitable error occurs*, are important. Our experience in discussing this with the technology developers and company management is that the question is left in a grey zone. We understand their reticence; admitting to inevitable errors is admitting to an imperfect product. Yet we want to know how the operator should deal with an error when it does happen. Not preparing for such situations in advance is another - unethical - way of developing Internal Logistics 4.0 by putting the plan and ideal of the technology first, and leaving the uncertainties to be absorbed by human operators who have little or no control over the implementation of the technology. It amounts to: 'here you go, make the best of it.'

4.7 Physiological and Psychological Issues

Finally, physiological and psychological issues round up the issues encountered. We observed the work environment of the Use Case in the control room and the factory to be collegial and smooth running. Introducing the new technology will certainly change operator team interactions. Even assuming the technology eventually works very well, the trip to verify the bag label will be eliminated. This is a loss of opportunities for physical movement, and a loss of human contact with co-workers. The change is only a small one, but many such small changes - making up part of the larger vision of Internal Logistics 4.0 - will tend to deaden collegiality and make human interaction less smooth. If the technology works intermittently or poorly, the situation could be much worse, with the frustrations of correcting for the errors of the application, and the stresses of clarifying the situation and delays to a distant work colleague. Plant noises may also cause difficulties in using the voice activated assisted reality helmet. The latter may also add to visual or balance issues, typical of such products and typically included in product specifications, and here aggravated by the fact that the work environment is both dangerous and more expressly 'three dimensional.'

5 Our Ethical Recommendations and Responses to Them

5.1 Clarifying and Questioning

The ethical issue of adopting new technologies for public relations purposes is built into such projects, thus difficult to address. Our approach was to continually try to clarify what was being developed, how long different stages would take, and whether the device was actually in use. We only achieved partial success. It was difficult to get clear and timely information from the developers and industrial partner. We recommended considering whether the AI service should be used at all, reasoning that the ultimate goal was to reduce the occasional time delay of operator verification of bag labels which sometimes resulted in missing the fact that a wrong bag had been selected. Estimating time delay due to AI error and comparing this to original time delay occurrences which caused the problem, was suggested. If the former were just as great, then the installation of the

AI service *at all* should be reconsidered. This conflicts with one of the project goals however - to test and research new AI technologies for industry - and also with a goal of the lead developers: showcasing new technology developments. The recommendation has not been taken up so far.

5.2   Reviewing Product Specifications Thoroughly

Overlooking product specifications was addressed by reviewing the product documentation of proposed off the shelf devices. We noted product warnings and potential issues and raised them in meetings. On this basis one planned off the shelf fully virtual reality wearable was removed from consideration, and replaced by an assisted reality helmet wearable. Hence this approach achieved some success. The tendency of the developer to fall back upon product specifications as reasons for not adapting the device in certain ways is more difficult to address. We did not succeed in convincing the lead developers to change off the shelf products that they had already committed to. This highlights one of the weaknesses of internal logistics 4.0: that - as in the notion of Industry 4.0 - it assumes fixed components linked by various processes. If components such as off the shelf products have ethically problematic aspects built into them already from their design phase, e.g., default language assumptions, then these problems bleed into other areas of the internal logistics process. The assumption is that the human user will adapt to the problems already built into the system, accordingly; those built in problems are simply accepted rather than addressed.

5.3   Direct and Early Contact with Operators

The issue of actual use on the factory floor was engaged variously. We kept asking this question explicitly at different stages, but answers remain vague. We specifically recommended that the lead developer works directly with the operators from an early stage. It would have the benefit of making the service and app better and testing HMI appropriateness in real conditions, so that if actual factory floor use occurred, the groundwork would already be done. A related recommendation was to gather feedback from the operators early. A further recommendation was to retain the role of colleagues in the control room in verifying the bag labels, but in a modified way, so as to avoid redundantly retaining old tasks and technologies together. The first recommendation has been only partially taken up and the second and third not at all so far.

5.4   Preparing for Inevitable Errors

We addressed the lowering of the bar with regard to reliability in a roundabout way. Since we only had estimates of the reliability/error rate of the OCR/AI services at the beginning, and also later, we made a number of recommendations based on the assuming that at some point the operator would have to deal with an error, no matter if the likelihood was only 0.5%. These recommendations included: adopting a formal protocol of the steps the operator would take in case of error, undertaking a logical conceptual analysis of the points at which error might occur, and early recommendations for the lead developer to work directly with the operators in designing the HMI. These recommendations have not yet been adopted. As noted earlier, the tendency has been to avoid the issue, by saying that the technology "will almost always work."

5.5 Pushing for Open Operator Developer Interaction

The difficulty of not having regular access to the operators is partly a result of the necessities of work conditions. Introducing outside observers into an ongoing production process is also tricky. We visited the factory and observed the workers in their day to day activity, but the ideal of observing the workers *using* the new AI services and discussing their satisfaction has not yet been possible. We receive second hand affirmations from company managers that the workers have tested the new technology, but we have no simple means of verifying this. We remain unsure how to address this issue better. A recommendation that the ethics team have regular access to speak with operators, and participate in deliberately scheduled meetings between operators and developers would seem an obvious path. This would depend upon operator willingness.

5.6 Clarifying *Who* is Responsible and AI as a Tool

The issue of responsibility for error was especially important and resulted in a number of recommendations. We advised formal clarifications regarding *who* was to be responsible for checking the accuracy rate of the OCR/AI service and what type of operator feedback the AI training would require and for how long. We recommended not taking the control room operator out of the label verifying process. Trial stages were recommended, in cooperation with the operators. Finally, we presented reasoning and a recommendation to urge viewing the AI services as a *tool* to augment operator capabilities rather than viewing the operator as a safeguard, or 'AI supervisor.' Regarding the first recommendation, partial clarifications were made. On the other hand, the recommendation regarding retaining the control room operator's role was not adopted, and we remain unsure of trial stages implementation.

5.7 Testing Under Real Conditions and Monitoring Effects

Physiological and psychological issues were engaged in various practical recommendations. We recommended providing a holster to the operator to offset the weight of the OCR scanning tablet. Testing the HMIs in actual conditions with work gloves on was also recommended. To address the changes in role and human interaction, we recommended directly asking the operator team about their satisfaction, and regular monitoring of the cohesion of the larger operator team in trials or after actual introduction of the new technology. Commitments to adopt the first two recommendations were made in deliverables. The other recommendations have not yet been adopted.

6 General Discussion of Insights Gained and Pathways Forward

The causes of the ethical issues encountered are interwoven in a larger theme, which is the very notion of Industry 4.0 itself. The idea of Industry 4.0 and Internal Logistics

4.0 depends upon accepting *a conceptual level social engineering of the workplace.* Instead of a process of industry slowly acting from the actual situation in the workplace to improve the human work experience, Industry 4.0 is laid out as a conceptual vision of a future workplace. The technology is supposed to develop to fill out the vision. The authors of [17], for example, speak of implementation strategies where the point of reference is Industry 4.0; a company's internal logistics must be ready *for* the latter. To some extent the technology can fill out the vision. Engineering the human to fill it is more difficult. At best, this approach is unethical, in not addressing humans needs. Instead, the approach *creates* those needs. It syncs very well with a public relation driven development of technology, however. This is so much the case now that simply coining buzzwords generate hype. The iterative buzzword, e.g., Industry 5.0, and now Industry 6.0 [10], can be advanced as a goal, before Industry 4.0 is even close to a 'first draft' in terms of actual implementation. The authors of [3] note the technology focus of Industry 4.0, and advocate moving on to Industry 5.0, in which humans are presumed to be present. But this does not prevent the ideal of Industry 5.0, from shifting the emphasis of development away from where it arguably ought to be. It is still development *toward/for* Industry 5.0, rather than development *from* the human worker.

Our own project uses the Industry 4.0 notion thus helping normalize this vague 'technology first' approach, which creeps into the development process in various ways, e.g., in accepting the limitations of off the shelf technologies - the latter allow the development to proceed at the speed which is sought. The rush to get things done in technology development is hype and public relations driven to a large degree, and inherently contrary to ethics. Again, the 'technology first' approach facilitates an 'if you can make it work we'll take it' attitude to technology development, which is the driver of change, on the part of manufacturing management. It also gives power to technology developers to do whatever it takes to get the technology working.

Recognizing this emphasis, it becomes easier to see why KPIs related to human worker satisfaction are not typically considered. Human participation is messy and not easily compartmentalized under generalized goals. Avoiding such KPIs, or gradually watering down the more quantitative measurements that could be attached to them, guarantees tidy results. The open and continual access to workers that such human sided measurements would take, does not fit within what [8] has called "corporate logics and incentives," so it is easy to brush aside. Difficulties in admitting errors also arise here. Since Industry 4.0 or Internal Logistics 4.0 masterplans are laid out in advance in a generalized way, proceeding by trial and error runs counter to them. The latter approach would require that technology developers admit in advance and by default that products are often unreliable, which damages the marketability of the product. In particular, this creates an incentive to overlook types of error which only inconvenience the worker.

The notion of human centredness is sometimes raised as a sort of 'proto-ethical foundation.' The authors of [5] mention human-centredness, but not ethics. But without sustained reflection this concept does not advance us toward an ethical outcome. Authors of [11], for example, develop a third axis representing the Human-Technology Relation in a three-dimensional taxonomy scheme for Internal Logistics. Ostensibly this is a good conceptual base to develop human-centredness upon. The delineations of the axis however - automation and support - show that the human technology relation is viewed

M. M. Anderson and K. Fort

from the technology side. The technology automates some human tasks and supports others. The direction is: *what can the technology do for the human?* This is precisely not a human-centred approach, i.e., one proceeding from the human to the technology. This approach proceeds from the technology. An ethically amenable approach would reverse it: *what does the human seek, need, or desire, in the technology* and, taking a step further: *what can the human contribute to the community developing the technology* [2]. The current 'for the human' emphasis leads easily to declarations and assumptions about the necessity of the technology: 'we will build it and we will find someone to use it.' What if the human doesn't want the technology on offer?

It is not enough that the human be statically somewhere within - ostensibly at the centre of - the development process. There has to be a process 'to or from,' with, we suggest, a 'from the human,' approach, being the most promising. Viewed as a component, even though 'centred' within the great plan, the human becomes secondary, being moulded to the plan, rather than the plan being adapted to the needs issuing out of the human. Aspects of [16] illustrate the problem. In their framework for a logistics maturity model, they outline five main dimensions: *manipulation, storage, supply, packaging,* and *material identification.* They outline then the characteristics of the subdimension of *manipulation technology,* by level, through six levels, i.e., from no application of Internal Logistics 4.0 elements to a full application of such elements. Yet in all of these levels - including the lowest - there is only one explicit mention of human involvement "Human loads and unloads material," repeated at five levels, and verbatim in three of those, while at the sixth and highest level of logistics automation we get suddenly: "One only oversees here." The impression is that the changes in automation occur all around the human operator without affecting the latter's role, until abruptly the operator transforms into a mere supervisor of the system. Not only is this internally inconsistent, since an operator truly disconnected from the development and implementation of Internal Logistics 4.0 in the production process is in no position to appreciate the system he must suddenly supervise, but it provides no room for a movement 'from' the operator regarding the changes occurring all around.

Given the above issues, we suggest that Internal Logistics 4.0 should be advanced along the following lines, in order to be ethical.

- Internal logistics 4.0 - and Industry 4.0 - should be re-envisioned as a set of improvements relative to a thorough analysis of a given particular industrial context. Instead of a high-level generalization of where we must get to, we would work from the question: "given this particular human work context, how can we improve it for the worker involved?"
- The first stage of this process toward an ethical internal logistics would be a sustained and patient interaction between workers and developers, where the latter would outline their needs relative to what could improve their work experience.
- That first stage requires the normalization and acceptance of regular access to the workers by developers, ethicists, and related specialists.
- KPIs that measure what is satisfying to the human worker relative to their particular work context would need to be developed and implemented.
- Off the shelf products should be modified to the needs of the humans involved and abandoned if they cannot be, or if they cannot be brought up to a sufficient level of

quality, as envisioned for example, by the EU Commission's High-Level Guidelines for Trustworthy AI, or others.

- The need for the technology should only be built upon the needs of efficiency, insofar as those needs can also be integrated with the needs of the humans who will use the technology, i.e., helping the human do a better job should be the first consideration in adding new logistics technologies to the workplace; the push for efficiency should be built upon and serve the former goal.

7 Conclusion

Our goal was to describe the ethical issues in developing internal logistics solutions in an actual working context, discuss what we have encountered, and offer a number of general but practical suggestions toward making internal logistics ethical. We have presented various issues encountered in a heavy industry Use Case where we act as ethical advisors. They include: developing technologies primarily for public relations, overlooking product specifications or leaning upon them to avoid better adaptations to worker needs, ambiguity regarding real world use of the technology and worker responsibility for error, lack of access to workers, and various physiological and psychological issues. We have argued that all of these issues arise from an outlook where Internal Logistics 4.0 (as a sub-field of Industry 4.0) is conceptualized as a general goal and then inexorably advanced toward without considering particular industrial contexts more deeply as beginnings. The conditions for that deeper consideration, beginning with the worker's point of view, are the conditions for an ethical Internal Logistics 4.0, and we present some of those conditions.

Future research pathways include: developing methods to facilitate worker and tech developer interaction, showing how strong KPIs proceeding from worker assessments of their own needs can be developed, and exploring ways to change the outlook of industry with regard to beginning from the worker. We think it will require a change in the idea of Industry 4.0, and a rethink of the need to iterate new Industry *X.0* buzzwords ultimately, to develop the former ethically. We are optimistic that it could be done.

Acknowledgements. AI-PROFICIENT has received funding from the EU's Horizon 2020 research and innovation program under grant agreement No. 957391.

References

1. Anderson, M.M., Fort, K.: From the ground up: developing a practical ethical methodology for integrating AI into industry. AI & Soc. (2022). https://doi.org/10.1007/s00146-022-01531-x
2. Anderson, M., Fort, K.: Human where? A new scale defining human involvement in technology communities from an ethical standpoint. Int. Rev. Inf. Ethics. **31**, 1 (2022)
3. Berrah, L., Cliville, V., Trentesaux, D., Chapel, C.: Industrial performance: an evolution incorporating ethics in the context of industry 4.0. Sustainability **13**, 9209 (2021)
4. Trentesaux, D., Caillaud, E.: Ethical stakes of Industry 4.0. 21st IFAC World Congress, Berlin, Germany (2020). https://doi.org/10.1016/j.ifacol.2020.12.1486

5. Cimini, C., Lagorio, A., Romero, D., Cavalieri, S., Stahre, J.: Smart logistics and the logistics operator 4.0. IFAC-PapersOnLine **53**, 10615–10620 (2020)
6. EU High-Level Expert Group on Artificial Intelligence. Ethics Guidelines for Trustworthy AI (2019). https://digital-strategy.ec.europa.eu/en/library/ethics-guidelines-trustworthy-ai
7. Ferretti, T.: An institutionalist approach to AI ethics: justifying the priority of government regulation over self-regulation. Moral Philos. Polit. **9**(2) (2021). https://doi.org/10.1515/mopp-2020-0056
8. Green, B.: The contestation of tech ethics: a sociotechnical approach to ethics and technology in action. ArXiv, abs/2106.01784
9. Jakobsson, P., Kaun, A., Stiernstedt, F.: Machine Intelligences: An Introduction. Culture Machine. vol. 20 (2021)
10. Annanperä, E., et al.: From industry x to industry 6.0: antifragile manufacturing for people, planet, and profit with passion. In: Kuosmanen, P., Villman, T. (eds.), Allied ICT Finland (AIF). Business Finland AIF White Paper No. 5 (2021)
11. Lagorio, A., Cimini, C., Pirola, F., Pinto, R.: A taxonomy of technologies for human-centred logistics 4.0. Appl. Sci. 11, 9661 (2021). https://doi.org/10.3390/app11209661
12. Radzikowski, K., Wang, L., Yoshie, O., Nowak, R.: Accent modification for speech recognition of non-native speakers using neural style transfer. EURASIP J. Audio Speech Music Proc. **2021**(1), 1 (2021). https://doi.org/10.1186/s13636-021-00199-3
13. Rossi, D., Zhang, L.: Landing AI on networks: an equipment vendor viewpoint on autonomous driving networks (2022). arXiv:2205.08347, https://doi.org/10.48550/arXiv.2205.08347
14. Schmidtke, N., Behrendt, F., Thater, L., Meixner, S.: Technical potentials and challenges within internal logistics 4.0. In: 4th International Conference on Logistics Operations Management (GOL), pp. 1–10 (2018)
15. Smil, V.: Grand Transitions: How the Modern World Was Made. Oxford University Press, Oxford (2021). ISBN: 978-0190060664
16. Zoubek, M., Simon, M.: A framework for a logistics 4.0 maturity model with a specification for internal logistics. MM Sci. J. **1**, 4264–4274 (2021). https://doi.org/10.17973/MMSJ.2021_03_2020073
17. Zoubek, M., Simon, M.: Evaluation of the level and readiness of internal logistics for industry 4.0 in industrial companies. Appl. Sci. **1**(13), 6130 (2021). https://doi.org/10.3390/app111 36130

Predicting Medicine Demand Fluctuations Through Markov Chain

Daniel Vélez, Siao-Leu Phouratsamay, Zakaria Yahouni$^{(\boxtimes)}$, and Gülgün Alpan

Université Grenoble Alpes, CNRS, Grenoble INP G-SCOP, 38000 Grenoble, France
{daniel.velez,siao.phouratsamay,zakaria.yahouni,
gulgun.alpan}@grenoble-inp.fr

Abstract. Nowadays, the healthcare sector is rapidly changing. Hospitals are facing limited budgets and high costs. The logistics activities of the hospitals in France (stock management, delivery, etc.) represent one of the highest cost components. The logistics costs can be reduced through an optimized inventory management system. The inventory optimization is strongly dependent on the accuracy of the demand prediction of medicines. Many factors influence this demand, such as seasonality, hospital size and location, etc. The objective of the paper is to propose a Markov chain model to estimate medicine demand fluctuations for hospital logistics in France. An analysis of the first experimental results is proposed to assess the effectiveness of the method. This preliminary result could contribute to the management of hospital inventories.

Keywords: Inventory management · Demand forecasting · Hospital pharmacy · Stochastic models · Markov chain

1 Introduction

The healthcare sector is rapidly changing, especially since the 90s. Organizations in the sector are promoting projects in areas such as care-related logistics, information systems and quality of care to cope with characteristics such as increased competition, patient influence and the need for a more efficient and effective service [1].

The logistics-related activities can account for about 46% of a hospital's operating budget in the United States, for instance [2]. The average health expenditure in France in 2019 was 11% of gross domestic products. The main logistics activities include planning, designing, implementing, and managing material [3]. These activities seek to support functions such as inventory management, procurement, and distribution [4]. One of the highest hospital costs is related to these logistical activities, for example in OECD[1] it represents 30% and is the second largest cost according to Volland et al. [5]. More precisely, in the health sector, inventory costs are estimated at 10%–18% [1].

---

[1] The OECD's 38 members are: Austria, Australia, Belgium, Canada, Chile, Colombia, Costa Rica, Czech Republic, Denmark, Estonia, Finland, France, Germany, Greece, Hungary, Iceland, Ireland, Israel, Italy, Japan, Korea, Latvia, Lithuania, Luxembourg, Mexico, the Netherlands, New Zealand, Norway, Poland, Portugal, and Slovakia.

© The Author(s), under exclusive license to Springer Nature Switzerland AG 2023
T. Borangiu et al. (Eds.): SOHOMA 2022, SCI 1083, pp. 329–340, 2023.
https://doi.org/10.1007/978-3-031-24291-5_26

The logistics costs in hospitals can be reduced, in some cases even at half, by implementing efficient logistics management [5]. That is why hospitals have to generate better returns in their internal service. In this case, inventory management becomes more important [1]. In order to decrease the costs associated to inventories, French hospitals are interested in the reduction of the inventory supplies [6]. Since the inventory optimization depends highly on the quality of the demand prediction, in this study we focus on the estimation of medicines demand fluctuations for hospital logistics.

The study begins with an overview of existing methods for forecasting medicine demand in hospital pharmacies. Many factors influence this demand, such as seasonality, size and location of hospitals, etc. In particular, demand can be stationary if its behaviour does not follow an up or down trend over time. Therefore, a stochastic method may capture fluctuations in demand from period to period. Markov chain is a particular type of no-memory stochastic process with discrete or continuous time and states [22] where the prediction of the future state only depends on the current state [25].

It is commonly used in different fields including inventory management to forecast demand. Indeed, depending on the definition of the states of the model (for instance, a state can be the demand level during one season), the forecast of the demand fluctuation in the current state in next season does not depend on the previous state (previous season). Moreover, compared to machine learning methods, we do not need to identify the factors impacting medicine consumption to predict the consumption fluctuation.

The objective is to use monthly historical data from French hospitals to develop a Markov chain model in order to evaluate its performance to predict monthly demand fluctuations. A state of the proposed Markov chain represents an interval of consumption for one month.

The paper is structured as followed. Section 2 presents a literature review about inventory management in hospitals and stochastic models. Section 3 presents the methodology followed to develop the Markov chain model. Section 4 describes a case study in the hospital of Montpellier to exemplify the application of the proposed methodology. The results obtained from the application of the proposed method to selected medicines are given in Sect. 5. Section 6 discusses some of the findings in this work. Section 6 presents the conclusion and the perspectives.

2   State of the Art

There is considerable amount of research on inventory management and demand prediction in the literature. In this section, we first focus on papers related to factors that influence the demand in hospitals (to understand how the demand can be modelled), and then on existing forecasting methods.

2.1   Influencing Factors in Hospitals

The inventory management system is an important factor that affects the performance of the operations in hospital systems. Inventory management in this context means controlling and managing a variety and quantity of items stored in a hospital pharmacy [7].

Healthcare systems are complex systems linked with several influencing factors such as the demand which is by itself related to other factors: the size of the hospital, the type of population, etc. These factors influence the performance of inventory systems and must be considered when modelling and analysing these systems [8].

An important factor in inventory management is the demand. In hospitals, the demand for health supplies is strongly linked to the doctor's recommendations based on the patient's condition [9]. The consumption of healthcare supplies is non-stationary (follows an up or down trend); the demand may depend on aspects such as the number of patients, the patient's conditions, the stage of treatment, among others. Some studies assume demand as independent and constant. However, to perform an analysis in a more real scenario, it is necessary to consider a stochastic demand [10].

The classification of inventory problems in healthcare sectors depends on the nature of the associated factors: the problems can be either deterministic or uncertain [11]. In the first case, the influencing factors are known over the period. This kind of inventory problems are not frequent in hospitals but may be applied to general medical items like syringes, gloves, intravenous fluids, and vaccines [11]. The other case is uncertainty; healthcare faces conditions of uncertainty such as the patient's clinical conditions and response to treatment, availability of providers, demand for medicines and lead time [12]. The demand character distinguishes stochastic inventory models from deterministic models. Demand is fixed in deterministic models while in stochastic models it is a random variable with a probability distribution [13]. Several factors that fluctuate at random, such as the patient's condition, his uncertain reaction to therapy, period of stay at hospital and the transition from one type of hospital care unit to another at various phases of treatment can all have a substantial impact on demand [11]. These random factors change with time, which explain why demand of required medicines is uncertain [14].

2.2 Forecasting Methods

Traditional methods for predicting demand in a hospital exist, and several of them can also anticipate the stationary demand for healthcare elements. Several works consider time series to forecast the demand. Ramírez et al. [16] apply different traditional methods such as simple moving mean, exponential smoothing and ARIMA models on no changing demand. Causal models like simple and multiple linear regression are applied when a cause-effect relation between a dependent variable and independent one exists. A forecasting exercise is performed by Varghese et al. [17] in two hospitals using, in most cases, ARIMA models for several medicines with no significant trends or non-stationary patterns in the demand. However, healthcare item consumption is usually non-stationary and uncertain [18]. Kadri et al. [19] successfully apply the ARIMA method to forecast the daily attendances in an emergency department using one year of data from the paediatric emergency department in Lille regional hospital centre in France.

In recent years, machine learning approaches have gained popularity, especially when it comes to integrate parameter estimation like the demand and inventory optimization [20]. Research in pharmacology identifies accurate methods for constructing a predictive model: linear regression, random forest method, neural network, and support vector regression [20]. The demand for most medicines depends on the periods of increased risk of spreading diseases; in this case, seasonal trends need to be considered when developing the model. A study using real drug consumption data in Rwanda was conducted by Mbonyinshuti et al. [21]. The authors applied machine learning methods to predict the demand for 10 selected medicines. According to their findings, machine learning methods can be used to predict the demand for medicines and Random Forest algorithm has the best performance. Machine learning can also be used to compare and to select the best prediction model among a set of predictive models for non-stationary demand [22].

Stochastic approaches can be used to solve problems when information on the probability distribution on some parameters such as demand and patient conditions are available. For instance, Markov decision process which relies on functional stochastic dynamic programming equations is used to determine the inventory levels [8]. In this context, Saha and Ray [11] use a Markovian demand approach to incorporate external factors impacting demand. They model a patient condition-based medication demand process which allows to capture the uncertain patient conditions (in terms of treatment stages and type of care units). Hermosilla et al. [24] propose models to predict the demand for cardiovascular drugs using Markov chain. They consider models with 3 and 4 states, each state representing patient conditions. These methods can be used to prioritize patients with greater consumption levels in critical inventory situations. Lastly, another Markov chain model is defined by Uzunoglu Kocer [25] to model and estimate the intermittent demand in products.

The reviewed studies on methods that predict drug demand are summarized in Table 1. In this table, the objectives, the data characteristics, the performance measures, and the limitations and advantages of each method/paper are given. Abbreviations: MAE - Mean Absolute Error, MSE – Mean Squared Error, MASE – Mean Absolute Scaled Error, MAPE – Mean Absolute Percentage Error, RMSE – Root Mean Square Error.

Table 1. Methods used in inventory management to predict medicines demand.

Method classification	Reference	Method(s)	Data	Performance measures	Limitations	Advantages
Traditional	Ramírez et al. (2014)	Simple moving average	Monthly	Typical deviation	- Better results in non-variable demand - Cannot capture demand variation well	- Easy to apply - Easy to understand
		Exponential smoothing				
		Simple and Multiple linear regression				
	Hyndman and Athanasopoulos (2018)	Average	Monthly	MAE, MSE MASE, MAPE		

(*continued*)

Table 1. (*continued*)

Method classification	Reference	Method(s)	Data	Performance measures	Limitations	Advantages
		Naïve				
		Naïve seasonal				
	Varghese et al. (2012)	Simple moving average	Weekly	MAE		
		Cumulative average				
		Exponential smoothing				
		Naïve				
Machine learning	A G Kravets et al. (2018)	Linear regression	Monthly	MAPE MSE	- Require a large number of observations - Require good data - Not Easy to apply - Not Easy to understand	Powerful in Big data
		Random forest				
	Mbonyinshuti et al. (2021)	Linear Regression	Monthly	RMSE R-square		
		Artificial neural network				
		Random forest				
Stochastic processes	Hermosilla et al. (2020)	Markov chain- several numbers of states	Monthly	MAPE Inventory parameters MASE **R-square MSE MAE**	- Need large observations number - Good quality data	- Capture variation in demand well
	Saha and Ray (2019)	Markov chain based in stage of treatment				
	Uzunoglu Kocer (2013)	Modified Markov chain model				
	Present study	**Markov chain based on consumption intervals**				

In the present study, our objective is to capture medicine demand fluctuations as the demand is variable. In this preliminary study, we will focus on Markov chain using monthly data model to confirm the fluctuations of demand and predict them. In such model, we do not need to include factors such as patient conditions, hospital size, etc.

3 Methodology

The Markov chain technique is used to capture consumption fluctuations from one period (month) to another. The objective of this section is to propose a discrete-time (representing the months) Markov chain model that can predict the medicine demand fluctuations. The proposed steps of our Markov-Chain-based method are shown in Fig. 1.

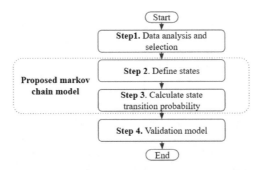

Fig. 1. Proposed methodology for forecasting consumption - Markov chain model.

Step 1: Data Analysis and Selection
We consider the historical data related to consumption of medicines aggregated by months in the Montpellier hospital for 2016, 2017 and 2018. The data set has 65450 observations (consumption of a medicine for one month), and 2396 medicines. For one observation, the quantity, the name, the year, and the month of the medicine consumed is available. In the following, for each year y, for each medicine m, we denote by M_m^y the set of observations. Note that an observation can be any type of medicine.

Preliminary results show that medicines which are not consumed frequently or constant are not interesting for our study. The following criteria were used to select the relevant medicines:

1. High level of consumption at each month (greater than 500).
2. Large variation of consumption value. The ratio between the minimum and the maximum value was calculated and medicines with a ratio less than 0.5 were taken.

After applying these criteria, the resulting data set has 156 medicines. For each of these medicines, steps 2, 3 and 4 are then applied.

Step 2: Define States of Markov Chain Model
We define the states as intervals of consumption. The number of states is denoted by n. Four methods are proposed to define the intervals.

Method 1. For each year y, for each medicine m, the states are determined dividing the demand data in quantiles. It is divided into equal sized groups. It means that the number of points in each state/interval is the same. The intervals for n states are represented as follow:

$$[0, Q_{i1}^y],]Q_{i1}^y, Q_{i2}^y], \ldots,]Q_{m,n-1}^y, Q_{mn}^y],]Q_{mn}^y, \infty[\tag{1}$$

where Q_{mn}^y is the n-th quartile of medicine m at year y.

Method 2. The grouping of the data is performed taking the average value of consumption $\overline{C_m^y} = \sum_{c \in M_m^y} c / |M_m^y|$ and applying a percentage tolerance α above and below this value. The method can only be used for odd number of states. For three states for instance, we have the following intervals:

$$[0, (1 - \alpha)\overline{C}],](1 - \alpha)\overline{C}, (1 + \alpha)\overline{C}], [(1 + \alpha)\overline{C}, \infty[\tag{2}$$

Method 3. In this method, the extreme of the intervals for each year y and each medicine m are calculated taking the maximum $\overline{l_{ym}}$ and minimum $\underline{l_{ym}}$ values of consumption in demand data set M_m^y and dividing by the number of groups desired. We denote by $L_m^y = \left(\overline{l_{ym}} - \underline{l_{ym}}\right)/n$ the size of the intervals for medicine m in year y. In contrary to method 1, the number of values in each interval/state is not necessarily the same. The intervals are given by:

$$[0, vL_m^y],]vL_m^y, (v + 1)L_m^y] \, for \, v = 1, \ldots, n - 1 \tag{3}$$

Method 4. According to Du et al. [26] a common machine learning technique can be used which is K-means. K-means clustering takes the data sequence and classifies each data point into a specific cluster. To define the Markov chain states, we use K-means to compute n clusters. Each cluster will be equal to a state. Now that the four methods are defined, the states are generated and the probability transition between each pair of states is defined in step3.

The number of states n depends on the method used. In the following, we set n to 3 and compare the performance of the four proposed methods.

Step 3: Calculate the Transition Probability Matrix

The probability transition P_{ij} between states S_i and S_j can be determined using the frequencies of transitions following expression [26]:

$$P_{ij} = \frac{T_{ij}}{\sum_{j=1}^{K} T_{ij}} = P_{ij}\{ X = S_j | X = S_i \} \tag{4}$$

where T_{ij} is the number of transitions between the state i (S_i) and the state j (S_j) in one period, and K is the total number of states. In the present study the state transition matrix is composed of historical data of consumption by months. The probability transition from one state to another describes the probability of changing the state (demand interval) in the next month.

Step 4: Validation of the Proposed Model

To validate our model, we choose to create it based only on historical consumption data between 2016 and 2017. We call it the prediction model. Ideally, this model should allow to predict the consumptions for every year such as 2018. Therefore, we apply it on 2018 data using the same consumption intervals/states (the probability transitions are updated). Then, we compare the prediction error/gap: the real probabilities of 2018 are compared with the probabilities of the prediction model.

In the literature, there is a variety of ways for evaluating predicting error metrics. For this purpose, coefficient of determination (R^2), Mean Square Error (*MSE*), and Mean Absolute Error (*MAE*) were selected for evaluating the predicted model with the real model constructed using historical data. The closer the R^2 is to 1, the greater the prediction value's fit to the observation value. The average squared difference (resp. mean of absolute errors) between the estimated values and the actual values are measured by the MSE (resp. MAE). An MSE/MAE value closer to zero reflects a smaller difference between the original values and the estimated values.

4   Case Study

A case study is presented to illustrate the proposed Markov chain model. Within the selected medicines in step 1, the consumption of a particular medicine MED1 at Montpellier hospital is selected and presented as an example of illustration. We assume that the Markov chain has three states for this preliminary work implying that we have three consumption intervals. For sake of clarify, we call the three intervals L "Low consumption", M "Medium consumption" and H: "High consumption". This simple assumption comes from the fact that a demand is usually normal; however, depending on multiple parameters, consumption can either be higher or less than usual.

Step 1: Define States of Markov Chain Model
Table 2 presents the results to define the states after applying the four methods on MED1 using the historical data between 2016 and 2017.

Table 2. Intervals of consumption for the four proposed methods

	States	Method 1	Method 2	Method 3	Method 4
Intervals	Low	[0–773]	[0–717]	[0–901]	[(0–825]
	Medium	]773–860]	]717–971]	]901–1139]	]825–1051]
	High	>860	>971	>1139	>1051

After applying each of the methods to group the data, it can be observed that the size of the intervals varies depending on the method implemented.

Step 2: Calculate the Transition Probability Matrix
For each method, the frequencies of transitions are computed using Eq. (4) on historical data of year 2016 and 2017. For validation purposes, the model of 2018 is generated

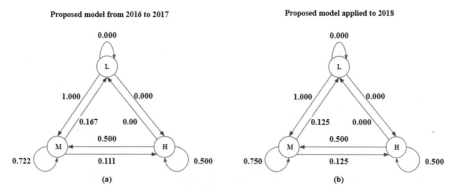

Fig. 2. Prediction Markov chain model from 2016 to 2017 (a) applied to 2018 (b) using method2.

using the same states calculated for 2016 and 2017. Figures 2(a) and (b) show these two models using method 2.

Step 3: Validating the Proposed Model

Table 3 shows the error metrics comparing the predicted model for 2018 using the historical data of 2016–2017 and the Markov chain model using the real consumption of 2018 for the four methods.

Table 3. Error metrics for the proposed model applying the 4 methods with 3 states.

Methods	R2	MSE	MAE
Method 1	−11.095	0.382	0.350
Method 2	0.997	0.017	0.009
Method 3	0.683	0.205	0.123
Method 4	0.784	0.159	0.115

The results show that the proposed model using method 2 to define states has the best value which is 0.997 for R2. Method 2 also has the best accuracy from the values of MSE and MAE that are 0.017 and 0.009, respectively. Method 1 has the worst results for R2, MSE, MAE.

5   General Results

We propose to evaluate the approach using three states model for the 156 selected medicines. We assume that a predicted Markov chain model is valid if the corresponding R^2 (resp. MSE/MAE) is higher (resp. smaller) than 0.9 (resp. 0.1). Table 4 shows the medicines for which the predicted Markov chain is valid for each method.

The results show that method 1 did not predict the consumption of any medicine. However, method 2 was able to perform better. MED1 presented in the previous sections

Table 4. Error metrics for the proposed model applying the 4 methods with 3 states.

Method	R^2 (>0.9)	MSE (<0.1)	MAE (<0.1)
Method 1	/	/	/
Method 2	MED1	MED1, MED3	MED1, MED3, MED4, MED6
Method 3	MED2	/	MED2, MED4, MED5, MED7
Method 4	/	/	MED4, MED5, MED8

has been predicted only by method 2, confirming a good R^2, MSE and MAE. MED2 only by method 3. Other medicines such as MED4 were predicted by the three methods but only using MAE. This shows that the metric evaluation affects the results. However, with this metric we compare values between 0 and 1 and it is easy to have an error less than 10% (MAE $< 0,1$). This can be observed in Fig. 2 where method 2 is applied. A low difference between the actual probability value and the predicted value makes the MAE always generating good result. The same is true for R^2 and MAE. Hence, these metrics do not allow to precisely validate the models. In such case, it is worth to mention that our prediction is not accurate because it predicts a range of values and not the exact value. Other metrics need to be analyzed and tested in future research to confirm the results.

6 Conclusion

Uncertainty about medicine demand in hospitals decreases the performance of the inventory management. Predicting medicine demand fluctuations can provide an important input for inventory decision making. In this study, we propose Markov chain models where a state represents an interval of consumption for one month to predict medicine demand fluctuations. Four methods are proposed to define the states using historical data from Montpellier hospital. The experimental study shows that the proposed Markov chain models can be used as a forecasting tool for some of the medicines.

Some limitations of this study are:

1. The present study used monthly data available for two years to propose the model. This means that the model was proposed with 24 observations per medicine. More data could better capture fluctuations in drug consumption. Cloud systems can be used to store and treat more data.
2. The proposed model can only model stationary consumptions and not trendy ones. However, if consumption behaviour changes significantly in the future (the minimum or maximum values changes for instance), the model will have to be adjusted to the new conditions. Furthermore, seasonality behaviour can be studied to better understand the fluctuations in demand; evaluation metrics can be adapted to confirm the accuracy of the proposed models.

For future research, new methods to group the data can be tested. Even the variation in the parameters used to calculate the intervals in each method can generate interesting results. In addition, the application of the proposed model in the other hospitals can generate suggestive to validate the model with more data. Models that include more variables about influencing factors such as patient condition and number of care units can be built to improve the accuracy. For example, the model developed by Saha and Ray [1] can be explored in depth to capture better the demand fluctuations. Furthermore, the proposed model can be compared with the ones presented in Table 1.

Acknowledgement. This work has been partially supported by the MIAI Multidisciplinary AI Institute at the Univ. Grenoble Alpes: (MIAI@Grenoble Alpes - ANR-19-P3IA-0003).

References

1. de Vries, J.: The shaping of inventory systems in health services: a stakeholder analysis. Int. J. Prod. Econ. **133**(1), 60–69 (2011)
2. Landry, S., Philippe, R.: How logistics can service healthcare. Supply Chain Forum Int. J. **5**(2), 24–30 (2004)
3. (Thomas) Pan, Z.X., Pokharel, S.: Logistics in hospitals: a case study of some Singapore hospitals. Leadersh. Health Serv. **20**(3), 195–207 (2007)
4. Pokharel, S.: Perception on information and communication technology perspectives in logistics: a study of transportation and warehouses sectors in Singapore. J. Enterp. Inf. Manag. **18**(2), 136–149 (2005)
5. Volland, J., Fügener, A., Schoenfelder, J., Brunner, J.O.: Material logistics in hospitals: a literature review. Omega **69**, 82–101 (2017)
6. Aptel, O., Pourjalali, H.: Improving activities and decreasing costs of logistics in hospitals: a comparison of U.S. and French hospitals. Int. J. Account. **36**(1), 65–90 (2001)
7. Gebicki, M., Mooney, E., Chen, S.-J., Mazur, L.M.: Evaluation of hospital medication inventory policies. Health Care Manag. Sci. **17**(3), 215–229 (2014)
8. Saha, E., Ray, P.K.: Modelling and analysis of inventory management systems in healthcare: a review and reflections. Comput. Ind. Eng. **137**, 106051 (2019)
9. Abdulsalam, Y., Gopalakrishnan, M., Maltz, A., Schneller, E.: The impact of physician-hospital integration on hospital supply management. J. Oper. Manag. **57**, 11–22 (2018)
10. Attanayake, N., Kashef, R.F., Andrea, T.: A simulation model for a continuous review inventory policy for healthcare systems. In: 2014 IEEE 27th Canadian Conference on Electrical and Computer Engineering (CCECE), pp. 1–6 (2014)
11. Saha, E., Ray, P.K.: Patient condition-based medicine inventory management in healthcare systems. IISE Trans. Healthc. Syst. Eng. **9**(3), 299–312 (2019)
12. Addis, B., Carello, G., Grosso, A., Lanzarone, E., Mattia, S., Tànfani, E.: Handling uncertainty in health care management using the cardinality-constrained approach: advantages and remarks. Oper. Res. Health Care **4**, 1–4 (2015)
13. Polanecký, L., Lukoszová, X.: Inventory management theory: a critical review. Littera Scr. **9**(2), 11 (2016)
14. Saha, E., Ray, P.K.: Inventory management and analysis of pharmaceuticals in a healthcare system. In: Ray, P.K., Maiti, J. (eds.) Healthcare Systems Management: Methodologies and Applications. MAC, pp. 71–95. Springer, Singapore (2018). https://doi.org/10.1007/978-981-10-5631-4_7
15. Roni, M.S., Eksioglu, S.D., Jin, M., Mamun, S.: A hybrid inventory policy with split delivery under regular and surge demand. Int. J. Prod. Econ. **172**, 126–136 (2016)

16. Lopez Ramirez, A.J., Jurado, I., Fernandez Garcia, M.I., Isla Tejera, B., Del Prado Llergo, J.R., Maestre Torreblanca, J.M.: Optimization of the demand estimation in hospital pharmacy. In: Proceedings of the 2014 IEEE Emerging Technology and Factory Automation (ETFA), pp. 1–6 (2014)

17. Varghese, V., Rossetti, M., Pohl, E., Apras, S., Marek, D.: Applying actual usage inventory management best practice in a health care supply chain. Int. J. Supply Chain Manage. 1(2), 10 (2012)

18. Vila-Parrish, A.R., Ivy, J.S., King, R.E.: A simulation-based approach for inventory modeling of perishable pharmaceuticals. In: Winter Simulation Conference, Miami, FL, USA, pp. 1532–1538 (2008)

19. Kadri, F., Harrou, F., Chaabane, S., Tahon, C.: Time series modelling and forecasting of emergency department overcrowding. J. Med. Syst. 38(9), 1–20 (2014). https://doi.org/10.1007/s10916-014-0107-0

20. Goltsos, T.E., Syntetos, A.A., Glock, C.H., Ioannou, G.: Inventory – forecasting: mind the gap. Eur. J. Oper. Res. 299, 397–419 (2021)

21. Mbonyinshuti, F., Nkurunziza, J., Niyobuhungiro, J., Kayitar, E.: The prediction of essential medicines demand: a machine learning approach using consumption data in Rwanda. Processes 10(1), 26 (2021)

22. Villegas, M.A., Pedregal, D.J., Trapero, J.R.: A support vector machine for model selection in demand forecasting applications. Comput. Ind. Eng. 121, 1–7 (2018)

23. Anderson, T., Goodman, L.: Statistical inference about Markov Chains. Ann. Math. Stat. 28, 89–110 (1957)

24. Hermosilla, A., Carmagnola, R., Sauer, C., Redondo, E., Centurion, L.: Demand forecasts for chronic cardiovascular diseases medication based on Markov chains, vol. 5, no. 2, p. 6 (2020)

25. Uzunoglu Kocer, U.: Forecasting intermittent demand by Markov chain model. Int. J. Comput. Inf. Control (IJICIC) 9, 3307–3318 (2013)

26. Hongyan, Du., Zhao, Z., Xue, H.: ARIMA-M: a new model for daily water consumption prediction based on the autoregressive integrated moving average model and the Markov chain error correction. Water 12(3), 760 (2020). https://doi.org/10.3390/w12030760

Industry 4.0 Technologies and Low-Cost Digitization

The Role of Low-Cost Digitalisation in Improving Operations Management

Jaime Macias-Aguayo$^{(\boxtimes)}$, Gokcen Yilmaz, Anandarup Mukherjee, and Duncan McFarlane

Institute for Manufacturing, University of Cambridge, Cambridge CB30FS, UK
{jem233,gy239,am2910,dm114}@cam.ac.uk

Abstract. Financial resource limitations and lack of technological awareness are two of the most common barriers to digitalisation in Small-and-Medium-sized enterprises (SMEs). In response to these challenges, recent research has focused on digital solution areas that can be deployed at a low cost in companies. In this article, we focus on how these solution areas benefit SME operations. In particular, we identify and categorise the value activities these solution areas enable and their potential benefits for the operations management function. Furthermore, this article proposes a conceptual model for the role of low-cost digital solutions in creating operational benefits and describes the opportunities that arise from it.

Keywords: Industry 4.0 · Digitalisation · Digital transformation · Operations management · Digital manufacturing on a shoestring

1 Introduction

This article is related to adopting digital solutions in SMEs to improve their operational processes. We are mainly focused on understanding the role of low-cost digitalisation as part of this improvement process. This role could include, for example, improving the company's ability to react to unexpected changes in demand or providing support to operators to perform manual activities, to name a few. In this section, we provide an overview of the topic. Then, we briefly explain the approach and describe how the article is organised.

1.1 The Problem

The operations management (OM) function is a critical part of companies' value chain; as a result, it is subject to a growing number of challenges. Some of these are pressure for shorter delivery times and higher dependability, rapid increase in the number of products offered coupled with low-demand volume, high demand variability, and environmental concerns, among others. Successfully meeting these challenges requires superior service and high efficiency in the OM function. However, achieving this in practice is difficult as it involves reconciling two conflicting objectives: increasing the level of service and

© The Author(s), under exclusive license to Springer Nature Switzerland AG 2023

T. Borangiu et al. (Eds.): SOHOMA 2022, SCI 1083, pp. 343–355, 2023.
https://doi.org/10.1007/978-3-031-24291-5_27

reducing costs. More specifically, the OM function needs to excel in cost, speed, quality, dependability, and flexibility [1], as well as in its environmental performance [2]. As a result, increasingly companies see digital transformation as an alternative to improve their performance. However, multiple barriers prevent technology adoption, especially in SMEs [3]. To obtain the benefits of digital transformation, a substantial change is required in companies [4], which could be less likely in SMEs due to barriers such as capital investment and lack of technological awareness. In this context, low-cost digitalisation arises as a viable option for SMEs willing to go digital; however, how this type of digitalisation delivers value to the OM function of these companies still needs to be analysed in more detail.

1.2 The Approach

In this article, we aim to empirically explore the role of low-cost digitalisation in improving the performance of the OM function. For this purpose, we identify the potential applications of different low-cost digital solutions and develop a conceptual model that links these applications to potential benefits for the OM function. We then explain some of the applications of this model for practitioners and researchers.

1.3 Article Structure

The article is organised as follows: in Sect. 2, we review previous related academic work. Section 3 introduces key concepts and describes the research methodology used. Then, in Sect. 4, we determine the value-added activities enabled by low-cost digitalisation and introduce a conceptual model relating these activities to operational benefits. Finally, Sect. 5 presents the research conclusions.

2 Background

2.1 Low-Cost Digitalisation

A digital solution can be defined as a system of digital technologies that facilitates an activity in an organisation [5]. A critical factor in adopting digital solutions in companies is their cost. In fact, the limitation of financial resources to access technology has been identified as one of the main barriers to digitalisation in manufacturing SMEs [3]. Consequently, low-cost digital solutions are a more viable alternative for this sector. In the context of digitalisation, the term low-cost solution refers to a system that requires a low capital investment, a low cost of development, and a low operating cost [5–7]. Although it is unlikely to reach a consensus on a threshold value to consider a digital solution as "low cost", there are some development considerations introduced in [6] that could help to characterise them. Overall, these solutions:

1. Make use of non-industrial, off-the-shelf technology components and software when developing the solution.
2. Make use of a design approach that facilitates the combination of components belonging to different technology groups.

3. Make use of an "incremental architecture" to ease the sharing of data and services among a company's initial and subsequent digital solutions.
4. Focus on facilitating the integration of components and the operation and maintenance of the digital solution to reduce costs.

Therefore, we could think of *low-cost digital solutions* in line with the above considerations. Furthermore, as a digital solution for a specific purpose can be developed using different technologies available in the market, the term *digital solution area* has been proposed in the literature as a way to capture the purpose of a digital solution, independent of the specific enabling technologies [5]. Digital solution areas could be deployed at low or high cost depending on whether or not they follow the previously mentioned development considerations. Those that can be deployed at a low cost are sometimes called *digital solution areas for SMEs*. The reader is referred to [5] for a good overview of this type of digital solution areas in the manufacturing industry. In this study, we will use the term *low-cost digitalisation* to refer collectively to low-cost digital solutions or digital solution areas that can be deployed at low cost.

2.2 Turning the Outputs of Digital Solutions into Benefits

The uncertainty about the benefits is a crucial barrier to technology adoption in SMEs [3]. Although previous work in the context of OM has identified the potential benefits (or impacts) of digitalisation (see, e.g., [4]), the systematic study of the way these benefits are achieved has received little attention in the literature.

Some authors suggest that the availability of digital technologies in a company leads to benefits by improving *capabilities*, which, if successfully deployed, could lead to superior performance [8]. Others, however, have found direct associations between the availability of digital technologies and benefits, bypassing capability formation (see, e.g., [9]). These direct associations are especially evident in analytic studies addressing digital solutions, given that identifying a cause-effect relationship between the output of a solution and a target metric is typical when formulating an analytical model (see, e.g., [10]). By *outputs* of a solution, we mean the "deliverables" of this, which could be a *message* (e.g., diagnostics, alerts, images), a *physical action* (e.g., moving an object), or *both*. The notion of the output of a solution is particularly useful for studying how digital solutions deliver benefits to operations from different perspectives. For instance, when the output of a digital solution is a message, the process can be studied through the lens of Decision science. In this context, the message is a *stimulus* that alters our current knowledge and helps us make better decisions [11]. Information is formed from the cognitive processing of a message (i.e., data) that is the *output* of an information source, which could be, for example, a digital solution. According to [11], the *knowledge* used for decision-making is seen as the result of information processing. However, it is essential to note that in practice, we could expect more applications of the knowledge gained from a digital solution than just decision making and more outputs than just a message (e.g., a physical action). Furthermore, understanding how the outputs of a digital solution translate into activities and these into benefits is interesting and important both from the perspective of technological awareness and the perspective of quantifying the benefits of digitalisation.

2.3 Summary and Opportunities

The critical points drawn after analysing the related scientific literature are:

1. Technological unattainability and lack of technology awareness are two critical barriers to SMEs' adoption of digital technologies. This contributes to companies not digitalising their operations or less than large companies.
2. How the outputs of digital solutions translate into operation benefits is not well understood, which may contribute to uncertainty in SMEs about the usefulness of digitalisation in their particular businesses.
3. Most research on the benefits of digitalisation in OM has not distinguished between those corresponding to low and high-cost solutions. Therefore, the literature is limited in its ability to drive technology adoption in SMEs, as these may focus on digital solutions that are unlikely to be deployed at a low cost.

From these conclusions, it can be inferred that there is an opportunity for research on how the availability of low-cost digital solutions translates into activities that create value (i.e., profit) for the operations function of a company. Just as important as this is a framework that categorises these actions and relates them to the expected benefits of digitalisation. Such a framework would facilitate future research efforts to quantify the benefits of digitalisation. It could also help guide these efforts toward digital solution areas most likely to be deployed at low cost.

3 Methodology

3.1 Value-Added Activity Areas

In Sect. 2, we reviewed some of the benefits of digitalisation in operations and argued for the need to understand how the outputs of digital solutions lead to them. For a digital solution to deliver benefits to the OM function, a necessary (although not sufficient) condition is that the company uses the outputs of the solution. We refer to these uses as *activities derived from digital solutions*, which can be categorised according to their purposes. We introduce the following working definition:

Value-added activity areas define the purpose of an activity derived from the output of a digital solution, which contributes to achieving a benefit for the customer and/or the organisation.

The aim of this definition is twofold. On the one hand, to help in the process of categorising the possible activities derived from digital solutions, focusing attention only on those that contribute to the achievement of benefits for the company; and, on the other hand, highlight the importance of the *output* of the solution in the process of obtaining value from it.

3.2 Approach to Developing the Conceptual Model

To understand how low-cost digitalisation leads to operational benefits, we conducted an inductive qualitative study based on interviews with seventeen SMEs in the UK. These companies were in the process of adopting low-cost digital solutions in their operations. The data collected was analysed through the lens of Grounded theory [12], which is a suitable approach to conceptualising or developing a theory from the data [13]. The latter attribute makes it very appealing for our study, given the limited research on the applications of low-cost digital solutions in operations management.

In Grounded theory, excerpts of participants' views regarding a topic are compared with those of others to identify patterns (i.e., underlying commonalities) and develop concepts or coded categories. Overall, three levels of coding (i.e., theoretical constructs) are used in this method to analyse the excerpts: *open code, axial code, and selective code* [14]. The resulting theoretical constructs or tentative theories are tested by theoretical sampling, which implies that, as new excerpts emerge, these are constantly compared with existing theoretical constructs, and if no further contributions to these are observed, the theory is considered "conceptually dense and grounded on the data" [15].

The interviews for data collection were conducted during in-person and online sessions with managers of SMEs. During these sessions, different low-cost digital solutions -of interest to the managers- were presented, and they were asked questions about the potential uses in their companies (i.e., user stories). Concretely, the points addressed during the interviews were:

- Users: Who in the company's operations function would need the solution?
- Needs: What kinds of applications would have the solution for each user?
- Benefits: For each application, what would be the benefit of deploying the solution?

The seventeen companies in the study were selected based on (1) being SMEs according to the criteria of the European Union [16] and (2) belonging to the manufacturing, logistics, or construction sectors. In contrast with other sampling methods (mainly probabilistic ones), in Grounded theory, sample size cannot be defined in advance but is the result of the 'theoretical saturation' process described before in this section.

In this study, the seventeen interviews led to data saturation, given that no new insights were obtained after ten interviews. Therefore, the sample size is considered reasonable based on the information redundancy criterion, which is key for sampling in qualitative research [17]. Following each interview, the data were divided into two groups: data on *applications* and data on *perceived benefits*. Each data group was processed using the coding and pattern analysis of the Grounded theory approach. This analysis involves three levels of coding. The first one is open coding, where the responses from interviewees (i.e., excerpts) are examined and assigned into suitable emergent categories. The second level is axial coding, where the pattern of similarities and connections between open codes are determined. The third level is selective coding, which aims to identify the core(s) category by identifying relationships among axial codes. The categories that emerged from the selective coding, both applications and benefits, represent the value-added activity areas of low-cost digital solutions and their benefits areas. Therefore, these are considered the theoretical elements of a model that describes how low-cost digitalisation delivers value to the OM function.

4 Findings

Seventeen companies participated in the study: 8 in the logistics sector, 5 in manufacturing, and 4 in off-site construction. Participants provided user stories for one or more digital solution areas during the data collection. In total, 28 solution areas were studied in this stage: 13 belonging to the data gathering-and-visualisation category, 6 to the data-analysis-and-visualisation category, 1 to the actuation category, and 8 to the support-systems category. The source of the digital solution areas considered were the works of [5] and [18]. This process yielded 103 user stories which were included in the coding and pattern analysis. Table 1 shows examples of the solution areas considered and the open coding process of the user stories.

Table 1. Open coding for excerpts from the interviews

Digital solution area	Solution output	Label	Description of the application	Open codes	Description of operational benefit	Open codes
Real-time tracking of internal jobs	Location and status of jobs in the process [manufacturing/construction]	L_1	Bring the next job to the shop floor right when the previous one is completed	Guide workers	Save time by not having to go to the office between jobs	Time-saving
		L_2	Inform customers of job progress	Communicate the status of a process	Shorter time to respond to a request for information	Fast response to customer requirements
Monitoring of lead time	Time spent for a job in the process [Logistics]	L_3	Reschedule operatives' activities in case part of the job is delayed	Execute case-based action to expedite an operation	React to unexpected changes at the right time to avoid a delay	Timely completion of the operation
Automated tool changer/part feeder	Parts required for the operation (e.g., nuts and bolts) [Manufacturing]	L_4	Access to parts required for the work using voice commands	Automate part of a process	Reduce time spent searching for materials	Time-saving
Monitoring and reporting on container unloading time	Average time required per container type [Logistics]	L_5	Decide on how to resource the team (e.g., size the team)	Expand inputs for making a decision	Increase the likelihood of being on track with the schedule	Timely completion of the operation

4.1 Value-Added Activities and Operational Benefits of Digital Solution Areas

Following the methodology in Sect. 3, similarities and connections between open codes were identified, and *axial codes* were proposed. Likewise, axial codes were grouped into *selective codes.* The resultant selective codes represent the value-added activity areas and potential operational benefit areas. We begin by describing the six resulting value-added activity areas and provide examples.

1. **Solve a decision problem. –** This is the case when the output of the digital solution is used to refine an input for a decision problem. For example, the probability of an event or the cost of possible action under different scenarios; alternatively, it might provide inputs which may be new to the decision-maker, for example, an alternative route for delivery not previously considered in the decision analysis.
2. **Perform a task new for the staff. –** This is the case when the output of the digital solution is used to learn how to do a task. An example is material in a digital format intended to train new staff. As a result, staff can perform new activities such as using a physical tool, driving a forklift, and performing a calculation, among others.
3. **Act in response to specific events. –** This is the case when the output of the digital solution is used to know when an event X -of interest to the company- occurs. For example, when an item in the warehouse has reached its reorder point, or a job has been in process longer than planned. As a result, the company can react in a timely manner by mitigating the event or taking advantage of it.
4. **Communicate the state of operations. –** In this category, the output of the digital solution is data about a process or resource. As a result, staff can promptly provide operational information to other areas of the company or the customer.
5. **Identify improvement initiatives for the operations. –** This is the case when the output of the digital solution is used to detect structural inefficiencies in a system. These inefficiencies include bottlenecks, congested zones, and suboptimal product placement, among others. As a result, the company can identify the most appropriate and convenient activities to improve its operations.
6. **Modify the usual way to perform an activity. –** This is the case when the result of the digital solution is a physical action or message that is used to automate a specific task in a manual handling process, or it is a message (i.e., data) that shows staff the correct way to perform a routinary task or that provides support to perform it.

The potential *operational benefits* of performing value-added activities were also identified and grouped into the seven benefit areas (BAs) below. It is essential to notice that all these BAs could contribute to lower operating costs and higher service levels.

1. **Utilisation. –** Although not recognised as an operations objective, capacity utilisation can significantly affect a company's unit cost. By utilisation, we mean the percentage of time of a resource that jobs or customers occupy. In particular, we found that low-cost digitalisation could impact utilisation by enabling a company to identify idle resources quickly (e.g., using sensors). These could make it easier to reallocate these resources and increase their utilisation.
2. **Speed. –** This area includes benefits associated with the time to satisfy a customer request, whether tangible (i.e., products) or intangible (i.e., information). The subareas identified were: (shorter) process time and (higher) productivity.

3. **Reliability.** – This area covers improvements in the ability of a process to operate under target parameters of availability and time. Subareas in this category are (higher) level of product availability and (higher) process dependability.
4. **Safety.** – Low-cost digitalisation could also impact the level of safety of a company. Although low-cost digital solutions are usually not implemented in safety-critical operations, these could contribute to a lower risk of accidents for staff by helping to create hazard awareness (e.g., through training material in digital format)
5. **Environment performance.** – It was also found that the operations function's environmental performance could be improved due to low-cost digitalisation. Overall, two benefit subareas were identified: (lower) energy consumption and (lower) material consumption.
6. **Flexibility.** – Although flexibility has multiple dimensions in the context of OM, we found that the one related to the capacity of a firm to react to unexpected changes in demand is the one that most benefits from low-cost digitalisation. Overall, the subareas identified were: (higher) workforce flexibility (e.g., by making it easier to train staff) and (shorter) changeover times between orders.
7. **Quality.** – We found that low-cost digitalisation could help reduce the likelihood of errors with the potential to affect the operations and the satisfaction of internal and external customers. The subareas identified were: reduction of information or estimation errors (e.g., by automating the process of collecting and processing data) and lower number of orders not according to specifications (e.g., by providing digital instructions to perform a task correctly).

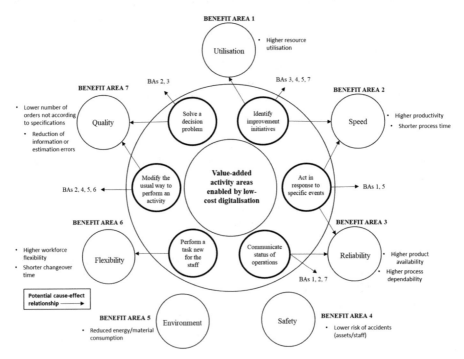

Fig. 1. A conceptual model for the role of low-cost digitalisation in improving operations

4.2 A Model of How Low-Cost Digitalisation Delivers Value to Operations

Combining the activities enabled by low-cost digitalisation with the operational benefit areas identified, Fig. 1 presents a simple conceptual model for the role of low-cost digitalisation in improving business operations. The model suggests that when a low-cost digital solution is available in an industrial setting, the organisation could use its output for one (or more) of six possible purposes, identified as value-added activity areas. Furthermore, the model suggests that the successful execution of these activities could lead to benefits that are grouped into seven potential benefit areas (BAs).

4.3 Case Study

To illustrate the use of the proposed model, in this section we present a case example for one of the digital solution areas considered in the study.

Job Tracking at a Construction SME
The job tracking solution is developed and deployed at an SME in the UK, which is a timber fire door set producer and Tier 2 supplier in the construction supply chain.

Problem: Managers/ employees working in the office had to visit the shop floor whenever they needed feedback or updates about the jobs.

Solution: The system is a standalone web-based application running on multiple Raspberry Pis and includes a barcode scanner at each workstation. The solution tracks jobs by scanning barcodes at each workstation and then capturing the process times of jobs at the main workstations. Figure 2 illustrates this process.

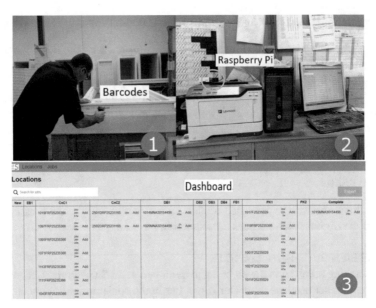

Fig. 2. Scanning of a job barcode (1). A Raspberry Pi receives and stores location and time data from workstations (2). A dashboard shows job tracking information (3)

Outputs of the Solution

- Location of jobs on the shopfloor (output type: message)
- Time spent for jobs in each workstation (output type: message)

Benefits Analysis for the Solution: The proposed model was used to infer the operational benefits of the digital solution under study *before* implementation. For this purpose, the value-added (VA) activity areas in Fig. 1 were used as an aid to reflect on all the potential uses of the digital solution's outputs. As a result, four specific activities (shown in Table 2) were identified, validated with the company, and classified within their corresponding VA activity areas. Then, the operational BAs of each VA activity area were identified using the potential *cause-effect relationships* in the model. These benefit areas are shown in Table 2. Finally, the benefit areas were used to assist the company in identifying specific benefits. These inferred benefits are shown in Table 2.

Table 2. Operational benefit pathways for the job tracking solution

Value-added activity areas [From the model]	Specific activities	Operational benefit areas [From the model]	Inferred operational benefits
Act in response to specific events	Management can take action (e.g., expedite) if a job has spent too much time in a workstation	• Utilisation • Speed • Reliability • Environment	• Speed: Higher productivity • Reliability: Higher process dependability
Communicate the state of operations	Management can inform customers and sales staff on the progress of jobs	• Utilisation • Speed • Reliability • Quality	• Speed: Shorter process time (to satisfy requests for information)
Identify improvement initiatives	Management can identify workstations where jobs are frequently stuck and propose improvements	• Utilisation • Speed • Reliability • Safety • Environment • Flexibility • Quality	• Speed: Shorter process time / higher productivity • Reliability: Higher process dependability
Modify the usual way to perform an activity	Management automates the collection of time and location data about the jobs, hence, saving time	• Speed • Safety • Environment • Flexibility • Quality	• Speed: Higher productivity • Safety: Lower risk of accidents (by avoiding the need to go to the shop floor) • Quality: Reduction of information errors (due to more precise data collection)

5 Conclusions

This article has analysed how low-cost digitalisation brings value to the OM function. By introducing a simple conceptual model and detailed descriptions of the enabled activities and their potential operational benefits, this work has attempted to scope the role of low-cost digitalisation in improving operational performance. We conclude this article by highlighting the potential applications of the model developed:

1. **Categorisation of digital solutions.** – The model discussed here could help categorise digital solutions from an application perspective (i.e., value-added activities). This could contribute to a more *structured study* of digital solutions, facilitating their comparison and report of findings. Also, as this categorisation is application-based, it might provide a more robust classification framework than one based on enabling technologies, which can be expected to vary over time.

2. **Quantification of the impact of digital solutions.** – The proposed model illustrates the potential mechanisms by which low-cost digitalisation influences operations performance. Concretely, the model identifies 22 potential cause-effect relationships between activity types enabled by digital solution areas and operational benefits. These *theoretical propositions* (indicated with arrows in Fig. 2) provide input for future studies aimed at quantifying the impact of digital solutions, as they provide a broad perspective of the possible impact types for different digital solution applications. Also, the sub-areas of the BAs could facilitate identifying KPIs likely to experience improvement due to low-cost digital solutions.

3. **Creation of technological awareness.** – The application of the model in this area is twofold. On the one hand, it contributes to understanding the role of low-cost digitalisation in operations. For example, for companies planning to start their technology adoption process, this model can provide insights on the types of activities and benefits they might expect from low-cost digital solutions, as well as on the type of solutions that could be more important to their companies. Furthermore, for companies that have already started their digitalisation journey, the model could be used as a diagnostic tool to identify value-added activity areas that are not currently being exploited by the company or expected benefits that are not yet achieved. On the other hand, the model allows us to understand better the applications and benefits of low-cost digitalisation that SMEs perceive. This provides a valuable insight into these companies' technological awareness levels. For instance, a digital solution area that assists staff in learning to perform a new task likely helps a company increase its range-of-service flexibility, but this was not mentioned during interviews. Similarly, it is possible that a solution that modifies the way to perform a task influences capacity utilisation, but this was not mentioned during the discussions either. Consequently, the model could help identify areas of weakness regarding technological awareness in SMEs and guide future research.

References

1. Slack, N., Chambers, S., Johnston, R.: Operations Management. Pearson Education (2010)
2. de Burgos Jiménez, J., Céspedes Lorente, J.J.: Environmental performance as an operations objective. Int. J. Oper. Prod. Manage. **21**(12), 1553–1572 (2001). https://doi.org/10.1108/014 43570110410900
3. Masood, T., Sonntag, P.: Industry 4.0: adoption challenges and benefits for SMEs. Comput. Industr. **121**, 103261 (2020). https://doi.org/10.1016/j.compind.2020.103261
4. Pousttchi, K., Gleiss, A., Buzzi, B., Kohlhagen, M.: Technology impact types for digital transformation. In: 2019 IEEE 21st Conference on Business Informatics (CBI), vol. 01, pp. 487–494 (2019). https://doi.org/10.1109/CBI.2019.00063
5. Schönfuß, B., McFarlane, D., Hawkridge, G., Salter, L., Athanassopoulou, N., de Silva, L.: A catalogue of digital solution areas for prioritising the needs of manufacturing SMEs. Comput. Industr. **133**, 103532 (2021). https://doi.org/10.1016/j.compind.2021.103532
6. McFarlane, D., et al.: Digital manufacturing on a shoestring: low cost digital solutions for SMEs. In: Borangiu, T., Trentesaux, D., Leitão, P., Giret Boggino, A., Botti, V. (eds.) SOHOMA 2019. SCI, vol. 853, pp. 40–51. Springer, Cham (2020). https://doi.org/10.1007/ 978-3-030-27477-1_4

7. Hawkridge, G., et al.: Monitoring on a shoestring: low cost solutions for digital manufacturing. Ann. Rev. Control **51**, 374–391 (2021). https://doi.org/10.1016/j.arcontrol.2021.04.007

8. Evangelista, P., Mogre, R., Perego, A., Raspagliesi, A., Sweeney, E.: A survey based analysis of IT adoption and 3PLs' performance. Supply Chain Manage. Int. J. **17**(2), 172–186 (2012). https://doi.org/10.1108/13598541211212906

9. Choy, K.L., et al.: Impact of information technology on the performance of logistics industry: the case of Hong Kong and Pearl Delta region. J. Oper. Res. Soc. **65**(6), 904–916 (2014). https://doi.org/10.1057/jors.2013.121

10. Kelepouris, T., McFarlane, D.: Determining the value of asset location information systems in a manufacturing environment. Int. J. Prod. Econ. **126**(2), 324–334 (2010). https://doi.org/10.1016/j.ijpe.2010.04.009

11. Lawrence, D.B.: The Economic Value of Information. Springer, New York (2012). https://doi.org/10.1007/978-1-4612-1460-1

12. Glaser, B.G., Strauss, A.L.: The Discovery of Grounded Theory: Strategies for Qualitative Research. Routledge, New York (2017). https://doi.org/10.4324/9780203793206

13. McGregor, S.L.T.: Understanding and Evaluating Research: A Critical Guide. SAGE Publications (2017)

14. Strauss, A.L.: Qualitative Analysis for Social Scientists. Cambridge University Press (1987)

15. Schwandt, T.A.: Dictionary of Qualitative Inquiry, p. 281. SAGE Publications (2001). ISBN 978-0761921660

16. Publications Office of the European Union: User Guide to the SME Definition, European Union [Online] (2015). http://ec.europa.eu/regional_policy/sources/conferences/stateaid/sme/smedefinitionguide_en.pdf

17. Ary, D., Jacobs, L.C., Irvine, C.K.S., Walker, D.: Introduction to Research in Education. Cengage Learning (2018)

18. Macias-Aguayo, J., McFarlane, D., Schönfuß, B., Salter, L.: A catalogue of digital solution areas for logistics SMEs. Presented at the 2022 10th IFAC Conference on Manufacturing Modelling, Management and Control, MIM 2022, Nantes, France (2022). https://doi.org/10.1016/j.ifacol.2022.09.664

Function-as-a-Service on Edge for Industrial Digitalization: An Off-the-Shelf Case Study

Pallav Kumar Deb[✉] and Himanshu Kumar Singh

Siemens Technology and Services Pvt. Ltd., Bengaluru, India
{pallav.deb,himanshu.singh}@siemens.com

Abstract. As Industry 4.0 gathers traction, smart manufacturers are exhibiting a significant interest in the digitalization of legacy infrastructure for realizing its potentials and benefits. Enabling digital data transmissions and processing on a platform such as cloud or edge is a common solution approach. However, due to limited capacity on the edge platforms there is a need to improve and optimize their utilization for multiple connected use cases. Building a serverless platform that develops, runs, and manages applications while avoiding infrastructure complexities on the edge is non-trivial. In this paper, we focus on enabling Functions-as-a-Service (FaaS) for edge devices using off-the-shelf solutions. We present a study on two popular IoT and FaaS-enablement platforms *viz.* AWS Greengrass V2 (proprietary) and OpenFaaS (open-source). We highlight their deployment mechanisms and indicate the pros and cons of each. Using an experimental setup, we also present the resource consumption (CPU and memory) for both and present the latency for executing operations as simple as the sum of two numbers and compute-intensive ones like making inferences using pre-trained machine learning models.

Keywords: Function-as-a-service · Edge computing · Real-time systems · Distributed computing · IIot

1 Introduction

The edge computing platform has the potential to automate factory floors and enhance supply chain and automation as IT in industries [13]. It allows processing the data close to where it is generated, which facilitates real-time decision-making. This is due not only to avoiding latencies for data to reach the cloud, process, and then get back to the originating device but also mitigating challenges due to intermittent connectivity. Apart from overcoming latency, the edge computing platform offers improved security and privacy, reliability, and scalability. Such features are beneficial across all industrial applications, particularly those that consist of connected assets, which typically includes process monitoring, predictive maintenance, manufacturing, oil and gas, energy, transportation, and others [14].

© The Author(s), under exclusive license to Springer Nature Switzerland AG 2023
T. Borangiu et al. (Eds.): SOHOMA 2022, SCI 1083, pp. 356–367, 2023.
https://doi.org/10.1007/978-3-031-24291-5_28

Considering the demand for platforms that can ease the application development, deployment and maintenance at edge, there has been a surge in number of platforms claiming to address these demands. In general, we expect the following properties on any edge computing solution, and the edge platforms to support most if not all of these.

- **Edge driven**: Perform operations on-site.
- **Event driven**: Execute functions on-demand.
- **Low latency**: Execute functions with near real-time possibilities.
- **Stateless**: Independent of previous executions.
- **Serverless**: Automated orchestration of edge infrastructures.
- **Lightweight**: Low software footprint on the edge devices.
- **Open architecture**: Seamless integrations of edge devices.
- **Modular**: Functions are independently created and maintained.
- **Self-sustaining**: Independent of cloud connectivity.
- **OTA updates**: Receive updates from the cloud.
- **Distributed computations and storage**: Collaborative operations.
- **Privacy and security**: Secured communications.
- **Support mobility**: Follow end user and optimize delivery.
- **Oblivious of heterogeneity**: Interoperability among the edge devices.
- **Reliable/Robust**: Fault tolerant and provide uniform Quality of Service.

Some of the popular available edge computing platforms in the current traction (while writing this manuscript) include EdgeX Foundry [5], IOFog [7], Coaty [3], Pub/Nub [10], Amazon Web Services (AWS) Greengrass (v2) [1], Azure IoT Edge [2], KNative [8], OpenFaaS [9], Sinumerick Edge [11], and many others. Among these platforms, some focus on facilitating interoperable communications and others on edge computing. EdgeX Foundry allows vendor independent interoperable communications in the form of middleware. Similarly, Coaty is a middleware that allows building and deploying distributed (loosely coupled) applications. It allows prosumers to act in an autonomous, collaborative, and ad-hoc fashion. PubNub is another such platform that specializes in real-time streaming, which is inclusive of protocols such as WebSockets, Socket.IO, SignalR, WebRTC Data Channel, and others. Additionally, PubNub has functions that allows developers to add code and deploy features in real-time in the form of customizable microservices.

AWS and Azure: Platforms such as AWS Greengrass and Azure IoT Edge specializes in offering compute power at the edge, near to where the data is located. They satisfy most of the features expected from a typical edge deployment, in addition both offer similar QoS and developer experience at the edge as their well-established cloud services. Azure provides better support for using OPC-UA for data collection and communications whereas AWS takes a more open approach by focusing on more open pub - sub protocols like MQTT. OPC-UA is a resource-intensive protocol in addition to being a client-server model, which makes it unsuitable for resource-constrained devices and distributed setup common on edge. In summary, adopting Azure means getting it as the ecosystem and

AWS helps in achieving true modularity. Packaged solutions may not always be beneficial, particularly in edge computing. One of the reasons is that both AWS and Azure may result in vendor lock in, which is undesirable for edge environments with heterogeneous devices communicating with one another. Refer [12] for detailed comparison.

Kubernetes: Also known as K8, Kubernetes is a system that helps containerized applications in automating deployment, scaling, and management. It attracts more attention as it is an open-source system that opens multiple possibilities, especially for distributed system orchestration such as the edge. Additionally, it has the following features that are worth mentioning.

- Automated rollouts and rollbacks
- Service discovery and load balance
- Storage orchestration
- Secret and configuration mgt.
- Automatic bin packing
- Batch execution
- IPv4/IPv6 dual-stack
- Horizontal scaling
- Self-healing
- Designed for extensibility

Kubernetes also has straightforward methods for application deployments. KNative (pronounced kay-nay-tiv), an open-source community project, aids in deploying, running, and managing serverless, cloud-native applications to Kubernetes. Such projects help developers to be oblivious to complex details on infrastructures and focus on the code. Such practices create space for offering Backend-as-a-Service (BaaS). Other platforms also use Kubernetes to offer their services for the edge. For instance, ioFog (an Eclipse Foundation project based on Kubernetes) helps in deploying and managing microservices at the edge and it works on top of Kubectl [4]. Kubectl is a command line tool used to run commands against Kubernetes clusters. It helps in operations such as deploying applications, inspecting, and managing cluster resources, and viewing logs. OpenFaaS is another framework that simplifies function and existing code deployment to Kubernetes, while offering a unified experience. As it runs on top of Kubernetes, it removes the need for developers to manage servers and focus on building and deploying code instead.

Industrial Focus Edge Platforms: Lastly, Sinumerick Edge and Industrial Edge are frameworks by Siemens that facilitate edge computing, with focus on industrial applications and their digitalization. It may be noted that the platforms mentioned in this section have support for most of the available programming languages like C, C++, C#, Java, NodeJS, Go, Python, JavaScript, and others to name a few. However, we recommend going through the individual documentations for more details.

We primarily focus on two of the mentioned platforms, one commercial and the other open source, and their ability to support serverless aspects on edge *viz.* AWS Greengrass v2 and OpenFaaS, respectively. AWS provides a more modular approach towards deployment (as mentioned earlier) as compared to its other proprietary counterparts. Additionally, the communications (both inter process and network) are based on the popularly accepted publish-subscribe model. OpenFaaS runs on top of Kubernetes and its adoption at the backend for resource orchestration is common for other open platforms.

It may be noted that AWS provides a serverless framework like Lambda and managed using Kubernetes environments (refer to EKS 1). Lambda is also extended to be used on edge through AWS Greengrass, even though this is not evolved into completely robust solution. OpenFaaS eases the deployment of event driven functions and microservices to Kubernetes. It extensively supports using docker images, which helps in scaling easily while avoiding repetitive and boiler-plate coding. The two platforms have different architecture modes and strategies, with their own set of merits and demerits. We attempt to highlight the key differentiating factors of each while deploying on resource-constrained environments to ease the decisions of developers and other practitioners working on a similar domain. For the ease of better comparison, we deploy both the platforms on the AWS cloud. The rationale behind the selection of the two frameworks and their deployment strategy is primarily to be able to update the functions through the cloud infrastructure and not by accessing the local edge device. While OpenFaaS on the edge may be accessed by making the device public, it will require modifications to cater to security compliance.

Organization: We highlight some of the existing platforms in this Section. We explain the deployment scenarios for the two platforms in Sect. 2 followed by our observations in terms of resource consumption and delays in Sect. 3. Finally, we conclude and provide our key insights in Sect. 5.

2   Deployment Architectures

In this section, we discuss the deployment architectures for AWS Greengrass v2 and OpenFaaS. AWS Greengrass v2 conforms with the definition of edge computing as being a medium to compliment the cloud. Accordingly, there are two major elements: 1) the AWS IoT Core at the cloud device and 2) the AWS IoT Greengrass Core at the edge (refer Fig. 1). The edge device may connect to the other nearby IoT devices through both wired and wireless media to receive data. It hosts several functions (called components) that contain routines for local execution. Based on the scenario, these components may run in a prolonged or on-demand manner. Further, depending on the requirements, these components may be both custom and default. AWS also provides several templates that cover most of the commonly used functionalities.

The Greengrass core communicates with the AWS IoT core using the MQTT protocol. Consequently, topics are strategically chosen for logical exclusion and security. AWS provides a few topics by default for sharing CloudWatch metrics and other administrative data that are particularly useful for edge administrators and debugging. As mentioned earlier, AWS services are highly modular, which makes the AWS IoT Core flexible to connect to other AWS cloud services, which in turn connects the IoT edge device to a myriad range of services. As shown in Fig. 1, a developer may deploy functions on both edge and cloud devices. At the

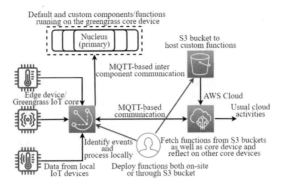

Fig. 1. Deployment architecture for AWS Greengrass v2.

cloud level, AWS provides multiple methods through lambda functions or Simple Storage Server (S3) buckets coupled with recipes in either JSON or YAML format. To conform with the definitions of AWS, for each component, they refer to the files containing the functions as artifacts and the file that identifies the dependencies to deploy to the edge device as a recipe. The recipe plays an important role in deploying platform-aware components as it allows defining the unique dependencies and lifecycles with ease. The local deployment of the components is made in a similar manner. It may be noted that, at the time of writing this manuscript, AWS suggests sticking to deploying native functions, either directly on the edge device or through S3 buckets. There are a few deployment challenges that exist in the IoT lambda functions, which we believe the AWS team will address soon. It will be interesting to see how the lambda function will perform as in our opinion, it operates in a serverless fashion. Components deployed on the devices either locally or through the cloud get transcended on the other connected edge devices in the same group too. This happens because of a special component running on the edge device as part of the Greengrass service called Nucleus. Nucleus may be thought of as a background function that is responsible for keeping all the other components alive as well as reporting to the cloud the events occurring locally. The components are deployed locally to communicate with one another through an MQTT-based Inter-Process Communication (IPC) method. Permissions to access the topics as well as hardware resources are supplied through the recipes.

Finally, the Greengrass core service generates log files that may be pushed to the IoT core as CloudWatch metrics, which may later be viewed in a browser. For further details on the AWS IoT core service, please refer to [1].

In the case of OpenFaaS, no daemon needs to run on the edge device. A server runs on the cloud which receives HTTP/HTTPS requests from the edge device and executes the relevant routines on-demand. However, OpenFaaS runs on top of Kubernetes, which implies that the server requires a minimum of 2 active cores to operate. As shown in Fig. 2, the edge device receives data from sensors through wired/wireless media. It detects the events and then sends a request

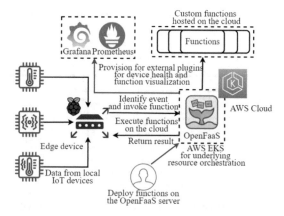

Fig. 2. Deployment architecture for OpenFaaS.

to the cloud server to invoke appropriate functions. The server executes the requested routine and then sends the result to the edge device. The OpenFaaS server hosts all the functions and similarly executes them. From the developer's perspective, in contrast with AWS Greengrass v2, they have only one method to deploy the functions. These functions need to be built as Docker images and then pushed to the OpenFaaS server. It has support for almost all the popular programming languages and the official links provide detailed templates for easy adoption. OpenFaaS also has provisions for attaching external plugins such as Prometheus and Grafana for extracting the metrics from the deployed functions and visualizing them, respectively. They can display information on the hardware usage and some details about the functions like execution time, number of invocations, and others.

OpenFaaS server may also be deployed locally on an edge device. However, the functions will be limited only to the local network. Accessing the edge devices over the Internet may not be favourable for most industries as it may require violating security norms. Additionally, a global sync on the functions is challenging if OpenFaaS servers are hosted directly on the edge.

Table 1 summarizes our findings relevant to the deployment strategies for AWS Greengrass and OpenFaaS. We also highlight some lacunae in terms of features. For instance, both components in AWS and functions in OpenFaaS are bound to the processor architecture in which they are deployed. They are not flexible enough to be deployed on heterogeneous processors with a single deployment. While there are platforms that are subject to constant development, we believe solutions to problems like device discovery (its presence and configuration) and device-aware deployment are imminent.

Table 1. Deployment techniques in AWS IoT and OpenFaaS.

Parent	Service	On-demand	OTA updates	Group deployment	Device discovery	Device config discovery	Flexibility
AWS IoT	Local	✗	✓	✓	✗	✗	✗
	S3	✗	✓	✓	✗	✗	✓
	Lambda	✓	✓	✓	✗	✗	✗
OpenFaaS	Cloud	✓	✗	✗	✗	✗	✓
	Local	✓	✗	✗	✗	✗	✗

3   Experiments, Observations, and Discussion

We present our observations on using the two edge computing platforms. We focus on resource usage in terms of computing and memory, followed by the latency in retrieving the results on-demand. In this work, for uniform comparison, we deploy both platforms on the AWS cloud. While the Greengrass v2 deployment is native to AWS, they also have EKS, which is a managed container service. It aids in running and scaling Kubernetes applications in the cloud. We deploy our OpenFaaS server in a serverless fashion on top of EKS.

3.1   Experiment Setup

We use a Raspberry Pi 4 (RPi) as edge node. In the case of AWS Greengrass v2, we use the standard infrastructure mentioned in their documentation. To have a similar experimental setup, we use the AWS EKS service for OpenFaaS. The EKS worker nodes act as the compute nodes and the RPi sends HTTP/HTTPS requests to the server along with the necessary parameters for computations. In summary, our experimental setup conforms with those depicted in Figs. 1 and 2, respectively. We perform two types of executions viz. simple sum and You Only Look Once Version 3 (YOLOv3) on them. In the former case, we take two integer variables as input and return the sum. In the latter case, we take the URL of an image as input and perform object detection using the pre-trained YOLOv3 model.

3.2   Resource Consumption (CPU and Memory)

As mentioned in Sect. 2, in addition to the user components, the Nucleus component runs as a background service in the case of Greengrass v2. In this section, we have a look at the CPU and memory (RAM) consumption when the Nucleus is turned on for some time and then turned off (refer Fig. 3). Before turning the Greengrass service on, we record how the processor acts when the RPi is in an idle state. As shown in Fig. 3a, the dashed line is stable at 75%. However, in the case of turning on the service at time instant 18, we notice a significant increase in CPU usage. The sudden change in behaviour shows a rising trend for the next

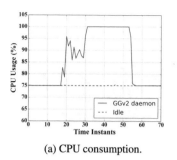

(a) CPU consumption.

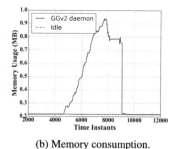

(b) Memory consumption.

Fig. 3. Resource consumption by the Greengrass v2 service (Nucleus).

12 time instants until it stabilizes at 100%. We notice that the service keeps the CPU utilization at 100% without any deviation. At a time instant little over 50, we turn off the service and notice the CPU trend coming down to 75%, which was what we observed in the ideal state. We comment that for using the Greengrass service at the edge, a single-core processor board may not suffice. For a device like RPi, it may not be alarming to keep one of its four cores at 100% as the others may help in compensating for the same. In summary, the choice of a device as edge node and its configuration needs to be determined strategically to avoid over-stressing its processor.

We notice a similar trend in the memory consumption (refer Fig. 3b). It may be noted that the time instants in Figs. 3a and 3b are not comparable as they are recorded at different sampling rates. In Fig. 3b, we turn on the Nucleus at a time instant somewhere close to after 4000. We notice a sharp increase in memory usage from 0.2 Mb to 0.9 Mb. At around time instant 8000, the memory usage stabilizes to 0.78 Mb, which is a jump by almost 56% from the idle state. As we turn off the Nucleus at time instant 9000, we notice that the memory usage comes down to its original position. We comment that the Nucleus has a higher CPU footprint compared to memory usage. This may be because the Nucleus is responsible for orchestrating the component life cycles, and their inter-process communications and ensures constant contact with the AWS IoT core for updates as well as message transfers.

Apart from the background processes, in the case of custom components running on the edge device, they also consume explicit CPU and memory (refer Fig. 4). We deploy multiple Python programs on the edge device and capture the hardware resource consumption after allowing them to startup completely. These programs range from adding two numbers, sending a message to the IoT core, listening to GPIO pins for turning on a LED bulb, running pre-trained machine learning models, and others. Interestingly, we notice that the CPU usage for these components individually is typically more than 95%. Additionally, unlike the Nucleus, the components have a high memory footprint. This is true for both Virtual Memory Size (VSZ) and Resident Set Size (RSS). It may be noted that VSZ is the total memory that the component may access during its lifetime. Consequently, VMZ includes the size of the component's binary and its linked

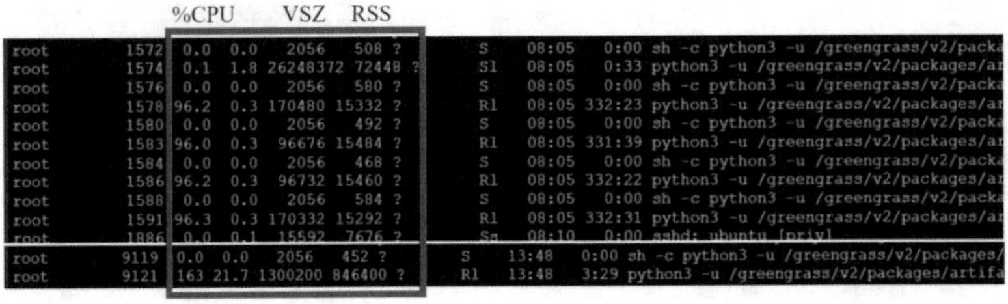

Fig. 4. Hardware consumption by components deployed on Greengrass v2.

libraries, along with its stack/heap allocations after it has started. On the other hand, RSS is the RAM allocated during the component's execution. Interestingly, the components are always active, irrespective of their invocation. This plays a major role in overcoming the slow-start problem. However, this happens at the cost of high dependency on potentially significant resource consumption. In the case of executing YOLOv3 (bottom of Fig. 4), which is typically computationally intensive, we observe a CPU usage of 163%. While this is acceptable for edge devices with multicore processor boards, it may slow down the usual performance of the device. In summary, the Greengrass v2 has the potential to offer high performance in delivering Function-as-a-Service on the edge device, while imposing a sizeable overhead on it. This leads to the challenge of selecting the appropriate edge device for achieving the desired Quality of Service (QoS).

In contrast to the AWS Greengrass v2, OpenFaaS does not put any load on the edge device. It only depends on making HTTP/HTTPS requests to the AWS EKS service and receiving the corresponding replies. It also does not require installing additional packages or libraries. However, it mandates the need for an Internet connection, which is not the case for AWS Greengrass v2. It is designed to operate even in the absence of any connection with the cloud. It may be noted that the OpenFaaS server may also be hosted on the edge device in the case of the absence of an Internet connection. However, its footprint on the device's CPU and memory is beyond the scope of this manuscript and may be taken up as part of our extended work.

3.3   Execution Latencies

As mentioned in Sect. 3.1, we perform our experiments on two sets of routines. The first is to generate a simple sum by adding two integer values. The second is to take the URL of an image and perform object detection using a pre-trained machine learning model. We use the YOLOv3 model to perform the inferences. We select these two routines to exhibit how the two platforms perform in the case of minimum and maximum compute load and get a gauge on how they might perform for others. In this section, we present the latencies that we observed on running the routines on-demand for 500 times, each in the case of AWS Greengrass v2 and OpenFaaS, respectively.

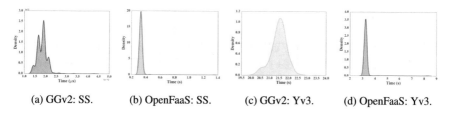

| (a) GGv2: SS. | (b) OpenFaaS: SS. | (c) GGv2: Yv3. | (d) OpenFaaS: Yv3. |

Fig. 5. Latencies on executing Simple Sum (SS) and YOLOv3 (Yv3) routines.

Simple Sum: In the case of AWS Greengrass v2, the execution occurs on the edge device directly, implying that the delay only involves the time required for execution. Consequently, we observe the delays in the range of s (refer Fig. 5a). It may be noted that the representation in Fig. 5 is the Kernel Density Estimations (KDE) and not a probability distribution. The delays are spread out in the range of 1.5–2.2 µs. The variation in the delays may be due to context switching and other Linux-based housekeeping operations within the RPi. On the other hand, in the case of OpenFaaS, the request travels from the edge device to the EKS server and then performs the execution before sending back the results. This implies that the delays in the case of OpenFaaS involve the round-trip time for the information to travel over the Internet and the execution time. Consequently, we observe delays typically in the range of 0.2–0.4 s (refer Fig. 5b). Please note that the densities in Fig. 5 represent the likelihood of the occurrence of the metric on the x-axis. The skewed nature is because of the low standard deviation and the mandate that the integral under the curve (probability) needs to be equal to 1.

YOLOv3: In the case of executing pre-trained machine learning models, compute processors need to fetch libraries, the model, and its weights among other things. This adds load to the processor and consumes time. Lower the configuration, higher will be the delay. We observe this in the case of AWS Greengrass v2 in Fig. 5c, where the delays are in the range of 20–23 s. The delays are evenly spread in the mentioned range, with most of the delays as 21.5 s. Such a delay is significantly high, which is not suitable for real-time applications. Interestingly, OpenFaaS outperforms AWS Greengrass v2 by almost 84%. It demonstrates delays in the range of 3–3.5 s. This is because AWS EKS hosts the OpenFaaS servers on t2.medium EC2 machines (2 cores of 3.3 GHz) in contrast to the 1.5 GHz processor on the RPi, which compensates for the network travel time. In our opinion, hosting the OpenFaaS server on the RPi will generate similar delays as AWS Greengrass v2.

From the observations on the delays, we comment that it is beneficial to use the AWS Greengrass v2 for routines that put low compute load and for nonreal-time applications. Services such as OpenFaaS on EKS mandate the need for an Internet connection, which may be a challenge for some deployments.

4   Key Insights

While both AWS Greengrass v2 and OpenFaaS are promising, they have their advantages and disadvantages. Table 2 summarizes our observations and comments.

Table 2. Summary of the case study.

Platform	Pros	Cons	Comment
Greengrass	• Internet connection not mandatory • Flexible deployment	• Nucleus consumes device resources	• Suitable for intermittent connectivity • Application-specific device selection recommended
OpenFaaS	• Low/No resource consumption	• Requires internet	• Suitable for complex operations

On one hand, Greengrass works irrespective of the presence of the Internet, on the other hand, OpenFaaS mandates the requirement (unless deployed on the edge device). This can be overcome by: 1) deploying Kubernetes on the edge device, 2) adding the local edge nodes as cluster nodes for EKS, or 3) adopting faasd [6] as an alternative. However, the Greengrass component Nucleus consumes significant resources on the device, in addition to the individual functions, also sizeable requirements, implying the need for a strategic decision on the device configuration, in contrast to OpenFaaS.

In summary, we comment that Greengrass is more promising with rich security compliance available in AWS cloud and the edge devices. Further, the setup helps in keeping the data on-premises.

5   Conclusion

In this work, we presented a study on enabling Function-as-a-Service toward the digitalization of industrial process flows. We focused on a proprietary and an open-source framework, namely AWS Greengrass v2 and OpenFaaS, respectively. We highlighted their process architectures and deployment strategies. With a real-world experimental setup using a Raspberry Pi as an edge device, we present the resource consumption in terms of CPU and memory usage. We also present the latency involved in performing simple operations such as simple sums and complex computations such as making inferences using pre-trained ML models like YOLOv3. Finally, we presented our insights on the two to help making real-world deployment decisions with ease.

In the future, we plan to extend our study on FaaS on the edge by deploying the OpenFaaS server on the edge device. We also plan to include lightweight middleware for establishing low latency inter-process communications.

References

1. AWS IoT Greengrass. https://docs.aws.amazon.com/greengrass/v2/developer guide/what-is-iot-greengrass.html. Accessed 02 June 2022
2. Azure IoT Edge. https://azure.microsoft.com/en-in/services/iot-edge/. Accessed 02 June 2022
3. Coaty. https://coaty.io/. Accessed 02 June 2022
4. Command line tool (kubectl). https://kubernetes.io/docs/reference/kubectl/. Accessed 02 June 2022
5. EdgeX Foundry. https://www.edgexfoundry.org/. Accessed 02 June 2022
6. Faasd Deployment - A Lightweight & Portable FaaS Engine. https://docs.openfaas.com/deployment/faasd/. Accessed 26 August 2022
7. IOFog. https://iofog.org/. Accessed 02 June 2022
8. Knative. https://knative.dev/docs/. Accessed 02 June 2022
9. OpenFaaS. https://www.openfaas.com/. Accessed 02 June 2022
10. PubNub. https://www.pubnub.com/. Accessed 02 June 2022
11. SINUMERIK Edge. https://new.siemens.com/global/en/markets/machine building/machine-tools/cnc4you/fokus-digitalisierung/sinumerik-edge.html. Accessed 02 June 2022
12. Das, A., Patterson, S., Wittie, M.: Edgebench: benchmarking edge computing platforms. In IEEE/ACM International Conference on Utility and Cloud Computing Companion (UCC Companion), pp. 175–180. IEEE (2018)
13. Nain, G., Pattanaik, K., Sharma, G.: Towards edge computing in intelligent manufacturing: past, present and future. J. Manuf. Syst. **62**, 588–611 (2022)
14. Singh, A., Kumar, A., Chauhan, B.K.: A comprehensive study of edge computing and the impact of distributed computing on industrial automation. In: Mallick, P.K., Bhoi, A.K., Barsocchi, P., de Albuquerque, V.H.C. (eds.) Cognitive Informatics and Soft Computing. LNNS, vol. 375, pp. 215–225. Springer, Singapore (2022). https://doi.org/10.1007/978-981-16-8763-1_19

Control of Open Mobile Robotic Platform Using Deep Reinforcement Learning

Mihai-Daniel Pavel[1], Sabin Roșioru[2], Nicoleta Arghira[2],
and Grigore Stamatescu[2(✉)]

[1] Asti Automation, Calea Plevnei 139, 060011 Bucharest, Romania
daniel.pavel@astiautomation.com
[2] Department of Automation and Industrial Informatics, University Politehnica
of Bucharest, 313 Splaiul Independentei, 060042 Bucharest, Romania
sabin.rosioru@stud.acs.upb.ro,
{nicoleta.arghira,grigore.stamatescu}@upb.ro

Abstract. Advanced control for mobile robotic platforms allows efficient real-time navigation in structured and unstructured environments in various industry applications. Deep reinforcement learning is an emerging control strategy where an agent is trained iteratively according to an optimisation objective by using reward and penalty actions. The agent generates the neural network weights used for computing the robot command towards the reference set point. We present an application for an open hardware mobile robotic platform navigation that integrates the sensing, communication, computing and control functions into a single system for navigation in unstructured environments. Implementation is performed through a dedicated software and communication layer that integrates the hardware platform with the MATLAB environment using standardized Robot Operating System (ROS) libraries. Quantitative testing results are presented, in order to prove the viability of the solution, by defining both simulation and laboratory setting scenarios.

Keywords: Open mobile robot platform · Deep reinforcement learning · ROS

1 Introduction

Navigation of mobile robots in dynamic, unstructured, environments is relevant for many industrial, operational and emergency recovery use cases. The main challenge lays in the orchestration of the sensing, computing and control functions that allow real-time object detection and avoidance while accounting for the tracking error against the control objectives, [1].

Several recent works describe development of algorithms used for mobile robot platforms and include new approaches that are based on several types of neural networks, [2]. These range from fully connected networks up to convolutional deep networks. In [9], the authors state that most of the applications

© The Author(s), under exclusive license to Springer Nature Switzerland AG 2023
T. Borangiu et al. (Eds.): SOHOMA 2022, SCI 1083, pp. 368–379, 2023.
https://doi.org/10.1007/978-3-031-24291-5_29

built for navigation and control of mobile robots, are based on computer vision, laser sensors or a combination of both. In [10], the authors present one of the most commonly used methods for interior navigation, which maps the environment and computes the distance between the start and finish points. In general, the trajectory computation is done using recursive algorithms applied on two levels: one that is global and one local, in order to avoid object collision. This implementation does not use neural networks and has a main drawback, that the platform has difficulties with environmental changes that can occur. A novel approach consists of deploying a neural network-based algorithm that will gather data for the sensors. Based on this information, the network will produce the best commands for the mobile platform to reach desired location, avoiding any perturbation that can appear in its path. This method does not need any prior knowledge of the surrounding space. The network is built using Reinforcement Learning for finding the optimal route taking into account any obstruction. Digital twin-type simulation can contribute to the time and cost effective modelling of the algorithm's performance using a realistically model of the robotic platform, [8]. Industrial communication protocols, described in [5] and in [6] serve as a support technology with which the advanced control layer is implemented.

In this context, the main contributions of the work consist of: formulation and implementation of a deep reinforcement learning (DRL) control methodology for a open hardware mobile robotic platform using standardized software and communication components; testing and evaluation of the control performance of the DRL technique in a dedicated simulation environment and through implementation on the physical platform in a real scenario.

The rest of the paper is structured as follows. Section 2 presents the methodology of our work which includes both a theoretical background of the reinforcement learning technique combined with deep neural network architecture training for parameter prediction, and a description of the open hardware mobile robotic platform used for experiments. The in-depth implementation and analysis of results are described in Sect. 3. The focus is on the stepwise implementation of the RL methodology, the heuristics used for the improvement of the control performance and the evaluation in both simulation and real scenarios. Section 4 concludes the paper and lists potential relevant improvements for future work.

2   Methodology

2.1   Reinforcement Learning

Reinforcement learning (RL) can be applied on general learning problems that optimize a metric in a sequential way. Thus, reinforcement learning is suited for optimal control and operation in robotic systems. It has close ties with statistics, optimization, game theory etc. and can be used in many scientific scenarios [4]. RL is built by implementing a logic policy by simulating different case studies in which an agent optimizes the cost function. The positive actions are rewarded while the ones that have no benefit to the global goal are penalized. The objective of the optimization problem is to minimize a cost function in order to produce a desirable control policy:

$$G = R_0 + R_1 + \ldots + R_k = \sum_{t=0}^{\infty} R_t \tag{1}$$

where G is the total reward at a moment of time. In a practical example, the infinite horizon of steps becomes a finite one since we want the algorithm to run for a fixed number of steps. The cost function can be written as $R_t = \gamma^t r_t$, where γ^t is the discount factor.

$$\pi : A \times S \to [0, 1]; \pi(a|s) = P(a_t = a|s_t = s) \tag{2}$$

Function π represents the logic in which the probability P will implement the action a as a state s. According to [7] and [3] and introducing two new performance evaluation functions, the state evaluation function described in Eq. 3, and the action evaluation function described in Eq. 4, it results that the critical network is trained according to Bellman model described in Eq. 5, and the actor network is updated according to Bellman model described in Eq. 6.

$$V_\pi(s) = E_\pi\{G|s = s_t\} = E_\pi\{R_t|s = s_t\} + E_\pi\{\gamma V_\pi(s)|s = s_{t+1}\} \tag{3}$$

where $E\{\}$ is the statistical averaging operator, $E_\pi\{R_t|s = s_t\}$ is the reward of state s at the current t or "immediate" time, denoted by R_{im}, and $E_\pi\{\gamma V_\pi(s)|s = s_{t+1}\}$ is the expected reward at the immediate next step, scaled by the discount factor γ,

$$Q_\pi(s, a) = E_\pi\{R_t|s = s_t, a = a_t\} + E_\pi\{\gamma Q_\pi(s, a)|s = s_{t+1}, a = a_{t+1}\} \tag{4}$$

with the notations mentioned in 3,

$$V_\pi^*(s) = \sum_{a \in A} \pi(a|s)(R_{im} + \gamma \sum_{s' \in S} P(s \to s', a)V_\pi^*(s')) \tag{5}$$

where $P(s \to s', a)$ is the probability of reaching state s' from the current state s by action/transition a, and $V_\pi^*(s)$ is the optimal value of the state evaluation function, considering the optimal $\pi(a|s)$ policy found,

$$Q_\pi^*(s, a) = R_{im} + \gamma \sum_{s' \in S} P(s \to s', a) \sum_{a' \in A} \pi(a'|s')Q_\pi^*(s', a') \tag{6}$$

with the same notations as in the Eq. 5.

Figure 1 illustrates how the RL algorithm integrates the two main components: the agent and the training environment and also the interactions between the network and the hardware equipment.

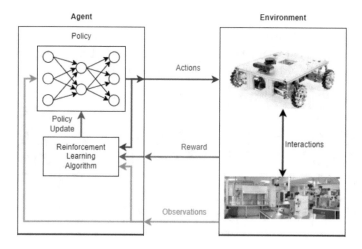

Fig. 1. Reinforcement learning architecture

2.2   Open Hardware Mobile Robotic Platform

This section presents the design of the robotic platform used for the study and the implementation of the deep reinforcement algorithm. The mobile robotic platform consists of:

- *Omnidirectional wheels*: based on their complex geometrical arrangement, the platform has a high degree of freedom, allowing for complex movements;
- *Suspension system*: perturbations caused by the vibration of the platform are reduced for precise positioning to the desired location;
- *Lidar*: used for scanning in order to accurately create a 3D map of the surrounding environment;
- *Depth camera*: used for complex tasks as object identification or spatial orienting method; in combination with the integrated NVIDIA Jetson module, the platform is capable of handling some of the latest and most complex optimization scenarios;
- *Drive system*: the mobile robot is equipped with two drivers needed for motor synchronization.

In addition to the physical equipment, a complex software architecture runs in parallel. ROS (Robot Operating System) is the open-source programming language used for mobile robot programming. Three main component compose a robot system: the perception of the surrounding, the logic used for calculations and the output used by the physical equipment, as shown in Fig. 2. In order to evaluate the environment the mobile platform is equipped with multiple sensors, including the depth camera enabling a large field of possible applications. In the logic part of the structure, either C++ or Python programming languages can be used. In general, a mobile robot uses DC or stepper motors to execute movement commands. In our case, the robot is using 4 DC motors. The operation of the DC motors is done by the integrated CANOpen module that receives the computed commands from the NVIDIA board and transforms them into usable information for the drivers. A summary of the main relevant technical specifications of the mobile robotic platform components is listed in Table 1.

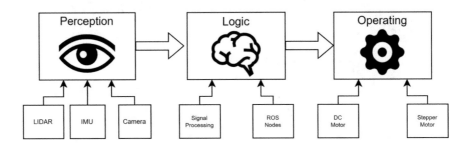

Fig. 2. Conceptual control system pipeline

Table 1. List of the hardware components

Component	Model	Specification
Main board	NVIDIA Jetson Tegra X2	Maxwell graphics processing unit and ARM A57 processor
Lidar	RP LIDAR A2M8 360	Resolution: 0.5 mm–1.5 m at a maximum range: 12 m
Camera	ASTRA PRO Depth	Distance: 0.6 m–8 m 1280 × 720 @30 fps
Motor	MD60 100W DC	Speed:175 rpm/67 rpm Power: 100 W
Wheel	Mecanum omnidirectional	Weight: 700 g Rolls number: 16
Driver	DFR0601 Motor Driver	Motor type: Brushed DC
Screen	SPI OLED LCD	Resolution: 128 px × 64 px

3   Results

3.1   Implementation

The first step in developing our application is building a neural network that will represent the brain of the robot. Using MATLAB Deep Network Designer we created an architecture for both networks, the actor (Fig. 3a) and the critic (Fig. 3b). The input data to those two networks is the same and it is distributed on two channels that will serve for a different purpose. The first channel will use all the data needed to avoid the obstacles. The data is received from the LiDAR sensor available as a vector of 720 values representing the distance in meters from the robot to the obstacle. The second channel serves for guiding the robot to the destination point and uses the information about the current state of the robot (current position, current velocity, target position and distance to the target) also available as an vector of 12 values. To help the agent develop a policy based on minimizing the distance between the robot and the target, we chose to also feed the data from the last three sample times.

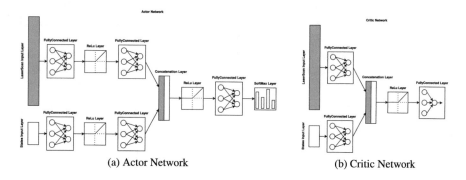

(a) Actor Network   (b) Critic Network

Fig. 3. Agent neural networks

After successfully building the network, we have created a simulation where the agent could start the training. Having access to the MATLAB Reinforcement Learning Toolbox, we have created a function-based environment to have maximum flexibility in the parametrization of training scenarios. The simulated agent is controlled by the neural network and it can interact with the simulated environment just like the real robot does, reading data and moving around according to the inverse kinematics of the real robot. We created more scenarios for the robot to train (Fig. 4), each scenario representing a different occupancy grid map and obstacles scattered around.

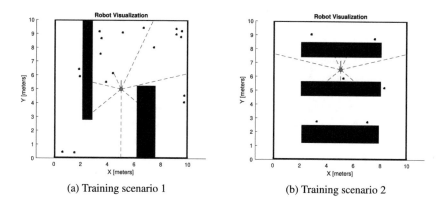

(a) Training scenario 1   (b) Training scenario 2

Fig. 4. Examples of training scenarios

The last step before starting the training is represented by the parametrization of the learning algorithm. We can set all common parameters of a learning algorithm like the learning rate, the gradient threshold, the discount factor and even the algorithm used for training. In this application we chose the Adam (Adaptive Movement Estimation) algorithm using an exploration policy for the agent training with a sample time of 0.05 s the learning rate is set to 10^{-4} and the discount factor is set to 0.99.

To accelerate the training process have we used the parallel training feature available in MATLAB. After activating this option, the learning process will create agents up to the NVIDIA multiprocessor count value; in our case we had six available processors. When using the parallel training we need to choose between the synchronous or asynchronous training. The synchronous training will make the agents pause their execution until all others are finished. This option is often used for a gradient-based parallelization where the main process updates the actor and critic weights according to the results of all agents. The asynchronous training uses the experiences sent by each agent to update the weights and as soon as a parallel process ends its episode it will receive a new set of updated parameters to start a new episode. We used the asynchronous method to speed up the training process because in our application an agent ends its episode if it reaches the destination or if it hits an obstacle. At first, it is not a problem for the agents to wait for each other to finish their episode since the chances of hitting an obstacle are higher than reaching the destination; however thinking about the future, when the agents end up making their way to the destination more and more often, it would be a waste if an agent hits an obstacle early and is forced to wait for the others to finish.

For the mathematical modelling of the desired behaviour we have used the branched reward function shown in the Eqs. 7 and 8. Each term is assigned an experimentally determined weight to encourage the good behaviours like "keeping a safe distance from the objects" and "approaching the target". Another way to help the agent develop a good policy is to feed a positive constant reward for reaching the target and a negative reward for hitting the obstacles.

$$R_t = Af_t^2 - 2Ar_t^2 + 0.5min(scans) + 0.1[dist(pos_t, target_t)]^{-1} \qquad (7)$$

$$R_t = \begin{cases} 1 & for\ dist(pos_t, target_t) < 0.1[m] \\ -10 & for\ min(scans) < 0.3[m] \end{cases} \qquad (8)$$

Once the training process begins, the developed environment will show the progress of the agents highlighting the reward accumulated by an agent on each episode, the average reward and the estimate of the discounted long-term reward Q0. We can also stop the training process if something goes wrong or if the agents do not make any progress, using this graphical interface. The training algorithm allows one setting the flags to stop the process automatically, when the average reward or even the episode reward exceeds some value. Another useful flag that can be set is the condition to auto-save an agent based on the average reward or the episode count, which will save all agents after that set value (Table 2).

Table 2. Training results

Episode	Ep. reward	Avg. reward	Ep. Q0	Elapsed time
2500	−112.13	−2392.29	−24.32	2 h
5000	−216.77	−1107.12	−33.18	4.5 h
7500	304.19	−974.02	−19.12	7 h
10000	417.71	−1109.16	−42.51	9.2 h

To deploy the network on the real-world robot, we needed to adapt the MATLAB environment to communicate with the ROS environment. The ROS Master must have access to the raw data of the sensors, i.e., the main process needs to run directly on the robot. However, the embedded system does not support the latest versions of MATLAB and that means we need the robot to send the ROS messages with the data from the sensors to the external station running the neural network and send back the ROS messages containing the commands for the movement. Using the MATLAB ROS toolbox and Robotics System Toolbox adapted our application for the real-world robot deployment. As long as the two stations are in the same wireless network, the ROS Master and the MATLAB ROS package can exchange messages without any trouble. Because our simulation was built around the real-world robot model, we had no problem adapting the code for the ROS environment. Instead of reading the data from the simulated environment we created the subscribers to access this information from ROS messages and instead of using the inverse kinematics to update the position of the agent in the simulation we created the ROS message that are sent directly to the robot to execute the real action.

3.2 Evaluation

As mentioned before, we used parallel computation to run the training simulations. At the end of each training session we obtain the evolution graph from which it can be seen how the agent started to accumulate more and more reward, meaning the policy developed becomes more accurate and the agent behaviour is approaching the one we want. To improve our network, we saved the agents with the most prominent result and used them as starting point of the new training sessions.

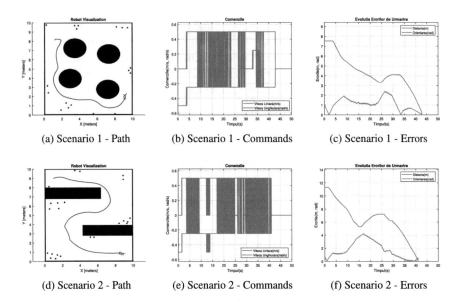

(a) Scenario 1 - Path   (b) Scenario 1 - Commands   (c) Scenario 1 - Errors

(d) Scenario 2 - Path   (e) Scenario 2 - Commands   (f) Scenario 2 - Errors

Fig. 5. Validation experiments on simulator

The final results are shown in the Fig. 5 in which it can be noticed that the validation scenarios are different from the training scenarios. We can see that even with no previous experiences the agent can find a way to the destination, keeping a safe distance from the obstacles scattered around the map.

The first position of the agent in each of the two scenarios was set to look at the upper left corner of the map. We see that the first action of the robot was to rotate until it would look in the direction of the target then proceed to search for the path to reach the destination. Every time it gets too close to the obstacles the policy developed chooses to stop in place and move around the walls until it finds a clear path to the target. When the robot arrives at the destination the simulation stops and, in real-world, the robot receives a "stop" command.

With this final agent we started testing in real-world environment and the results are shown in Fig. 6 and Fig. 7 along with the graphs for the executed commands and the graphs of the tracking errors. The commands are sent with a frequency 20 Hz and the data is also received 20 Hz. As we can see from these two sets of experiments, the agent is able to find a path to the destination even in the real environment. The sudden jump illustrated in Fig. 6d is caused by the change of orientation from 180° to −179.9° when it is moving backwards.

The wheels of the real robot are subjected to friction and at the same time the system gains inertia when it moves in relation to the simulator where the agent moves at a constant speed. These effects can be seen in Fig. 6c and Fig. 7c. The target of the robot is set manually at a laptop where we run the robot visualization program RViz included in ROS packages. It is interesting to see how the agent reacts to the real-life scenarios, since the LiDAR sensor has an

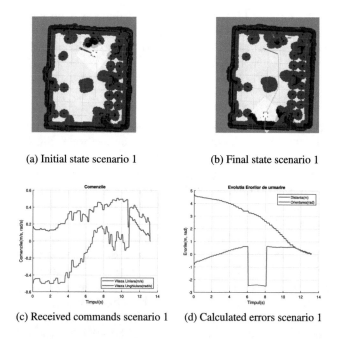

(a) Initial state scenario 1   (b) Final state scenario 1

(c) Received commands scenario 1   (d) Calculated errors scenario 1

Fig. 6. Validation experiment 1 on real-world environment

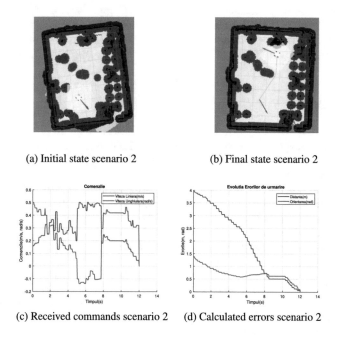

(a) Initial state scenario 2   (b) Final state scenario 2

(c) Received commands scenario 2   (d) Calculated errors scenario 2

Fig. 7. Validation experiment 2 on real-world environment

error dispersion of about 1 cm. Even if the environment is no longer as accurate as on the simulator, the agent still manages to comply with its policy, keeping a distance of about 10 cm from the nearest obstacle and looking for a way to the set target.

4   Conclusion

The article presented a reinforcement learning approach to train a fully connected neural network for the control of an open hardware mobile robotic platform. The implementation includes the design of the control and the communication between the MATLAB environment using ROS functions the robot hardware. The evaluation of the results was performed in a dedicated simulation environment as well as through implementation on the hardware robot system in a laboratory scenario.

Future work will be dedicated to comparing the DRL scheme to other classical (PID) and intelligent control methodologies such as genetic algorithms, nature-inspired heuristics fuzzy logic. The potential for deploying an online training algorithm directly on the robotic platform using the onboard computing resources will be explored. This will enable the robotic edge computing paradigm with or without an online cloud support system which should allow the continuous improvement of the control performance based real-time acquisition and perception of the changing environment.

Acknowledgements. Financial support from the Competitiveness Operational Program 2014- 2020, Action 1.1.3: Creating synergies with RDI actions of the EU's HORIZON 2020 framework program and other international RDI programs, MySMIS Code 108792, project acronym "UPB4H", financed by contract: 250/11.05.2020 is gratefully acknowledged.

References

1. Choi, J., Park, K., Kim, M., Seok, S.: Deep reinforcement learning of navigation in a complex and crowded environment with a limited field of view. In: 2019 International Conference on Robotics and Automation (ICRA), pp. 5993–6000 (2019). https://doi.org/10.1109/ICRA.2019.8793979
2. Fu, Y., Jha, D.K., Zhang, Z., Yuan, Z., Ray, A.: Neural network-based learning from demonstration of an autonomous ground robot. Machines **7**(2), 24 (2019). https://doi.org/10.3390/machines7020024, http://dx.doi.org/10.3390/machines7020024
3. Han, X.: A mathematical introduction to reinforcement learning (2018)
4. Li, Y.: Deep reinforcement learning: Opportunities and challenges. arXiv preprint arXiv:2202.11296 (2022)
5. Luchian, R.A., Rosioru, S., Stamatescu, I., Fagarasan, I., Stamatescu, G.: Enabling industrial motion control through IIoT multi-agent communication. In: IECON 2021–47th Annual Conference of the IEEE Industrial Electronics Society (2021)
6. Luchian, R.A., Stamatescu, G., Stamatescu, I., Fagarasan, I., Popescu, D.: IIoT decentralized system monitoring for smart industry applications. In: 2021 29th Mediterranean Conference on Control and Automation (MED), pp. 1161–1166 (2021)

7. Peng, Y., Zhang, X., Jiang, Y., Xu, X., Liu, J.: Leader-follower formation control for indoor wheeled robots via dual heuristic programming. In: 2020 3rd International Conference on Unmanned Systems (ICUS), pp. 600–605 (2020). https://doi.org/10.1109/ICUS50048.2020.9274823

8. Rosioru, S., Mihai, V., Neghina, M., Craciunean, D., Stamatescu, G.: PROSIM in the cloud: remote automation training platform with virtualized infrastructure. Appl. Sci. **12**(6), 3038 (2022)

9. Shabbir, J., Anwer, T.: A survey of deep learning techniques for mobile robot applications. CoRR abs/1803.07608 (2018)

10. Wang, B.: Path planning of mobile robot based on a algorithm. In: 2021 IEEE International Conference on Electronic Technology, Communication and Information (ICETCI), pp. 524–528 (2021). https://doi.org/10.1109/ICETCI53161.2021.9563354

Investigations on Real-Time Image Recognition with Convolutional Neural Networks on Industrial Controllers

Rando Raßmann[1]([✉]), Christoph Wree[1], Fabian Bause[2], and Brian Hansen[1]

[1] Faculty of Computer Science and Electrical Engineering, University of Applied Sciences Kiel, 24149 Kiel, Germany
rando.rassmann@fh-kiel.de

[2] Beckhoff Automation GmbH & Co. KG, 33415 Verl, Germany

Abstract. Modern manufacturing systems are collecting more data as a result of progressive digitization. In the application area of individualized production, the collected data can be used, e.g., with the help of image processing systems, to control required workflows by classifying individual objects. Machine learning (ML) methods can be used to evaluate and classify the captured images. As soon as the acquired objects shift rotationally or translationally within the acquisition area, a reliable classification of the objects with a simple multilayer perceptron (MLP) becomes difficult. Convolutional Neural Networks (CNNs) use methods that can offer a solution to these problems. This paper investigates whether it is possible to use computation intensive CNNs for image recognition, in real-time and in coordination with machine and motion control tasks. These CNNs are compared to less computation intensive MLPs. In this paper, an innovative approach is presented to show how trained CNNs can be integrated into the real-time environment of a programmable logic controller (PLC). Based on two application examples, several CNNs with different network architectures are integrated into a PLC runtime environment on a standard industrial PC (soft-PLC). The execution times of the different networks are measured and compared.

Keywords: Machine Learning · Individualized Production · Convolutional Neural Networks

1 Introduction

As the digitization of manufacturing systems increases, more data can be collected. The collected data can then be used in a variety of ways. Possible areas of application are, for example, the optimization of processes in quality management or the implementation of an individualized production. Individualized production refers to a production system that can manufacture products for a single customer at low cost [1].

Image acquisition, processing and analysis (machine vision) is an essential component here. Machine learning (ML) methods can help to reduce the programming efforts required to implement complex solution approaches. In addition, they can also improve the efficiency of machine vision solutions [2].

© The Author(s), under exclusive license to Springer Nature Switzerland AG 2023
T. Borangiu et al. (Eds.): SOHOMA 2022, SCI 1083, pp. 380–391, 2023.
https://doi.org/10.1007/978-3-031-24291-5_30

There are several ways to integrate a machine vision system into a production environment. For example, machine vision systems can be implemented with the help of an external computer [3], on an intelligent camera [3], or within a PLC runtime environment that is hosted on an industrial PC (soft-PLC) [4]. If the image processing is implemented within the PLC runtime environment, it is executed centrally on the same system as the machine control and motion control. This allows the different tasks to be synchronized with each other more efficiently in terms of time and reduces latency and additional interfaces. For real-time applications, models with a deterministic runtime complexity are required. The integration of all tasks into one central system avoids separate subsystems, guarantees full transparency overall processing steps (especially compared to smart cameras) and reduces the complexity to implement IT-security.

In [5] it is shown that it is possible to integrate Neural Networks for classification problems, in the form of Multilayer Perceptrons (MLPs), into the real-time environment of a PLC and to execute them in the sub-millisecond range.

For simple problems or in case of reproducible conditions, MLPs are sufficient for image classification. However, as soon as objects shift rotationally and translationally within the region of the image, a reliable classification of the images is difficult [2]. Convolutional Neural Networks (CNNs) are therefore used for more complex pattern and image recognition problems. CNNs have many advantages over an MLP in terms of image classification. However, they require more computational effort, which makes a direct integration of a CNN into a PLC runtime environment more challenging.

Another approach to executing Neural Networks (e.g., MLPs or CNNs) in conjunction with a PLC is to run the networks on separate processors. For example, the manufacturer Siemens offers the extension module TM-NPU [6] for the PLC S7-1500. Through this additional module it is possible to couple the execution of Neural Networks with the PLC runtime environment. This approach is, however, cost-intensive and introduces additional latency.

This paper demonstrates that it is possible to execute CNNs for image recognition in real-time directly on a industrial soft-PLC, i.e. no additional hardware and no special new controller design is necessary. The execution times of different network architectures with varying input image size are compared. The PLC used also performs machine and motion control tasks simultaneously.

This paper is structured as follows. Section 2 describes the differences between an MLP and a CNN. Then, Section 3 describes the network architectures used for the training of the networks and for the integration of the networks into the real-time environment of the PLC. Different Neuronal Networks are executed for two application examples of individualized production systems and presented in Sect. 4. Section 5 discusses how the networks are integrated into the runtime environment of the PLC. In addition, the required execution times and the accuracy of the different networks are measured. The results of the measurements are discussed in Sect. 6.

2   Neural Networks for Image Classification

Conventional image processing systems use feature extraction to identify specific objects within an image [3]. If such image processing systems are to be used for products with a high variance (e.g., due to changes in shape, colour, size, or surface texture), the complexity of the system increases. The amount of engineering work and expertise required for such a solution increases.

An alternative approach to solving such a complex problem is ML. In [7, 8], and [9] it is shown that with the help of ML methods solutions for image processing systems can be implemented. Possible models to deploy ML solutions are MLPs and CNNs.

Deep Multilayer Perceptron

The deep MLP is often considered to be a "fully-connected" or a "Dense Neural Network". It has an input layer, one or more hidden layers, and an output layer. Each layer comprises a certain number of neurons, which in turn consist of weights and an activation function. Fully connected implies that each neuron of one layer is connected to all neurons of the following layer. Deep MLPs have multiple hidden layers.

Convolutional Neural Networks

If the objects that need to be classified deviate in any way from the training data, the accuracy of MLPs can be greatly reduced. Due to their design, MLPs are effective at recognizing spatial features compared to CNNs. For this reason, MLPs are more sensitive to objects that shift within the image translationally or rotationally or to changes in scale of the object. The network architecture of CNNs allows to solve these problems more accurately.

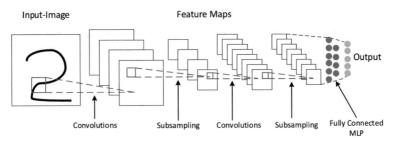

Fig. 1. Schematic structure of a Convolutional Neural Network

After the input data (the input image) has been transferred to the network in the form of a matrix, the input data is not passed directly to a dense layer, as in an MLP. Instead it is passed to a convolutional layer. This layer uses the mathematical operation of a convolution onto the image. Here, a convolutional matrix, which is called kernel, is placed over the matrix of the image, and gradually shifted over the entire image matrix. The result of a convolutional layer is called a feature map (see Fig. 1). The values within the kernel are the learnable parameters (weights) that are determined during training. The calculation of the values of a feature map is done as follows. Each element within the kernel is overlaid on an element of the matrix. For each of the elements overlying each other, the product is created and then summed up. The result is an element in the matrix of the feature map. Within a convolutional layer, there can be multiple kernels, each with its own feature map.

Another frequently used layer in CNNs is the so-called pooling layer. This layer is used to summarize and to generalize information. There are different pooling functions;

a frequently used pooling layer is MaxPooling. Here, the feature map is divided into many small squares of size z × z. Then the feature map is reduced by taking only the largest value within each square. The size of these squares should not be too large, otherwise too much information will be lost. Typical values for z are 2, 3, and 4.

In general, convolutional layers and pooling layers are repeated several times in series. Several convolutional layers can also be connected in series. The end of a CNN forms a (deep) MLP. The architecture of a CNN depends on its application and must be reinvestigated for each new application.

3 Application Example in Real-Time: The Production System

The trained Neuronal Networks are integrated into a flexible production system for the investigation of their execution time. The production system is operating with a TwinCAT runtime hosted on a standard Beckhoff C6920 industrial PC with an Intel® Celeron® T3100 1.9 GHz 2-core processor. It hosts the control of the machine itself, the control of the transport system, the control of the delta robot, the image acquisition and pre-processing, and the CNN respectively the MLP. The runtime environment used is TwinCAT 3.1 Build 4024. TwinCAT is an automation software from Beckhoff that enables to integrate PLC and motion control applications on PC-based control systems. Since the TwinCAT runtime environment can be executed on an industrial PC, the soft-PLC has sufficient processing power. It can be used to manage more complex tasks besides the regular control of the PLC program. The authors have chosen TwinCAT 3 as the PLC runtime environment because to the best of their knowledge it is the only PLC runtime environment that allows the integration of CNNs.

The production system is structured as follows:

Workpieces are located in an input storage. A delta robot removes the workpieces from the input storage and places them on a transport system. The transport system can move the workpieces to different stations and allows to determine the position of the movers at any time. The production system has a total of three stations, two processing machines and one camera system (see Fig. 2). The camera system has a LED ring light to reduce the influence of changing light conditions. The machine control, the control of the robot and the transport system as well as the image processing are executed on different tasks but in the same runtime environment. The cycle times of the different tasks vary between 250 μs for the delta robot and transport system and 1 s for the task on which the Neuronal Networks are executed.

Application example 1 "digit classification" is used to demonstrate the behaviour for products with a high variation within a product class. Depending on the classification result, the workpieces labelled with handwritten digits are sorted into the corresponding area of the output storage (0 to 4 in the upper area and 5 to 9 in the lower area of the storage). A video of this demonstrator can be seen in [10].

In application example 2, cuboids that represent hydraulic blocks must be classified. The hydraulic blocks vary in terms of the number, position, depth, diameter, and type of their boreholes. The hydraulic blocks are stored one above the other in the input storage. Depending on how the hydraulic blocks are stacked, they can also be rotated 180° around the z-axis in the input storage. Application example 2 is intended to represent a more realistic production example compared to the digit classification.

For both application examples, it is aimed that the networks are executed in less than 1 s.

a)   Setup for application example 1   b)   Setup for application example 2

Fig. 2. Application examples for individualized production using ML systems

4   General Process for Using Machine Learning

4.1   Data Collection

For application example 1, the free available MNIST dataset [11] is used. For application example 2, a self-defined dataset is generated. Here, ten different hydraulic blocks are created using a 3D CAD program. In order to have a sufficiently large dataset for training the networks, a synthetic dataset with slightly varying features is created.

4.2   Preparation of the Data Sets for Training

The data from the two existing data sets is prepared for the purpose of training. For both application examples, 8-bit grayscale images are used. The pixel values are normalized to a value range from 0 to 1 before training.

In application example 1, the images have a size of 28×28 pixels. The dataset consists of 70000 images in total, where 60000 images are used for training and 10000 images are used for network validation. For further investigations, the accuracy of the networks is tested with an own test data set, which contains only captured images from the production system (see Table 4).

For the application example 2, the images have a size of 300×200 pixels. The synthetic dataset consists of 20000 images in total. 18000 images are used for the training and 2000 images are used for the validation.

4.3   Training of the Models

The way a network is converted into machine-readable code using a compiler significantly affects the execution time of the network on the industrial controller [12].

In this research, MATLAB's "Deep Learning Toolbox" (R2022a) is used for training all networks. For both application examples, different network architectures with varying complexity are built. Figure 3 and Fig. 4 show an example of four different networks for application example 1. The input and output layers are identical for all four networks. The input layer reads the 28 × 28 pixel grayscale image. The output layer has 10 neurons (one neuron for each class). Using the activation function softmax, the output layer outputs return the probability for all 10 classes.

The first network (MLP_1) is an MLP which includes two hidden layers with 1024 and 512 neurons. The execution time of the MLP is later compared to those of the CNNs. The other three networks shown are CNNs. The network CNN_1 is a simple CNN, which consists of two convolutional layers and then goes directly to the output layer.

The structure of the two networks M_7 and C_3 is taken from [13]. In [13], an architecture for a CNN for the MNIST dataset has been developed for a very high classification accuracy. In the later investigation of the execution times, further network architectures from [13] are used.

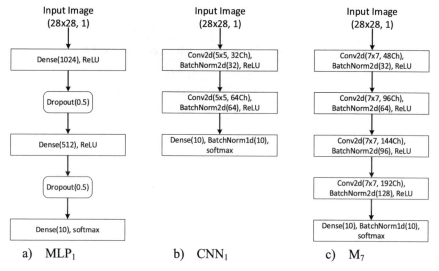

Fig. 3. Network architecture of selected MLP and CNNs for application example 1

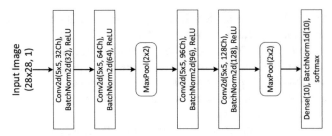

Fig. 4. Network architecture of CNN C₃ for application example 1

The number of learnable parameters for each layer of the networks in Fig. 3 and Fig. 4 are listed in Table 2. It can be seen that the number of parameters depend a lot on the architecture of the networks.

Table 1. Comparison of the learnable parameters for the individual layers for the networks from Fig. 3 and Fig. 4

Layer	MLP₁		CNN₁	
	Layer type	Parameters	Layer type	Parameters
1	Dense	803,840	conv2d	320
2	Dense	524,800	conv2d	18,496
3	Dense	5,120	Dense	501,760
	Total:	≈ 1,300,000	Total:	≈ 520,000
Layer	M₇		C₃	
	Layer type	Parameters	Layer type	Parameters
1	conv2d	2,852	conv2d	832
2	conv2d	22,592	conv2d	51,264
3	conv2d	677,520	Maxpool	–
4	conv2d	1,354,944	conv2d	153,696
5	Dense	1,505,280	conv2d	307,328
6	–		Maxpool	–
7	–		Dense	1,003,530
	Total:	≈ 3,600,000	Total:	≈ 1,500,000

For application example 2, a custom network architecture is chosen to solve the classification problem. The architecture of the network is built in such a way that the execution time is short and at the same time a high classification accuracy is achieved for the validation data. MATLAB's Experiment Manager is used for this optimization. The network architecture can be seen in Fig. 5.

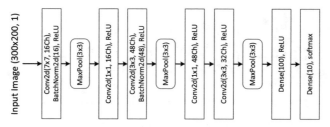

Fig. 5. Network architecture of CNN$_2$ for application example 2

5 Network Execution on the Production System

The following subsections describe how the Neuronal Networks are integrated into the program of the PLC runtime environment. Subsequently, the execution times of the networks on the production system are measured.

5.1 Integration of the Networks into the Runtime Environment of the PLC

A compiler is needed to convert the trained networks into machine-readable code. There are several ways to integrate MLPs into the PLC real-time environment [5]. Since CNNs are much more computationally intensive, incorporating CNNs into the PLC runtime environment is much more challenging. The trained Neural Networks are placed in MATLAB functions that call the predict method. The MATLAB Coder together [14] with the TE1401 TwinCAT Target for MATLAB [15] is used to convert the MATLAB functions into a PLC Library that can be used in TwinCAT Engineering to integrate those generated function blocks in the PLC project.

5.2 Pre-processing of Captured Images on the Production System

This section describes how the process from the acquisition of the image to the classification of the acquired image in the production system is implemented.

An external trigger for the camera is triggered by a virtual cam, which captures an 8-bit grayscale image of the workpiece with which the mover is loaded. The image is captured without the mover having to stop. Then, depending on the application example, the region of interest is determined. Thus, the captured image is reduced to the relevant image section (see Fig. 6a).

The size of the image segment must have the same dimensions as the input layer of the respective network. For this purpose, a bilinear filtering is chosen, which reduces the image segment to the corresponding number of pixels by a linear interpolation, as shown in Fig. 6b. For application example 1, this corresponds to a size of 28 × 28 pixels and for application example 2, to a size of 300 × 200 pixels.

The data type from the image cannot be passed directly to the networks. The image must be converted into a two-dimensional array beforehand.

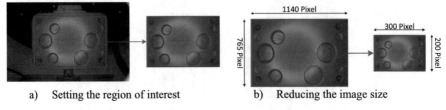

a)   Setting the region of interest   b)   Reducing the image size

Fig. 6. Defining the region of interest and scaling the image size

5.3   Comparison of the Trained Networks on the Production System

After integrating the networks into the PLC program the execution times of the different Neuronal Networks are measured. The measured execution times of the different network architectures are shown in Table 2 (application example 1) and Table 3 (application example 2). Many studies found that higher accuracies and more robust networks for classification problems can be achieved using CNNs. Especially for the MNIST dataset, there are numerous investigations in this area. In [13] it is shown, with the network architectures C_1, C_2, C_3 and M_7 accuracies of up to 99.8% are achieved for the validation data.

Table 2. Execution times of the networks for application example 1

Network	Minimum execution time [ms]	Maximum execution time [ms]	Average execution time [ms]	Number of executions
C_1	103.882	105.222	104.200	100
C_2	113.503	114.366	113.776	100
C_3	296.898	298.328	297.151	100
M_7	983.850	986.390	984.954	100
CNN_1	10.730	10.883	10.774	100
MLP_1	1.955	2.815	2.526	100

The six networks examined for application example 1 have significant differences in their execution times. The network M_7 takes by far the longest with 987 ms. The second slowest is the network C_3 with 299 ms, followed by C_2 and C_1 with 115 ms and 106 ms, respectively. The fastest CNN is the network CNN_1 with 11 ms. MLP_1 needs 3 ms for execution.

Table 3. Execution times of the networks for application example 2

Network	Minimum execution time [ms]	Maximum execution time [ms]	Average execution time [ms]	Number of executions
CNN_2	126.047	126.833	126.328	100

Finally, the trained networks were analyzed for accuracy using a self-produced test dataset. The test dataset contains 531 captured images of the production system. The test dataset and the used networks will be made publicly available[1]. Figure 7 shows example images for each digit from the test data set.

Fig. 7. An example of digits from the test dataset

The accuracy of the trained networks on the validation dataset and on the test dataset is shown in Table 4. It can be seen that all CNNs (C_1, C_2, C_3, M_7 and CNN_1) achieve an accuracy above 0.99 with the validation dataset. The MLP MLP_1 achieves an accuracy of 0.9691 with the validation data set.

The captured images from the test data set have significant deviations to the data from the original MNIST dataset (e.g., smudged background, blurred writing). For this reason, the accuracy of the analyzed networks for the test dataset is clearly below the accuracy for the validation dataset. It can be seen that all CNNs achieve a higher accuracy than the MLP (0.4840). The highest accuracy is achieved by the network C_2 (0.8023), followed by the networks C_1 (0.7740) and M_7 (0.6328).

Table 4. Accuracy of the networks from application example 1

Network	Validation dataset		Test dataset	
	Loss	Accuracy	Loss	Accuracy
C_1	0.0215	0.9941	0.7877	0.7740
C_2	0.0579	0.9947	0.7803	0.8023
C_3	0.0178	0.9959	1.4431	0.5235
M_7	0.0136	0.9968	1.0870	0.6328
MLP_1	0.1092	0.9691	1.8546	0.4840
CNN_1	0.0257	0.9923	1.6924	0.5254

6 Discussion

All created networks could be successfully integrated and executed in the runtime environment of the PLC. The reason for the different execution times of the analyzed networks for classification of the images in application example 1 is that the networks are varying in their size and use different layers in their architectures (compare Table 1). The comparison of the execution times of the networks MLP_1 and CNN_1 and the comparison of

---

[1] https://github.com/rrassmann/HandwrittenDigitRecognition.

the learnable parameters demonstrates once again how computationally complex CNN are.

When evaluating the execution times, it becomes clear that with increasing complexity the required execution time increases. The most complex network for application example 1 is the M_7. The average execution time for this is 985 ms.

The architecture of the CNNs in application example 1 is sufficient to successfully use them for quality control tasks or classification tasks in a production system. In [16] and [17], CNNs with a similar architecture are used to solve complex image classification problems.

All presented Neural Networks achieve a very high accuracy on the validation dataset during the training. However, the investigation shows that as soon as the networks are applied to a more challenging dataset, the accuracy of the networks decreases significantly. In this case, the CNNs are more robust to the variations and are able to achieve higher accuracy.

With application example 2, a more realistic production example is shown. The average execution time of CNN_2 is 127 ms for an input image size of 300×200 pixel.

The two application examples show that the classification results are available in a sufficient time (in less than 1 s). Thus, computationally complex CNNs can be integrated directly on a PLC, also in the case of dynamic processes. This can replace conventional image processing systems that need to be interfaced with the PLC and potentially other subsystems.

7 Conclusion

This investigation demonstrates that it is possible to integrate CNNs directly into the PLC runtime environment.

Neural Networks can be trained in MATLAB or they can be imported via MATLAB's import capabilities. A MATLAB function can be coded to call the predict method of the trained Neuronal Networks. The TwinCAT Target for MATLAB can be used to convert these functions into a standard PLC library for the integration into a TwinCAT PLC project. Thus, the Neural Networks, once compiled on the runtime system, are executed in the soft-PLC runtime on a standard industrial controller. The authors are convinced that the presented procedure for the use of CNNs in the real-time PLC runtime environment has not yet been described in this way and thus it represents an innovative approach for the integration of CNNs.

In the context of automation technology, this means that MLPs respectively CNNs do not necessarily have to be executed on a separate controller. This avoids interfaces to subsystems which often reduce execution time, increase the overall system complexity and make IT-security more challenging (e.g., multiple systems to maintain or open ports for communication).

Depending on the use case, machine learning applications can be integrated on existing systems without the need to purchase new hardware. The ability to run CNNs in addition to MLPs in the PLC runtime environment significantly expands the use of machine learning within an industrial application.

References

1. Koren, Y.: The local factory of the future for producing individualized products. Bridge Nat. Acad. Eng. **51**(1), 20–26 (2021)
2. Wani, M.A., et al.: Advances in Deep Learning. Springer, Singapore (2021). https://doi.org/10.1007/978981-13-6794-6
3. Hornberg, A.: Handbook of Machine and Computer Vision. Wiley-VHC (2017)
4. Beckhoff: TwinCAT 3 Engineering (2022). https://www.beckhoff.com/en-us/products/automation/twincat/te1xxx-twincat-3-engineering/te1000.html. Accessed 14 Jul 2022
5. Wree, C., Raßmann, R., Daâs, J., Bause, F., Schönfeld, T.: Real-time image analysis with neural networks on industrial controllers for individualized production. In: Borangiu, T., Trentesaux, D., Leitão, P., Cardin, O., Joblot, L. (eds.) Service Oriented, Holonic and Multi-agent Manufacturing Systems for Industry of the Future: Proceedings of SOHOMA 2021, pp. 471–481. Springer, Cham (2022). https://doi.org/10.1007/978-3-030-99108-1_34
6. Siemens, SIMATIC S7-1500 TM NPU. https://new.siemens.com/global/de/produkte/automatisierung/systeme/industrie/sps/simatic-s7-1500/simatic-s7-1500-tm-npu.html. Accessed 14 Jul 2022
7. Wang, J., et al.: Deep learning for smart manufacturing: methods and applications. J. Manuf. Syst. **48**(2018), 144–156 (2018)
8. Tercan, H., Meisen, T.: Machine learning and deep learning based predictive quality in manufacturing: a systematic review. J. Intell. Manuf. **2022**, 1–27 (2022)
9. Weimer, D., et al.: Design of deep convolutional neural network architectures for automated feature extraction in industrial inspection. CIRP Ann. **65**(1), 417–420 (2016)
10. University of Applied Sciences Kiel: Real-time object recognition with neural networks for industrial control systems. https://www.youtube.com/watch?v=KmPGQ2w_Peg&t=3s. Accessed 14 Jul 2022
11. LeCunn, Y., et al.: THE MNIST Database of handwritten digits. http://yann.lecun.com/exdb/mnist/. Accessed 14 Jul 2022
12. Li, M., et al.: The deep learning compiler: a comprehensive survey. IEEE Trans. Parallel Distrib. Syst. **32**(3), 708–727 (2021). https://doi.org/10.1109/TPDS.2020.3030548
13. An, S., et al.: An ensemble of simple convolutional neural network models for MNIST digit recognition. arXiv preprint arXiv:2008.10400 (2020). https://doi.org/10.48550/arXiv.2008.10400
14. MathWorks, MATLAB. https://de.mathworks.com/help/coder/release-notes.html. Accessed 14 Jul 2022
15. Beckhoff, TE1401 | TwinCAT 3 Target for MATLAB®. https://www.beckhoff.com/de-de/produkte/automation/twincat/texxxx-twincat-3-engineering/te1401.html. Accessed 29 Aug 2022
16. Liu, Q., Huang, C.: A fault diagnosis method based on transfer convolutional neural networks. IEEE Access **7**(2019), 171423–171430 (2019)
17. Du, B., et al.: Intelligent classification of silicon photovoltaic cell defects based on eddy current thermography and convolution neural network. IEEE Trans. Industr. Inf. **16**(10), 6242–6251 (2019)

Reconfigurable Manufacturing Systems

A Reinforcement Learning Approach for Solving Integrated Mass Customization Process Planning and Job-Shop Scheduling Problem in a Reconfigurable Manufacturing System

Sini Gao[(⊠)], Joanna Daaboul, and Julien Le Duigou

Université de Technologie de Compiègne Roberval (Mechanics, Energy and Electricity),
Centre de recherche Royallieu - CS 60319, 60203 Compiègne Cedex, France
sini.gao@utc.fr

Abstract. This paper addresses the integrated process planning and job-shop scheduling problem for mass customization in a reconfigurable manufacturing system. A bi-objective mixed-integer non-linear programming mathematical model for minimizing the total tardiness penalty of products and the total cost covering setup, machine reconfiguration as well as processing activities is built to formulate the problem. A Q-learning based reinforcement learning solution approach is presented to solve the formulated problem. Numerical experiments were carried out to validate the mathematical model and the solution approach. The computational results of the numerical examples show the great efficiency of the proposed solution approach in the aspect of computation time, compared with NSGA-II and the exhaustive search. The effectiveness of the problem-specific designed policies is also discussed.

Keywords: Reconfigurable manufacturing system · Mass-customized products · Process planning · Job-shop scheduling · Q-learning

1 Introduction

The fourth industrial revolution has profoundly transformed the world and described new data-driven manufacturing in mass customization (MC) and personalization [1]. Manufacturers have to accept small-scale orders due to the limited profit margins and huge costs caused by the vacancy of production lines [2]. For both Small to Medium Enterprises and Large Manufacturing Enterprises, Reconfigurable Manufacturing Systems (RMSs) have the potential to provide new agility, scaling beyond traditional design methodologies, adaptability and resilience to future disruptive crises [3]. Utilizing RMS enables building a "live" factory that structures its changes cost-effectively in response to markets and customers' needs, so it can keep supplying products at competitive price for many years after the factory design [4].

Indeed, the reconfiguration capability makes planning and scheduling processes even more complex [5]. Not surprisingly, meta-heuristics are the most used solution

© The Author(s), under exclusive license to Springer Nature Switzerland AG 2023
T. Borangiu et al. (Eds.): SOHOMA 2022, SCI 1083, pp. 395–406, 2023.
https://doi.org/10.1007/978-3-031-24291-5_31

approaches and the reconfigurability seems to arouse more interest among researchers working on optimization for RMS [6]. With the advent of the Industry 4.0, copious availability of data, high-computing power and large storage capacity have made of machine learning approaches an appealing solution to tackle manufacturing challenges [7]. A statistic of studies related to manufacturing and machine learning from 2000 to 2020 shows that the number of studies has been increasing rapidly, with the proportion of deep learning increasing in recent years and the number of reinforcement learning (RL) increasing in the last three years [8].

As the complexity of the production environment increases, the application of RL for dynamic scheduling becomes more beneficial [9]. Some researchers have already done this research in RMS. A job releasing schedule agent is trained using the double Deep Q-learning method to optimize the completion of the assigned order lists while minimizing the reconfiguration actions [10]. A dynamic control policy is found via a group of deep RL agents for the scheduling problem of RMS on multiple products aiming at minimizing the makespan of set of continuously updated customer orders by guiding a group of automated guided vehicles to move machine components, raw materials and finished products inside the system [11]. The total tardiness cost of dynamically arriving jobs in a reconfigurable flow line is minimized by adopting the *advantage actor-critic procedure*, a relatively new method of RL [12].

However, to the best of our knowledge, there is no study applying RL approach to concurrently solve multi-unit process planning and scheduling for MC in RMS, especially in the form of multi-objective optimization. Aiming to fill this research gap and explore the effectiveness of RL for MC operation management when considering the reconfiguration of RMS, a bi-objective mathematical model for MC integrated process planning and job-shop scheduling in RMS and a Q-learning based RL solution approach are demonstrated in this paper. Owing to the lack of practical data, the mathematical model and the solution approach are validated by the numerical experiments. Results are tested by the exact Pareto-optimal solutions obtained from the exhaustive search method.

The rest of this paper is organized as follows: Sect. 2 introduces the formulation of the integrated process planning and job-shop scheduling problem for MC in RMS. Section 3 presents the Q-learning based solution approach. Section 4 discusses the performance of numerical experiments for validation. Section 5 concludes the contribution of this work and the perspectives for further research.

2   Problem Formulation

The process planning and scheduling literature in RMS describes how operations are assigned to machines, tools and configurations of resources [13]. This paper settles the joint optimization problem. As shown in Fig. 1, an MC order consists of some mass-customized products. Each product can be decomposed into a number of parts belonging to several categories (part variants). Each type of part is made executing a set of operations scheduled according to a graph of precedencies. The task to produce a part is defined as a *job* in this paper. All the jobs are realized in an RMS having a few *machines* with multiple *configurations* that can be changed.

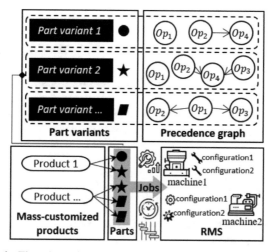

Fig. 1. The schematic diagram of producing an MC order in RMS.

Below is the mixed-integer non-linear programming mathematical model:

Sets

- $I = \{1, \ldots, i, \ldots, i', \ldots, |I|\}$ Set of mass-customized products.
- $V = \{1, \ldots, v, \ldots, v', \ldots, |V|\}$ Set of part variants.
- $J_{i,v} = \{1, \ldots, j, \ldots, j', \ldots, |J_{i,v}|\}$ Set of jobs belonging to part variant v for product i, $\varnothing$ if there is no job belonging to part variant v for product i.
- $OP = \{1, \ldots, e, \ldots, e', \ldots, |VOP|\}$ Set of operations.
- $VP_v = \{\ldots, e, \ldots, e', \ldots\}$ Set of operations to process part variant v.
- $K_{v,e} = \{\ldots, e', \ldots\}$ Set of operations precedent to operation e when processing part variant v, $\varnothing$ if there is no operation precedent to operation e.
- $M = \{1, \ldots, m, \ldots, m', \ldots, |M|\}$ Set of machines.
- $G_m = \{1, \ldots, g, \ldots, g', \ldots, |G_m|\}$ Set of configurations on machine m

Parameters

- D_i: the due date of product $i \in I$.
- W_i: the tardiness penalty of product $i \in I$ per time unit.
- $MG_{e,m,g} = \begin{cases} 1, & \text{if machine } m \text{ with configuration } g \text{ can perform operation } e. \\ 0, & \text{otherwise.} \end{cases}$
- $PT_{e,v,m,g}$: the processing time of operation e for a WIP belonging to part variant v on machine m with configuration g.
- $PC_{e,v,m,g}$: the processing cost of operation e for a WIP belonging to part variant v on machine m with configuration g.

- $ST_{e,v,m,g}$: the setup time of operation e for a WIP belonging to part variant v on machine m with configuration g.
- $SC_{e,v,m,g}$: the setup cost of operation e for a WIP belonging to part variant v on machine m with configuration g.
- $RT_{m,g,g'}$: the reconfiguration time from configuration g to configuration g' on machine m, if $g = g'$, $RT_{m,g,g'} = 0$.
- $RC_{m,g,g'}$: the reconfiguration cost from configuration g to configuration g' on machine m, if $g = g'$, $RC_{m,g,g'} = 0$.

Independent Decision Variables

- $\alpha_{i,v,j,e}$: the machine to perform operation e in the operation sequence of job j belonging to part variant v in product i.
- $\varphi_{i,v,j,e}$: the configuration on machine $\alpha_{i,v,j,e}$ to perform operation e in the operation sequence of job j belonging to part variant v in product i.
- $\beta_{i,v,j,e}$: the beginning time to perform operation e in the operation sequence of job j belonging to part variant v in product i.
- Auxiliary decision variables
- T_i: the tardiness time of product i.
- TPC: the total processing cost to produce all mass-customized products.
- TRC: the total reconfiguration cost during the production period for an MC order.
- TSC: the total setup cost during the production period for an MC order.
- $c_{i,v,j,e}$: the completion time of operation e in the operation sequence of job j belonging to part variant v product i.
- $i_{m,l}$: the product of the lth processed operation on machine m.
- $v_{m,l}$: the part variant of the lth processed operation on machine m.
- $j_{m,l}$: the job of the lth processed operation on machine m.
- $e_{m,l}$: the lth processed operation on machine m.

The first objective function to minimize the total tardiness penalty is as follow:

$$Min \sum_{i=1}^{|I|} T_i \times W_i \tag{1}$$

The second objective function to minimize the total cost for MC in RMS is as follow:

$$Min\, TPC + TRC + TSC \tag{2}$$

subject to:

$$c_{i,v,j,e} = \beta_{i,v,j,e} + PT_{e,v,\alpha_{i,v,j,e},\varphi_{i,v,j,e}}, \quad \forall i \in I, \forall v \in V, \forall j \in J_{i,v}, \forall e \in VP_v \tag{3}$$

$$T_i = max\left(\max_{\forall v \in V} \max_{\forall j \in J_{i,v}} \max_{\forall e \in VP_v} (c_{i,v,j,e}) - D_i, 0 \right), \quad \forall i \in I \tag{4}$$

$$TPC = \sum_{i \in I} \sum_{v \in V} \sum_{j \in J_{i,v}} \sum_{e \in VP_v} PC_{e,v,\alpha_{i,v,j,e},\varphi_{i,v,j,e}} \tag{5}$$

$$i_{m,1}, v_{m,1}, j_{m,1}, e_{m,1} = \underset{i \in I, v \in V, j \in J_{i,v}, e \in VP_v}{\arg \min} \left\{ \beta_{i,v,j,e} \middle| \forall \alpha_{i,v,j,e} = m \right\} \tag{6}$$

$$i_{m,l}, v_{m,l}, j_{m,l}, e_{m,l} = \underset{i \in I, v \in V, j \in J_{i,v}, e \in VP_v}{\arg \min} \left\{ \beta_{i,v,j,e} \middle| \forall \alpha_{i,v,j,e} = m, \beta_{i,v,j,e} > \beta_{m,l-1} \right\}$$
$$\forall l \in N+, \ l > 1 \tag{7}$$

$$TRC = \sum_{m \in M} \sum_{l \in N+} RC_{\varphi_{i_{m,l},v_{m,l}j_{m,l},e_{m,l}},\varphi_{i_{m,l+1},v_{m,l+1}j_{m,l+1},e_{m,l+1}}} \tag{8}$$

$$TSC = \sum_{m \in M} \left(SC_{e_{m,1},v_{m,1},m,\varphi_{i_{m,1},v_{m,1}j_{m,1},e_{m,1}}} + \sum_{l \in N+, l>1} SC_{e_{m,l},v_{m,l},m,\varphi_{i_{m,l},v_{m,l}j_{m,l},e_{m,l}}} \right)$$
$$e_{m,l} \neq e_{m,l-1} \vee v_{m,l} \neq v_{m,l-1} \vee \varphi_{i_{m,l},v_{m,l}j_{m,l},e_{m,l}} \neq \varphi_{i_{m,l-1},v_{m,l-1}j_{m,l-1},e_{m,l-1}} \tag{9}$$

$$MG_{e,\alpha_{i,v,j,e},\varphi_{i,v,j,e}} = 1, \quad \forall i \in I, \forall v \in V, \forall j \in J_{i,v}, \forall e \in VP_v \tag{10}$$

$$\beta_{i,v,j,e} \geq c_{i,v,j,e'}, i \in I, \quad \forall v \in V, \forall j \in J_{i,v}, \forall e \in VP_v, \forall e' \in K_{v,e} \tag{11}$$

$$\beta_{i,v,j,e} \neq \beta_{i,v,j,e'}, \quad \forall i \in I, \forall v \in V, \forall j \in J_{i,v}, \forall e, e' \in VP_v \tag{12}$$

$$\beta_{i,v,j,e} \geq c_{i,v,j,e'}, \quad \forall i \in I, \forall v \in V, \forall j \in J_{i,v}, \forall e, e' \in VP_v, \beta_{i,v,j,e} > \beta_{i,v,j,e'} \tag{13}$$

$$\beta_{i_{m,1},v_{m,1}j_{m,1},e_{m,1}} \geq ST_{e_{m,1},v_{m,1},m,\varphi_{i_{m,1},v_{m,1}j_{m,1},e_{m,1}}}, \quad \forall m \in M \tag{14}$$

$$\beta_{i_{m,l},v_{m,l}j_{m,l},e_{m,l}} \geq c_{i_{m,l-1},v_{m,l-1}j_{m,l-1},e_{m,l-1}}$$
$$\forall m \in M, \forall l \in N+, l > 1, v_{m,l} = v_{m,l-1} \wedge e_{m,l} = e_{m,l-1} \wedge \varphi_{i_{m,l},v_{m,l}j_{m,l},e_{m,l}} \tag{15}$$
$$= \varphi_{i_{m,l-1},v_{m,l-1}j_{m,l-1},e_{m,l-1}}$$

$$\beta_{i_{m,l},v_{m,l}j_{m,l},e_{m,l}} \geq c_{i_{m,l-1},v_{m,l-1}j_{m,l-1},e_{m,l-1}}$$
$$+RT_{m,\varphi_{i_{m,l-1},v_{m,l-1}j_{m,l-1},e_{m,l-1}},\varphi_{i_{m,l},v_{m,l}j_{m,l},e_{m,l}}}$$
$$+ST_{e_{m,l},v_{m,l},m,\varphi_{i_{m,l},v_{m,l}j_{m,l},e_{m,l}}} \tag{16}$$
$$\forall m \in M, \forall l \in N+, l > 1, v_{m,l} \neq v_{m,l-1} \vee e_{m,l} \neq e_{m,l-1} \vee \varphi_{i_{m,l},v_{m,l}j_{m,l},e_{m,l}}$$
$$\neq \varphi_{i_{m,l-1},v_{m,l-1}j_{m,l-1},e_{m,l-1}}$$

$$\alpha_{i,v,j,e} \in M$$
$$\forall i \in I, \forall v \in V, \forall j \in J_{i,v}, \forall e \in VP_v \tag{17}$$

$$\forall \varphi_{i,v,j,e} \in G_{\alpha_{i,v,j,e}}$$
$$\forall i \in I, \forall v \in V, \forall j \in J_{i,v}, \forall e \in VP_v \tag{18}$$

$$\beta_{i,v,j,e} \in R+$$
$$\forall i \in I, \forall v \in V, \forall j \in J_{i,v}, \forall e \in VP_v \tag{19}$$

Constraints (3) to (9) define the auxiliary decision variables derived from the independent decision variables. Constraint (10) ensures that every operation is available to be performed by the selected machine and configuration. Constraint (11) indicates the beginning time of every operation no earlier than the completion time of its precedent operations. Constraints (12) and (13) express that there is no overlap for any two operations. Constraint (14) states that the time to process the first operation on each machine should be after its setup. Constraints (15) and (16) guarantee the processing of two consecutive operations on the same machine will not overlap and respect the reconfiguration and setup conditions in case of any difference. Constraints (17) to (19) specify the domains of the independent decision variables.

3   Solution Approach

Problems that can be solved by RL must be described as a finite Markov decision process, and the rewards of all actions in different states must be clearly explained [14]. From this point of view, the formulated problem introduced in Sect. 2 is simulated by a sequential decision-making event where a central agent is employed to assign jobs' operations to machines. Assignments are chronological, depending on the state of mass-customized products and the state of machines. The number of assignments (n) is determined, since the mass-customized products and their components are given, and the operations to process every part variant are known. The state of product i contains the values of three independent decision variables for all assigned operations and the set of all unassigned operations in this product (NOT_DO_i). The state of machine m contains the current configuration (g_m), the part variant of the current processing job (v_m), the current processing operation (e_m) and its completion time (ct_m). $ct_m = 0$ For the initial state of machine m.

An assignment starts by selecting machine m^*:

$$m^* = \arg \min_{m \in M} ct_m \tag{20}$$

then select the mass-customized product i^*:

$$i^* = \arg \min_{i \in I} f(i) = \begin{cases} W_i \times |NOT\_DO_i| \big/ (D_i - ct_{m^*}), ct_{m^*} < D_i \\ W_i \times (ct_{m^*} - D_i), ct_{m^*} \geq D_i \end{cases} \tag{21}$$

If more than one machine/product meet the conditions, the selection of m^*/i^* will be stochastic among all those alternatives.

The final decision is to select an operation that can be performed on the selected machine m^* from $NOT\_DO_{i^*}$ with a proper configuration. To be more specific, only the operations with no precedent operation or for which all precedent operations are assigned are ready to be selected. The pending operations in job (i, v, j) are collected in Set $TO\_DO_{i,v,j}$. Concerning two objectives of optimizing time and cost respectively, two policies P_1 and P_2 are designed to find (v^*, j^*, e^*, g^*):

$$P_1 : (v^*, j^*, e^*, g^*) = \arg \min_{\forall v \in V, \forall j \in J_{i^*,v}, \forall e \in TO\_DO_{i^*,v,j}, g \in G_{m^*}} c_{i^*,v,j,e} \tag{22}$$

$$P_2 : (v^*, j^*, e^*, g^*) = \underset{\forall v \in V, \forall j \in J_{i^*, v}, \forall e \in TO\_DO_{i^*, v, j}, g \in G_{m^*}}{\arg \min} f(v, j, e, g) =$$

$$\begin{cases} SC_{e,v,m^*,g} + PC_{e,v,m^*,g}, \mid ct_{m^*} = 0 \\ PC_{e,v,m^*,g}, \mid ct_{m^*} > 0, v = v_{m^*} \wedge e = e_{m^*} \wedge g = g_{m^*} \\ RC_{m^*,g_{m^*},g} + SC_{e,v,m^*,g} + PC_{e,v,m^*,g}, \mid ct_{m^*} > 0, v \neq v_{m^*} \vee e \neq e_{m^*} \vee g \neq g_{m^*} \end{cases}$$

$$(23)$$

Q-learning is a form of model-free RL [15]. It is referred as the class of methods that identify an optimal decision strategy in a sequential decision setting to estimate the optimal Q-function which measures the expected cumulative utility of each currently available decision, given that the decision maker will follow the optimal decision strategy in the future [16]. This study draws on the idea of the Q-learning for finding an optimal decision strategy path of better obtaining the approximate Pareto-optimal solutions to the formulated problem. There are two Q-tables, with n rows and two columns, for the evaluation of the above two policies at each assignment. An entire n-assignment procedure is considered an episode. In each episode, a new Q-table and a solution s to the formulated problem are generated. The new Q-table helps to upgrade the old Q-table during iteration. Here comes the pseudocode for upgrading the old Q-table:

```
At the end of each episode:
   Compare the new solution s with all solutions saved in the Pa-
   reto-optimal solution set S;
   If s is dominated by a solution in S:
     Keep the old Q-table;
   If s dominates all the solutions in S:
     Reset S = ∅ and add s in S;
     The old Q-table = the new Q-table;
   If s dominates some of the solutions in S:
     Remove the dominated solutions in S add s in S;
     Update all the Q-values in the old Q-table with the corre-
     sponding Q-values in the new Q-table;
   If the two objectives' values of s are the same as those of
   any solution in S:
     Update all the Q-values in the old Q-table with the corre-
     sponding Q-values in the new Q-table.
```

At each assignment, there are only three pairs of Q-values $(1, 0)$, $(0.5, 0.5)$, and $(0, 1)$ for policies P_1 and P_2. Therefore the number of possible decision strategy paths is 3^n. The pseudocode for updating Q-values is as follows:

For any assignment ordinal $\leq n$:
  Find all $(v_{P_1}^*, j_{P_1}^*, e_{P_1}^*, g_{P_1}^*)$ with minimum completion time ct^{min} by P_1, calculate the corresponding total cost for each of them, collect all $(v_{P_1}^*, j_{P_1}^*, e_{P_1}^*, g_{P_1}^*)$ with minimum total cost $tc_{P_1}^{min}$ into Set A;
  Find all $(v_{P_2}^*, j_{P_2}^*, e_{P_2}^*, g_{P_2}^*)$ with minimum total cost tc^{min} by P_2, calculate the corresponding completion time for each of them, collect all $(v_{P_2}^*, j_{P_2}^*, e_{P_2}^*, g_{P_2}^*)$ with minimum completion time $ct_{P_2}^{min}$ into Set B;

  Calculate $\Delta tc = tc_{P_1}^{min} \big/ tc^{min}$, $\Delta ct = ct_{P_2}^{min} \big/ ct^{min}$;

  If $\Delta tc < \Delta ct$:
    The new Q-values = $(1, 0)$;
  If $\Delta tc > \Delta ct$:
    The new Q-values = $(0, 1)$;
  If $\Delta tc = \Delta ct$:
    If $|A| > |B|$:
      The new Q-values = $(1, 0)$;
    If $|A| < |B|$:
      The new Q-values = $(0, 1)$;
    Else:
      The new Q-values = $(0.5, 0.5)$;
  Get the old Q-values of this assignment ordinal from Q-table;
  If (the old Q-values = $(1, 0)$ and the new Q-values = $(1, 0)$) or (the old Q-values = $(1, 0)$ and the new Q-values = $(0.5, 0.5)$) or (the old Q-values = $(0.5, 0.5)$ and the new Q-values = $(1, 0)$):
    The updated Q-values = $(1, 0)$;
  If (the old Q-values = $(0, 1)$ and the new Q-values = $(0, 1)$) or (the old Q-values = $(0, 1)$ and the new Q-values = $(0.5, 0.5)$) or (the old Q-values = $(0.5, 0.5)$ and the new Q-values = $(0, 1)$):
    The updated Q-values = $(0, 1)$;
  Else:
    The updated Q-values = $(0.5, 0.5)$.

The (v^*, j^*, e^*, g^*) is arbitrarily selected from Set A if the updated Q-values = $(1, 0)$. If the updated Q-values = $(0, 1)$, the (v^*, j^*, e^*, g^*) is arbitrarily selected from Set B. If the updated Q-values = $(0.5, 0.5)$, the (v^*, j^*, e^*, g^*) is arbitrarily selected from Set $A \cup B$. The ε-greedy policy is adopted to decide whether performing the above (v^*, j^*, e^*, g^*) selected principle. A random number rn is generated from the unit interval $[0, 1]$ to control the probability of the ε-greedy policy occurring.

The flowchart of the solution approach is illustrated in Fig. 2. The old Q-table is initialized by selecting all the (v^*, j^*, e^*, g^*) in the first episode using Policy P_1 exclusively. The new Q-table is initialized by selecting all the (v^*, j^*, e^*, g^*) in the second episode using Policy P_2 exclusively. The old Q-table is first updated after the second episode. There are two conditions to stop the iteration. Besides the common episode limit, the number of continuously obtaining a solution with the same objective values as the solutions in the Pareto-optimal solution set also has a limit, nt.

4   Validation and Discussion

Unfortunately, with no access to real production data on RMS for MC, the mathematical model and the solution approach were validated with two numerical examples. The scale of the examples is represented by the total number of operations n and the total number of configurations $\sum_{m=1}^{|M|} |G_m|$. The fist numerical example is small in scale. Therefore, the exhaustive search method is used to obtain its exact Pareto-optimal solutions. The second numerical example is somewhat complicated. Thus, only approximate solution approaches, the proposed Q-learning based RL and NSGA-II, are used to solve it. Table 1 shows the mean values of the computation times for the exact and approximate solution approaches of the numerical examples. The likelihood of getting the exact Pareto-optimal solutions by these approximate solution approaches is defined as follows:

$$P_{ne} = \sum_{ne=1}^{10} \theta_{ne} \bigg/ 10 \tag{24}$$

The Q-learning based RL and NSGA-II solution approaches for each numerical example were carried out ten times. The exhaustive search for the first numerical example is carried out only once. At each run, θ_{ne} in expression (24) is equal to 1 if the approximate solution approaches obtain at least one exact Pareto-optimal solution.

The numerical experiments were carried out on a laptop computer with an Inter(R) Core i7-7600 CPU (2.80 GHz) and 16 GB of RAM. The Python language was used to implement the exhaustive search and the approximate solution approaches. The algorithms are programmed using the Pycharm software (version 2021.2.1 Professional Edition). Codes, numerical example data and the optimal solutions are available in the repository: https://github.com/Sini-GAO/SOHOMA22_Sini.

From Table 1, it is clear that the Q-learning based RL solution approach proposed in this paper is efficient in obtaining the approximate Pareto-optimal solutions. For the first small numerical example, it is probably to get some exact Pareto-optimal solutions in short computational time, reduced by 99% relative to that of the exhaustive search. The performance of this method is compared with that of a multi-objective heuristic, NSGA-II. By setting appropriate parameters to keep similar accuracy for the first small numerical example, the computation time of NSGA-II is more than five times longer than that of the proposed method. For the second larger numerical example, the computation time of NSGA-II is eight times longer than that of the proposed method.

The good performance of this solution approach is due to the appropriate problem-specific designed policies. The policy of selecting a machine by expression (20) helps to balance the production on machines in RMS, which is also good for minimizing the completion time of the entire MC production. The policy of selecting a product by expression (21) finds the most 'urgent' product. This helps to minimize the first objective of the total tardiness penalty. The Q-values denote the superiority of two policies P_1 and P_2 for selecting (v^*, j^*, e^*, g^*) at each assignment. If there are plenty (v^*, j^*, e^*, g^*) satisfying both policies P_1 and P_2, this means that these two policies are equally superior, thus the new Q-values $= (0.5, 0.5)$.

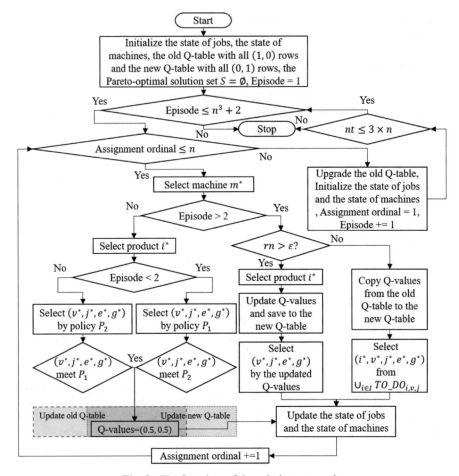

Fig. 2. The flowchart of the solution approach.

Table 1. The results of the numerical experiments

n	$\sum_{m=1}^{|M|} |G_m|$	Computation time (s)			P_{ne}	
		Exhaustive search	Q-learning	NSGA-II	Q-learning	NSGA-II
12	4	784.59	0.69	3.95	0.8	0.8
–	–	–	–	–	$\left(\overline{f_1}, \overline{f_2}\right)$	
126	7	–	9.92	79.39	(1670.12, 1289.56)	(6155.29, 1187.03)

5   Conclusion

This paper present our attempt to solve the multi-objective MC integrated process planning and job-shop scheduling problem in RMS by RL approach. Compared with the heuristic algorithms that are widely used in most studies to solve this research topic, the solution approach developed in this study is less parametric, because it requires less expertise on parameter setting.

The first contribution of this work is using RL to solve the mathematical model with multi-objective function. One limit of this paper is the absence of a real application. In future research, deep neural network may be applied to generate the action directly based on the production state information. RMSs require on-going analysis and monitoring of the environmental requirements to suggest change scenarios for the current configurations of the system, if necessary [17]. In this context, manufacturing firms need to shift their focus from linearly improving efficiency towards real time learning from big data and contextual decision making [18]. Furthermore, it is worthwhile to formulate the united problem and develop theory and techniques for the purpose of the green and intelligent industry [19].

Acknowledgments. The authors are grateful for the financial support (No. 201806280501) provided by China Scholarship Council (CSC).

References

1. Aheleroff, S., Philip, R., Zhong, R.Y., Xu, X.: The degree of mass personalisation under industry 4.0. Procedia CIRP **81**, 1394–1399 (2019)
2. Leng, J., et al.: Digital twin-driven rapid reconfiguration of the automated manufacturing system via an open architecture model. Robot. Comput. Integr. Manuf. **63** (2020).https://doi.org/10.1016/j.rcim.2019.101895
3. Morgan, J., Halton, M., Qiao, Y., Breslin, J.G.: Industry 4.0 smart reconfigurable manufacturing machines. J. Manuf. Syst. **59**, 481–506 (2021). https://doi.org/10.1016/j.jmsy.2021.03.001
4. Koren, Y., Gu, X., Guo, W.: Reconfigurable manufacturing systems: principles, design, and future trends. Front. Mech. Eng. **13**(2), 121–136 (2017). https://doi.org/10.1007/s11465-018-0483-0
5. Bruccoleri, M., Nigro, G.L., Perrone, G., Renna, P., Diega, S.N.L.: Production planning in reconfigurable enterprises and reconfigurable production systems. CIRP Ann. **54**(1), 433–436 (2005). https://doi.org/10.1016/S0007-8506(07)60138-3
6. Brahimi, N., Dolgui, A., Gurevsky, E., Yelles-Chaouche, A.R.: A literature review of optimization problems for reconfigurable manufacturing systems. IFAC-PapersOnLine **52**, 433–438 (2019). https://doi.org/10.1016/j.ifacol.2019.11.097
7. Usuga Cadavid, J.P., Lamouri, S., Grabot, B., Pellerin, R., Fortin, A.: Machine learning applied in production planning and control: a state-of-the-art in the era of industry 4.0. J. Intell. Manuf. **31**(6), 1531–1558 (2020). https://doi.org/10.1007/s10845-019-01531-7
8. Nassehi, A., Zhong, R.Y., Li, X., Epureanu, B.I.: Review of machine learning technologies and artificial intelligence in modern manufacturing systems. In: Design and Operation of Production Networks for Mass Personalization in the Era of Cloud Technology. pp. 317–348. Elsevier Inc. (2022)

9. Kardos, C., Laflamme, C., Gallina, V., Sihn, W.: Dynamic scheduling in a job-shop production system with reinforcement learning. Procedia CIRP **97**, 104–109 (2021)

10. Tang, J., Salonitis, K.: A deep reinforcement learning based scheduling policy for reconfigurable manufacturing systems. Procedia CIRP **103**, 1–7 (2021)

11. Tang, J., Haddad, Y., Salonitis, K.: Reconfigurable manufacturing system scheduling: a deep reinforcement learning approach. Procedia CIRP **107**, 1198–1203 (2022). https://doi.org/10.1016/j.procir.2022.05.131

12. Yang, S., Zhigang, Xu.: Intelligent scheduling and reconfiguration via deep reinforcement learning in smart manufacturing. Int. J. Prod. Res. **60**(16), 4936–4953 (2021). https://doi.org/10.1080/00207543.2021.1943037

13. Khan, A.S., Homri, L., Dantan, J.Y., Siadat, A.: An analysis of the theoretical and implementation aspects of process planning in a reconfigurable manufacturing system. Int. J. Adv. Manuf. Technol. **119**(9–10), 5615–5646 (2021). https://doi.org/10.1007/s00170-021-08522-0

14. He, Y., Xing, L., Chen, Y., Pedrycz, W., Wang, L., Wu, G.: A generic Markov decision process model and reinforcement learning method for scheduling agile earth observation satellites. IEEE Trans. Syst. Man Cybern. Syst. **52**, 1463–1474 (2022). https://doi.org/10.1109/TSMC.2020.3020732

15. Watkins, C.J.C.H., Dayan, P.: Q-learning. Mach. Learn. **8**, 279–292 (1992). https://doi.org/10.1007/BF00992698

16. Clifton, J., Laber, E.: Q-learning: theory and applications. Annu. Rev. Stat. Appl. **7**, 279–301 (2020). https://doi.org/10.1146/annurev-statistics-031219-041220

17. Azab, A., ElMaraghy, H., Nyhuis, P., Pachow-Frauenhofer, J., Schmidt, M.: Mechanics of change: a framework to reconfigure manufacturing systems. CIRP J. Manuf. Sci. Technol. **6**, 110–119 (2013). https://doi.org/10.1016/j.cirpj.2012.12.002

18. Morariu, C., Morariu, O., Răileanu, S., Borangiu, T.: Machine learning for predictive scheduling and resource allocation in large scale manufacturing systems. Comput. Ind. **120**, 103244 (2020). https://doi.org/10.1016/j.compind.2020.103244

19. Gao, S., Daaboul, J., Le Duigou, J.: Process planning, scheduling, and layout optimization for multi-unit mass-customized products in sustainable reconfigurable manufacturing system. Sustainability **13**, 13323 (2021). https://doi.org/10.3390/su132313323

Generic Aggregation Model for Reconfigurable Holonic Control Architecture – The GARCIA Framework

William Derigent[1](✉), Michael David[1], Pascal André[2], and Olivier Cardin[2]

[1] CRAN CNRS UMR 7039, Université de Lorraine, Campus Sciences, Boulevard des Aiguillettes, 54506 Vandœuvre-lès-Nancy, France
{william.derigent,michael.david}@univ-lorraine.fr
[2] Nantes Université, École Centrale Nantes, CNRS, LS2N, UMR 6004, 44000 Nantes, France
{pascal.andre,olivier.cardin}@ls2n.fr

Abstract. During the last twenty years, many innovative control architectures of manufacturing systems have been developed and promoted in literature. One of the main attributes, in correlation with the aims of Industry 4.0 paradigm, is to define control architectures where both the actors and the interactions between these actors could cope with an evolution of the environment. To do so, dynamic architectures are being recently developed, where the hierarchy of decision can be jeopardized at any time during the normal behaviour of the system. However, the deployment of such architectures faces major software development issues, that a proper initial modelling could help solving. The objective of this paper is to exhibit good practices in the modelling of dynamic architectures in order to enable an automatic reconfiguration when needed.

Keywords: Dynamic control architectures · Reconfigurable manufacturing systems · Digital twin

1 Introduction

The three-dimensional framework expressed in the RAMI 4.0 model (*Reference Architecture Model for Industry 4.0* – proposed by [1]), exhibits, all along the product lifecycle, the multi-scale aspects of the Industry 4.0 as well as the potential benefits of a coupling with dynamic vertical data integration (from assets to business process) more massive and intense than it is currently [2]. To allow a real-time adaptability of the whole company to external/internal changes and an efficient use of data available to manage these Industry 4.0-oriented processes, it becomes crucial to define enterprise information systems equipped with connected, interoperable, flexible and reactive control architectures [3].

This topic has been widely studied [4], and numerous control architectures, with multiple application fields (e.g., automotive industry [5], railway transportation systems [6] or radiopharmaceutical products [7]), were developed. The common idea shared by these control architectures is the distribution of intelligence throughout the actors of the

© The Author(s), under exclusive license to Springer Nature Switzerland AG 2023
T. Borangiu et al. (Eds.): SOHOMA 2022, SCI 1083, pp. 407–422, 2023.
https://doi.org/10.1007/978-3-031-24291-5_32

decision making process [8]. As the myopia phenomenon has been widely exhibited [9], these control architectures usually rely on the principle of heterarchy [10], where the actors can collaborate horizontally within the architecture, but can also be coordinated by an actor of superior level, often through an aggregation relationship [4].

The classical implementation of these control architectures is generally based on the concept of industrial agent [11], or holonic control architectures if the aggregation pattern is considered. Along the years, several reference architectures were developed [12], and led to the definition of dynamic control architectures, where the hierarchy of decision can be jeopardized at any time during the normal behaviour of the system. In those structures, the relationships between the agents/holons need to be redefined dynamically, without the intervention of a programmer. If the performance and dynamics of these architectures were studied and their efficiency proven, there is a lack of generic approach enabling easy reconfiguration of the structure. This paper proposes a generic definition of the reconfigurable architectures model and behaviour, aiming at specifying more clearly and thus developing more easily the dynamic architecture.

The paper is organized as follows: Sect. 2 presents the related works on dynamic Holonic Control Architectures and underlines the fact that few works describe how to apply reconfiguration in cyber-physical production systems. Section 3 describes the proposed generic and reconfigurable Holonic architecture. Section 4 exhibits the potential benefits on a simple manufacturing system case study.

2   Related Works

This section focuses on the recent dynamic Holonic Control Architectures (HCAs) available in literature. This presentation is not designed to be exhaustive but reflects the state of the art on the topic.

The initial purpose of HCAs has been to seek for reactivity, in opposition to classical centralized architectures. Since PROSA [13], the definition of the behaviour of each holon enables a quick reaction to disturbances. However, in case of high level of disturbances, the global behaviour of the system remains constant and can only cope with the initial flexibility provided to each holon to solve the issue. As a result, designing dynamic control architectures is probably one of the most promising current trends in literature in HCAs [14]. The idea is to design a behaviour of the architecture that can be modified dynamically in order to adapt to the changes of the environment and thus reduce the transient states and the associated loss of performance.

Several dynamic architectures have been proposed in the literature. An invariant of these architectures is that they are all characterized by a "switching mechanism", enabling them to transit from a structure of the holonic architecture to another. In [17], a mechanism is proposed to switch between a centralized and a decentralized holonic architecture in the presence of perturbations to ensure as long as possible both global optimization and agility to changes in batch orders during manufacturing. For more information, a survey of the different kind of switching mechanisms and their use in dynamic HCAs is available in [15].

In this section, three different dynamic HCAs are presented. ORCA [16] was one of the first dynamic architectures that was formalized in literature. In ORCA, a global optimizer Holon controls at a lower-level local optimizer Holons. The switching mechanism is activated if a perturbation happens and forbids the application of the schedule decided by the global optimizer. In this new mode, a cooperation between local optimizers is established in order to create a new distributed schedule. This evolution can be considered as discrete, between stable states. In opposition, an evolution of the ADACOR mechanism was introduced in [18] as ADACOR^2. The objective is to let the system evolve dynamically through configurations discovered online, and not only between a stationary and a transient state. The last dynamic HCA in date is denoted as POLLUX [15]. The main novelty is focused on the adaptation mechanism of the architecture, using governance parameters that enlarge or constraint the behaviour of the low level holons regarding the disturbances observed by the higher level. The simulation of the consequence on performances of several switching options ("what-if" scenarios) is proposed, since in POLLUX the number of possible switches at a given time is high compared to ORCA, for example. Each decided commutation is thus justified by an increase in the performances compared to other switching possibility.

What can be seen through these three examples is the lack of precise definition of the configuration/reconfiguration processes of the architectures. The objective of these studies was to exhibit the potential performance implied by switching rather than staying in a stable architecture, but no explanations are given on how to implement this switching in the HCA development process. Indeed, the switching process ends with a modification of the control architecture, resulting in a modified organization, exhibiting new aggregations or compositions of the different entities. All works describe when, why and what to modify in the architecture, but the implementation (the "how") is often let to the developers, in the sense that the process of applying the reconfiguration orders to the physical devices was never detailed. Methodologies and implementation guides are thus missing. As a result, to follow this objective, this paper details a framework that would complement existing dynamic HCAs, in order to solve these issues.

3   The GARCIA Framework

The following section presents the proposed generalized dynamic control architecture, by detailing its static and dynamic views using UML diagrams. Section 3.1 introduces the underlying principles supporting our proposal, Sect. 3.2 proposes a class diagram of our architecture and describes the main objects. Finally, the meta-model of reconfiguration process is presented Sect. 3.3.

3.1 Definition of the Generic Control Architecture

Before presenting the complete model of our architecture, some basic principles need to be introduced to understand its complexity.

Principle 1: Use of Aggregation Patterns. As described earlier, heterarchical architectures exhibit relations between entities belonging to different decision levels. In holonic architectures, these links are the result of aggregation relations, linking a decisional entity to its subordinates. An aggregation relation is a specific type of master/slave relation, with three different forms: *coercive, limitary* and *permissive* [15]. In a *coercive* relation, slaves follow the decision of their master. In a *permissive* relation, the decision is let to slaves with a control of the upper level. In a *limitary* relation, the upper level shares the decision with its slaves (e.g., it proposes a set of solutions to its slaves, which then elect together the best one). A previous work [19] shows that some basic and elegant modelling patterns, often used in holonic architecture design, may lead to impracticable programming concerns when implemented (Fig. 1a). As a result, our architecture relies on the aggregation pattern proposed in [19] and presented in Fig. 1b.

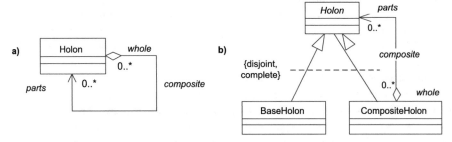

Fig. 1. Recursive composition patterns

Principle 2: Use of Recursion. This second principle is a consequence of the first one. A general control architecture should use recursive objects, since using recursion ensures a good system scalability. Indeed, scalability refers to the 'capacity to be changed in size or scale'. In HCAs, this notion is linked to the number of entities and to the number of decision levels operating in a complex system. A scalable HCA should be able to change its size, especially to face reconfiguration processes. Recursion combined with adequate aggregation patterns help to solve this challenge. Moreover, from an implementation perspective, recursion is also a way to reduce the programming efforts.

Principle 3: Integration of the "LOCAtion of the Intelligence". In a cyber-physical world, a part of the decision-making entities could be located directly on the physical device itself (product or resource) and the other part in a cloud-based system. There is thus a need to underline the link between both worlds in the HCA (see Figs. 1 and 2).

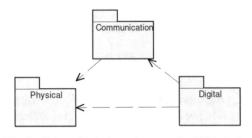

Fig. 2. Cyber-physical requirements for HCA design

Consequently, a correct modelling of a HCA should describe the digital and physical world, as well as the way to make connections between them (referred to as "communication" in the following), to link the different physical or virtual locations where decision algorithms are executed. Exhibiting the communication part is an example of the 'separation of concerns' principle. It means a better evolution of the system in practice because changing communication means (e.g., protocols) will have limited impact on the other parts.

As many decision-making mechanisms exist (heuristics, meta-heuristics, linear programming, fuzzy logic, etc.), the architecture should be "decision-free" in order to integrate every type of decision mechanism, by providing abstract functions which may be implemented during deployment. Note that our architecture is quite flexible by enabling local control for intelligent devices and global control for the centralized decision management (see Fig. 3).

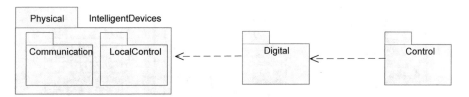

(a)   Organization with intelligence at object

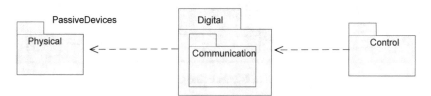

(b) Organization with intelligence in the Digital Twin

Fig. 3. Examples of intelligence location

Conditions of Application. The GARCIA framework is to be applied preferably on networked cyber-physical entities. In this context, two types of networks are coexisting: the virtual network, linking the cyber entities representing the physical devices (resources, products), as well as the physical network, which is the communication network linking the physical devices – physical devices that can speak directly with each other.

3.2   Meta-model of the Control Architecture

Applying these principles led to the following architecture (Fig. 4).

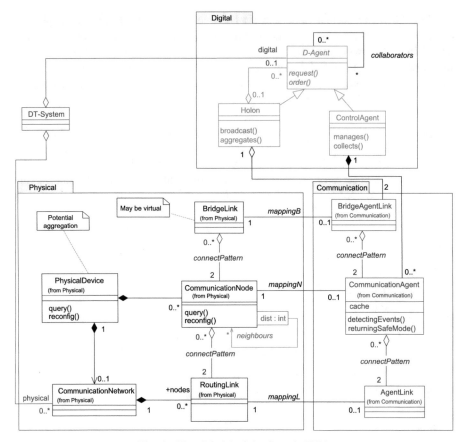

Fig. 4. Class Model of the Generic HCA

As introduced before, this model is divided into three domains: digital, physical and communication (the regions delimited in Fig. 4). In the following, the different objects included in each of these domains are described.

The Physical Domain: This domain gathers all the objects that could contain intelligent entities. Several objects are introduced:

- A PhysicalDevice describes an object able to sense and communicate data. It could be a complete system (fleet of objects) or a single object in which sensors are inserted (intelligent product).
- A CommunicatingNode describes the devices located inside the PhysicalDevice, generating data and taking decisions. It could be a computer unit, a network device or a sensor node.
- A CommunicationNetwork is a set of linked CommunicatingNodes that can exchange data/information or discuss decisions to take.
- A RoutingLink is a physical connection between CommunicatingNodes used for routing messages for one to the other. RoutingLinks is a subset of the NeighbourLinks

which represent the potential set of connections according to a communication technology; e.g., in a wireless support, all the nodes a given CommunicatingNode can discuss with, given a specific radio power level. They are also different from the BridgeLinks which are the links between two CommunicatingNodes that do not belong to the same PhysicalDevice.

The Digital Domain: This domain gathers all the decision-making entities. This is a complex domain that can be linked to the enterprise information system, e.g., stored in a cloud-based system. We only show here the low level classes of the digital part:

- A D-Agent contains all the needed information for describing a PhysicalDevice and managing its physical part. A composite pattern is introduced to define its two subtypes: Holon and ControlAgent.
- A Holon is a D-Agent composed of other D-Agents, according to the pattern of Fig. 4. Such a pattern enables both decentralised and centralised decisions.
- A ControlAgent is an 'atomic' D-Agent. It is directly linked to one or several CommunicationAgent of the communication domain.

The Communication Domain: This domain gathers all the entities representing the interface between the physical and the digital domain.

- A CommunicationAgent is an interface between the cyber and physical part. A CommunicationAgent is a digital twin of each CommunicatingNode existing as a PhysicalDevice. It monitors and controls its twin.
- An AgentLink represents a physical routing link.
- A BridgeAgentLink represents a physical bridge link.

3.3   Meta-modelling the Reconfiguration Process

The core static structure previously presented must be completed by a dynamic view, describing its evolution through time. This is done in the following section, in the form of diagrams. Each D-Agent or communication agent follows the same logical state-transition diagram (Fig. 5), composed of three steps:

- **Initialisation**: in this state, agents receive their original structure.
- **Working**: in this state, the agent accomplished its duties. In this paper, this state will not be detailed, since it depends on the application to which this framework is added. Moreover, it is not the subject of this paper.
- **Maintenance**: in this state, agents are collaborating with each other to decide if reconfiguration is needed and in case it is necessary, to apply it.

   In addition to this first model, some other agents execute a monitoring process (Fig. 6) to alert their hierarchy in case some type of events occurs at the physical device (breakdown, important delays, Work-In-Progress shortages, new products in the queue…). This alert is then repeated to the higher levels of the hierarchy.

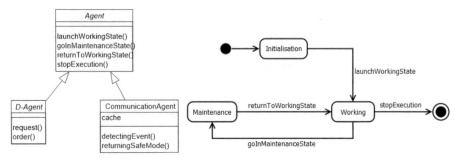

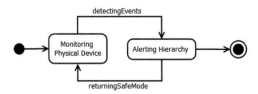

Fig. 5. Common agent process

Fig. 6. Event detection from the Communication Agent

These models are orchestrated in the next sequence diagram (Fig. 7), explaining how the reconfiguration process could be launched and applied in our architecture, decomposed into 8 steps.

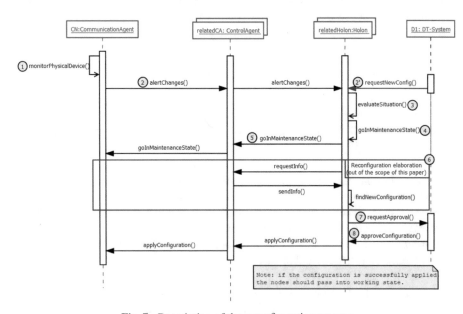

Fig. 7. Description of the reconfiguration process

In the working state, every `PhysicalDevice` is monitored by its related `Commu-nicationAgent` (*step 1*), querying data from the device `CommunicatingNo-de`. This event can be related to the physical device itself (breakdown, delay,...) and also to a modification of its local neighbourhood (i.e., a new resource appears in its communication range, because of a shop floor reconfiguration). In an alternative way, and following Principle 3, the monitoring can be ensured by the `Communica-tingNode`. Depending on the received events, an alert is then sent to the related `ControlAgent`, that can be conveyed to a `Holon` (*step 2*). In this case, the notification comes from the context of application (e.g., the shop floor in the case study exposed later). The overall digital control system (called DT-system) itself can also send a reconfiguration query to the Holon (*step 2*).

When receiving the alert, the corresponding `Holon` evaluates the necessity to reconfigure (*step 3*) and, if needed, first enters in maintenance state (*step 4*) and then asks its related sub-agents to enter in maintenance state (*step 5*). All agents are thus ready to cooperate to define the next configuration. The algorithmic mechanism needed to choose the new configuration to set up (*step 6*) depends on the types of the aggregation relations (*coercive, limitary* or *permissive* as explained Sect. 3.1) but is not in the scope of this work. Centralized or fully distributed solutions can be implemented. Once done, the chosen configuration is proposed to the DT-system and applied if validated (*steps 7 and 8*). *Note that the operations called by the message sends are omitted in Fig. 4 for sake of space.*

4   Instantiation on a Simplified Case Study

The remainder of this contribution will be illustrated by a simplified use case, aiming at exhibiting an application of this work. The main feature of this example is the necessity of the system to reconfigure dynamically its control due to its evolution. The global system is constituted of two subsystems each of five machines, able to execute various manufacturing activities (Fig. 8). In this case study, the shop floor is meant to be reconfigurable, in the sense that every machine and carts can be transferred from one subsystem to another according to production requirements. Each shop floor element is equipped with communication devices connected to a wireless communication network.

Inside those subsystems, the transfer of products is executed by autonomous carts, one in the first subsystem, two in the second one. The carts receive in real-time the transportation missions calculated by the control system, and regularly send in return various acknowledgments to the control system.

4.1   Static Description of the Initial Control Architecture

As a main hypothesis, it is considered that an independent control system is defined for each subsystem, both of them being aggregated in the global control system. The specificity of this case study is a possible transportation of products between the subsystems. Based on the decisions of the control system, one of the carts can be asked to transport products to the other subsystem and can stay in this other subsystem and help the other cart. In that case, the cart changes from one control system to another, and this change needs to be reflected in the control architecture.

GARCIA is "specialized" to the case study by adding subclasses to the generic HCA classes. For sake of space, we did not represent all the classes of the application, but an excerpt in Fig. 9.

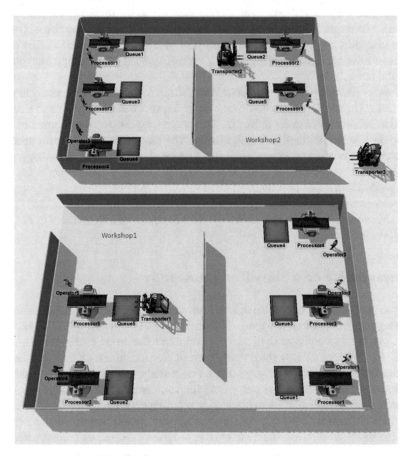

Fig. 8. Case study's workshop organization

The top of Fig. 9 represents the classes of the physical domain and the bottom represents the digital part; between the two there is the manufacturing part, closely related to the case study. The structuring resources of the manufacturing system are the physical devices while the individual resources are the communication nodes. In this case, we fixed arbitrarily the recursive (concrete) aggregation to three levels (workstation, workshop, factory) to make it clear for the reader. For sake of simplicity, we also omit the communication links and the constraints that are assessed by the case study (e.g., all resources are attached to workstations except the transporters). Also, the communication domain has been collapsed by the blue associations between physical elements and digital elements.

The digital part (in orange colour on the bottom of Fig. 9) is made of control agents that control the individual resources and three kinds of holons to control the processes (`ProductionHolon`), the product transfer by carts (`TransferHolon`) and the simplified manufacturing management (`MS-System`). Again, things have been simplified to be more readable.

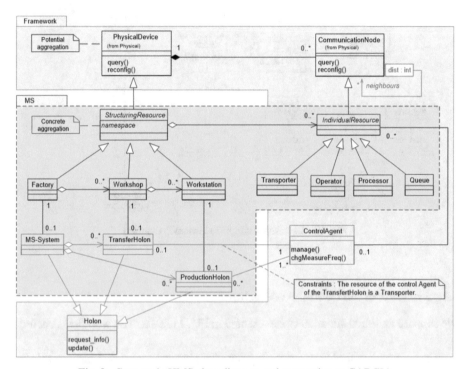

Fig. 9. Case study HMS class diagram and connection to GARCIA

The model of Fig. 9 has been instantiated on the case study. The physical domain includes at least 45 objects excluding the links, and the digital part includes at least 60 objects. We represented only a part of the MS layer's instances in Fig. 10: the physical devices and nodes of the `wsh1-ws4` workstation of the `wsh1` workshop, which is controlled by the `ph1-4` product holon. The transfer tasks are processed by `th1` and `th2`, two transfer holons. The interesting point here is that the transporters can be linked to different workshops by dynamic reconfiguration.

Each holon controls a set of `ControlAgents`, denoting digital twins for the physical elements. Each control agent controls a set of `CommunicationAgents`, one per `CommunicationNode`. The communication agents are implicitly denoted by the blue lines in Fig. 10. As a simplification, each control agent manages one resource in our case study i.e., the `ControlAgent` corresponding to the `wsh1-ws4-pr4` processor

is named `ca-wsh1-ws4-pr4` and its implicit `CommunicationAgent` is named
`co-wsh1-ws4-pr4`.

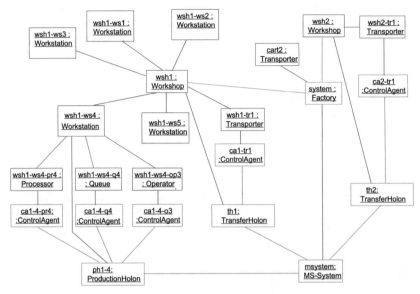

Fig. 10. Case study simplified instance diagram

4.2 Evolution of the Control Architecture

We suppose an initial situation as represented in Fig. 11. Two main workflows occur:

1. *Communication and interactions* between the physical and digital parts to achieve
   the manufacturing processes, i.e., the control part of the system and
2. *Reconfiguration* to evolve after disturbance events.

A centralized decision is not mandatory, both workflows can be controlled locally
depending on the disturbance scale. This initial situation is then perturbed, since the
transporter3 is available and detected by a processing resource. Figure 12 illustrates the
execution of the sequence diagram depicted in Fig. 7, from step 1 to 3.

Once the situation evaluated, the system will pass in maintenance state. Figure 13
describes the recursive use of the *GoInMaintenanceState* message (step 4–5 of the
sequence diagram). This framework allows to circumscribe this state to a given part
of the system, where the detection occurred.

As explained before, the reconfiguration of the decision-making process is not part
of this framework. However, decisions taken by reconfiguration are then applied recur-
sively, as in Fig. 12 for the transition to the maintenance state. In our case study, this
reconfiguration provokes changes in the collaboration structures, as shown in Fig. 13.

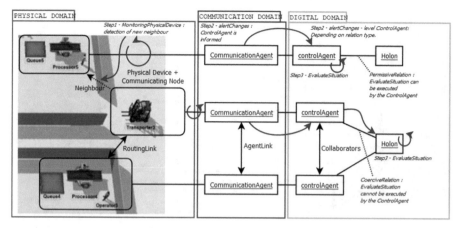

Fig. 11. Detection and alter triggering

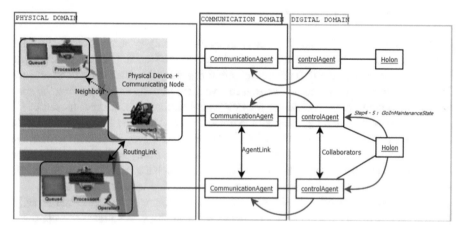

Fig. 12. Transition in MaintenanceState

This last figure shows the reconfigured architecture, with the different layers allowing the description of all components of a cyber-physical production system. This first simplified case study aims to show that GARCIA could support the reconfiguration process by implementing the different needed *pre* and *post* procedures. Because it is based on recursive patterns, GARCIA is believed to support scalability. Moreover, this first simplified case study explains how GARCIA can be very easily customized for a given domain. Of course, additional work is required to fully validate the GARCIA framework.

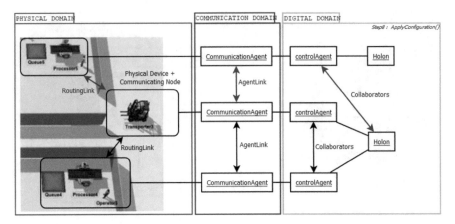

Fig. 13. State of the system at the end of the reconfiguration process

5   Conclusion and Perspectives

The literature review on dynamic control in manufacturing systems exhibits the need of reconfigurable architectures with a simple description on initial and future situations. But how to implement the reconfiguration process itself is missing most of time. Therefore, this work proposes a generic framework for guiding and improving the software development process related to reconfigurable architecture. It is composed of (1) a *generic class diagram* integrating all the objects needed to describe a dynamic HCA, (2) *state-transition diagrams* and a *global reconfiguration process* needed to manage the reconfiguration phase. A simple manufacturing case study with a reconfiguration scenario illustrates the potential benefits of this approach, but it is believed to be generic enough to be adapted to other contexts like the construction industry for example.

Indeed, a first perspective is to implement this framework with diverse Dynamic HCA (Pollux, ORCA or ADACOR) on simulated or real test cases, to validate this approach and study the reconfiguration strategies according to different criteria (time, cost,…). The domain of applications is not limited to manufacturing, and we already began to apply this framework to the context of the French ANR McBIM project, developing the communicating material concept for the construction industry. A third perspective is to filter the recursive path flows by means of contracts where each level of recursion denotes a particular context that influences the behaviour of control agents [20]. Other perspectives cover the design of new patterns for the control part, or the communication layers using multi-protocol brick [21].

Acknowledgments. Authors thank financial support from the French National Research Agency (ANR) under the McBIM project, grant number ANR-17-CE10-0014.

References

1. Zezulka, F., Marcon, P., Vesely, I., Sajdl, O.: Industry 4.0 – an Introduction in the phenomenon. IFAC-PapersOnLine **49**, 8–12 (2016). https://doi.org/10.1016/j.ifacol.2016.12.002

2. Barbosa, J., Leitao, P., Trentesaux, D., Colombo, A.W., Karnouskos, S.: Cross benefits from cyber-physical systems and intelligent products for future smart industries. In: 2016 IEEE International Conference on Industrial Informatics (INDIN), pp. 504–509 (2016)

3. Valckenaers, P.: Perspective on holonic manufacturing systems: PROSA becomes ARTI. Comput. Ind. **120**, 103226 (2020). https://doi.org/10.1016/j.compind.2020.103226

4. Derigent, W., Cardin, O., Trentesaux, D.: Industry 4.0: contributions of holonic manufacturing control architectures and future challenges. J. Intell. Manuf. **32**(7), 1797–1818 (2020). https://doi.org/10.1007/s10845-020-01532-x

5. Bussmann, S., Sieverding, J.: Holonic control of an engine assembly plant: an industrial evaluation. In: 2001 IEEE International Conference on Systems, Man and Cybernetics. e-Systems and e-Man for Cybernetics in Cyberspace (Cat. No. 01CH37236), pp. 169–174 (2001)

6. Le Mortellec, A., Clarhaut, J., Sallez, Y., Berger, T., Trentesaux, D.: Embedded holonic fault diagnosis of complex transportation systems. Eng. Appl. Artif. Intell. **26**, 227–240 (2013). https://doi.org/10.1016/j.engappai.2012.09.008

7. Borangiu, T., Răileanu, S., Oltean, E.V., Silişteanu, A.: Holonic hybrid supervised control of semi-continuous radiopharmaceutical production processes. In: Kondratenko, Y.P., Chikrii, A.A., Gubarev, V.F., Kacprzyk, J. (eds.) Advanced Control Techniques in Complex Engineering Systems: Theory and Applications. SSDC, vol. 203, pp. 229–258. Springer, Cham (2019). https://doi.org/10.1007/978-3-030-21927-7_11

8. Pujo, P., Broissin, N., Ounnar, F.: PROSIS: an isoarchic structure for HMS control. Eng. Appl. Art. Intell. **22**, 1034–1045 (2009). https://doi.org/10.1016/j.engappai.2009.01.011

9. Adam, E., Zambrano, G., Pach, C., Berger, T., Trentesaux, D.: Myopic Behaviour in holonic multiagent systems for distributed control of FMS. In: Corchado, J.M., Pérez, J.B., Hallenborg, K., Golinska, P., Corchuelo, R. (eds.) Trends in Practical Applications of Agents and Multiagent Systems, pp. 91–98. Springer, Berlin, Heidelberg (2011). https://doi.org/10.1007/978-3-642-19931-8_12

10. Zambrano Rey, G., Bonte, T., Prabhu, V., Trentesaux, D.: Reducing myopic behaviour in FMS control: a semi-heterarchical simulation–optimization approach. Simul. Model. Pract. Theor. **46**, 53–75 (2014). https://doi.org/10.1016/j.simpat.2014.01.005

11. Antzoulatos, N., Castro, E., Scrimieri, D., Ratchev, S.: A multi-agent architecture for plug and produce on an industrial assembly platform. Prod. Eng. Res. Devel. **8**(6), 773–781 (2014). https://doi.org/10.1007/s11740-014-0571-x

12. Cardin, O., Derigent, W., Trentesaux, D.: Evolution of holonic control architectures towards Industry 4.0: a short overview. IFAC-PapersOnLine **51**, 1243–1248 (2018). https://doi.org/10.1016/j.ifacol.2018.08.420

13. Van Brussel, H., Wyns, J., Valckenaers, P., Bongaerts, L., Peeters, P.: Reference architecture for holonic manufacturing systems: PROSA. Comput. Ind. **37**, 255–274 (1998). https://doi.org/10.1016/S0166-3615(98)00102-X

14. Cardin, O., Trentesaux, D., Thomas, A., Castagna, P., Berger, T., Bril El-Haouzi, H.: Coupling predictive scheduling and reactive control in manufacturing hybrid control architectures: state of the art and future challenges. J. Intell. Manuf. **28**(7), 1503–1517 (2015). https://doi.org/10.1007/s10845-015-1139-0

15. Jimenez, J.F., Bekrar, A., Zambrano-Rey, G., Trentesaux, D., Leitão, P.: Pollux: a dynamic hybrid control architecture for flexible job shop systems. Int. J. Prod. Res. **55**, 4229–4247 (2017). https://doi.org/10.1080/00207543.2016.1218087

16. Pach, C., Berger, T., Bonte, T., Trentesaux, D.: ORCA-FMS: a dynamic architecture for the optimized and reactive control of flexible manufacturing scheduling. Comput. Ind. **65**, 706–720 (2014). https://doi.org/10.1016/j.compind.2014.02.005

17. Borangiu, T., Răileanu, S., Berger, T., Trentesaux, D.: Switching mode control strategy in manufacturing execution systems. Int. J. Prod. Res. (2015). https://doi.org/10.1080/00207543.2014.935825

18. Barbosa, J., Leitão, P., Adam, E., Trentesaux, D.: Dynamic self-organization in holonic multi-agent manufacturing systems: the ADACOR evolution. Comput. Ind. **66**, 99–111 (2015). https://doi.org/10.1016/j.compind.2014.10.011

19. André, P., Cardin, O.: Aggregation patterns in holonic manufacturing systems. In: Borangiu, T., Trentesaux, D., Leitão, P., Cardin, O., Joblot, L. (eds.) Service Oriented, Holonic and Multi-agent Manufacturing Systems for Industry of the Future: Proceedings of SOHOMA 2021, pp. 3–15. Springer, Cham (2022). https://doi.org/10.1007/978-3-030-99108-1_1

20. André, P., Ardourel, G., Messabihi, M.: Component service promotion: contracts, mechanisms and safety. In: Barbosa, L.S., Lumpe, M. (eds.) FACS 2010. LNCS, vol. 6921, pp. 145–162. Springer, Heidelberg (2012). https://doi.org/10.1007/978-3-642-27269-1_9

21. André, P., Azzi, F., Cardin, O.: Heterogeneous communication middleware for digital twin based cyber manufacturing systems. In: Borangiu, T., Trentesaux, D., Leitão, P., Giret Boggino, A., Botti, V. (eds.) SOHOMA 2019. SCI, vol. 853, pp. 146–157. Springer, Cham (2020). https://doi.org/10.1007/978-3-030-27477-1_11

Safety Design and Verification in Reconfigurable Assembly Systems

Yassine Idel Mahjoub[✉], Thierry Berger, Thérèse Bonte, and Yves Sallez

LAMIH, CNRS, UMR 8201, Université Polytechnique Hauts-de-France, 59313 Valenciennes,
France
yassine.idelmahjoub@uphf.fr

Abstract. The reconfigurable manufacturing systems (RMS) have been proposed
in the last decades to deal with mass-customization problems and unpredictable
market conditions. Numerous studies deal with RMS design and control issues, but
very few studies explore RMS's inherent safety problems. To address these issues,
this paper aims to fill the current research gap of human safety in reconfigurable
assembly systems (RAS). The objective is to make standard industrial robot-
human safe platforms and assist safety managers in implementing flexible and
easy-to-deploy safety devices to efficiently ensure human safety in RAS. The result
of the developed approach will be tested and verified using a formal approach to
prove and guarantee the design's correctness of the safety bubble, i.e., no possible
violation of the safety rules.

Keywords: Reconfigurable assembly system · Reconfigurable manufacturing
system · Safety · Design · Verification · Petri nets

1 Introduction

Throughout the last decade, the manufacturing industry has been confronted with
dynamic and unpredictable market conditions. This latter was characterized by an
increasing variety of products along with demand fluctuation [1]. These issues have led
industrial companies to redesign their production practices, by shifting from traditional
systems, e.g., dedicated lines, and flexible manufacturing systems, to reconfigurable
manufacturing systems (RMS). Introduced by [2], RMS are considered to be one of
the most suitable solutions to cope with nowadays market conditions efficiently and
effectively. They are essentially designed with the necessary flexibility to handle various
products. Such systems have a modular structure; they are made up of reconfigurable
units (e.g., robotized unit, conveying unit) that can be easily added, removed, or reconfigured to meet market demands or to deal with external changes (new product introduction,
machine breakdown, etc.) [1, 2]. A reconfigurable assembly system (RAS) is considered
as a reconfigurable manufacturing system dedicated to assembly assignments [3].

In RMS, there is a constantly increasing need for robots to interact, collaborate and
assist humans [4]. Some dangerous robots (e.g., traditional industrial robots) require
heavy fence guarding (i.e., solid cages) equipment that reduce flexibility while increasing

© The Author(s), under exclusive license to Springer Nature Switzerland AG 2023
T. Borangiu et al. (Eds.): SOHOMA 2022, SCI 1083, pp. 423–433, 2023.
https://doi.org/10.1007/978-3-031-24291-5_33

costs. However, the current market asks for reduced lead times and mass customization [5]. As a result, human-robot interaction has to withdraw the traditional paradigm of robots living in a separated space inside safety cages, allowing humans and robots to work together in a more flexible, reconfigurable, and open space.

On one hand, several traditional safety approaches (i.e., risk analysis, risk assessment, and risk reduction) show various weaknesses and fail to consider RMS's versatility and reconfigurability [6]. On the other hand, various research studies have proposed promising results for maintaining human safety with a focus on new robotized unit design rather than existing industrial robotic platforms [3, 9]. To address all these issues, this paper aims to fill the current research gap of human safety in RMS, by building upon the human safety approach (i.e., safety bubble) developed by [3, 4]. The objective is to make standard industrial robots human-safe platforms and assist safety managers in implementing flexible and easy-to-deploy safety devices to efficiently ensure human safety in RMS. The developed approach will be tested and verified using a formal approach (Petri nets-based approach) to prove and guarantee the design's correctness of the safety bubble, i.e., no possible violation of the safety rules.

The rest of this paper is structured as follows. Section 2 focuses on reconfigurable assembly systems structure and presents safety considerations and the adopted methodology. Section 3 presents the safety bubble design and verification models. Finally, Sect. 4 summarizes the paper and defines directions for future works.

2 Reconfigurable Assembly Systems Safety

2.1 RAS Structure and Safety Consideration

As mentioned before, a RAS is a reconfigurable manufacturing system dedicated to assembly assignments. It is designed to promptly adjust its structure and layout to meet dynamic market demands and requirements. RAS layout is composed of one or many cells, as presented in Fig. 1. Each RAS cell consists of a set of basic connected and modular units (e.g., robotized units, conveying units), that can be easily added, removed, rearranged, or configured. The RAS layout changes every time a reconfiguration process occurs and a new arrangement of units is applied (Fig. 1).

In this article, four types of units are considered to build a RAS cell:

- *Robotized Unit* (RU) consists of some classical industrial robots (or cobots not used in cobotic mode). Since our study focuses on an "open-space" RAS, these robots are assumed dangerous and must be safeguarded when operators enter their workspace.
- *Conveying Unit* (CU) is in charge of product transport in a cell (intracell transport).
- *Manual Unit* (MU) on which operators work.
- *Other units*: this type includes all other harmless units, such as inspection, treatment, and storage units.

The flow of materials between different cells (inter-cellular flow) is ensured by a fleet of mobile units (e.g., material handling devices, autonomous mobile robots). These units are out of the present study scope.

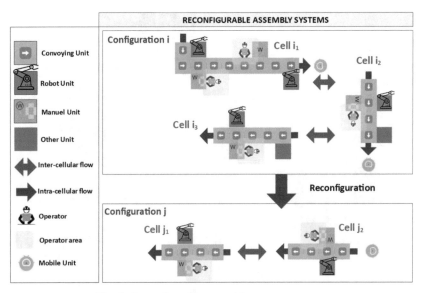

Fig. 1. Reconfigurable assembly system structure

2.2 Safety Bubble Concept and Methodology Adopted

The concept of "safety bubble" is developed by [3, 4], and is based on the cooperative behaviour of various RAS cell components, e.g., RU and safety devices. Once the safety bubble detects a human intrusion, it sends the information to the other units that can adapt their behaviours (e.g., speed reduction, halt). The objective is to ensure operator safety in an open dangerous space containing several RU. As a result, a safety margin (norm ISO 13855 [8]) is added to the robot workspace to prevent any collision between operators and the robot of RU. Furthermore, some classical safety devices, e.g., barriers and rigid domes are so difficult to move and take time to reconfigure. As a result, more flexible and easy-to-deploy devices (e.g., safety plans, safety laser scanners, removable barriers) need to be used for more flexibility and reconfiguration (Fig. 2).

In an open space RAS, three cases with different safety issues are considered:

- *Case 1.* This case is related to possible interference between the robot's workspace and the operator when a MU is located in the vicinity of a RU. It is mandatory to add a safety plan (SP) device. SP is used to restrict the robot's allowed workspace by forcing the robot to stay on the correct side of the defined plans and never cross them.
- *Case 2.* In this case, a safety issue arises. An operator tries to enter the robotized area. For this reason, a removable barrier is added to SLS to ensure the operator's safety.
- *Case 3.* This case is related to possible interference between the respective workspaces of the robot and the operator when a MU is located in the vicinity of a RU. It is mandatory to add the safety plan (SP) device. SP is used to restrict the allowed working space of the robot by forcing the robot to stay on the correct side of the defined plans and never cross them.

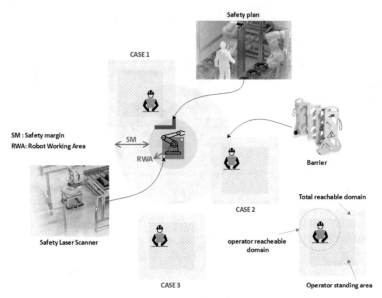

Fig. 2. Addressed safety issues

The adopted methodology considers two phases: "off-line" and "on-line" (Fig. 3). **Off-line phase:** This phase considers three main steps:

(i) Preliminary design and layout modelling

Firstly, a layout of the RAS cells composed of basic units is considered (Fig. 1). These layouts are afterwards modelled with a dedicated CAD tool. Layout modelling consists of positioning its representations of the robotized, conveying, and manual units on a polygonal grid composed of squares. At the end of this step, a precise layout of RAS cells is provided with the different basic units' locations. The next step is devoted to the safety study of the previously generated layout.

(ii) Safety design

The "safety bubble" approach is proposed to assist safety managers in implementing the safety devices (SLS, barriers, SP) and therefore to ensure operator safety. This approach is based on the detection of human intrusions in some hazardous areas such as robotic units. The "off-line" phase is supported by a software application based on the multi-agent Netlogo platform. Netlogo is a multi-agent programmable modelling environment for simulating complex systems created in 1999 by Uri Wilensky at Northwestern University [1].

The design of the safety bubble is structured in three steps:

1. First, an algorithm detects, for each MU, all possible cases as mentioned in Fig. 2.
2. Second, all safety devices' precise locations on the RAS layout map are determined.
3. Finally, a field file is automatically generated for each SLS.

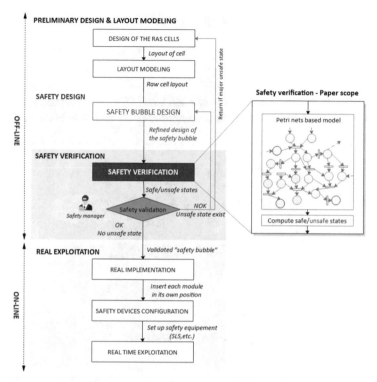

Fig. 3. The methodology adopted for each RAS cell

(iii) Safety verification

Once the refined design of the safety bubble is done, the verification step begins. In this regard, some models (Sect. 3) are developed to assist the safety manager and verify all the safety standards, namely the safety distances around each RU. If the verification result is not satisfactory, a new configuration of the safety bubble must be generated, or at worst (e.g., major problem unsolved by the safety bubble), a new RAS layout design must be generated. After having completed the safety verification, the on-line phase (i.e., real exploitation) can begin.

"On-line phase": In this phase, three main steps are considered:

(i) *Real Implementation.* Various units are placed in their respective locations. The units are coupled physically to allow the transfer of products.

(ii) *Safety devices configuration.* Then, all the safety devices are configured and placed in their locations.

(iii) *Real-time exploitation.* The RAS is fully operational in this phase. The safety devices are operational and share data to detect any human intrusion into the robotized areas.

In the next section, the case study is proposed to demonstrate our approach's applicability.

2.3 Case Study

In this section, we present a case study of a RAS cell (Fig. 4). The proposed approach is applied to each cell of a RAS which is composed of several cells in the general case.

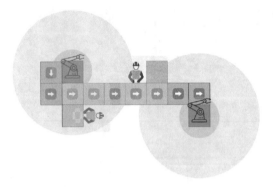

Fig. 4. RAS cell case study

Following the methodology described above (Fig. 3), the developed algorithms integrated in Netlogo have produced the results described in Fig. 5:

- The description of the safety limits for each RU,
- The precise locations of the barriers and the SLS,
- A field file for SLS (yellow zones) describing the geometry of the monitored areas.

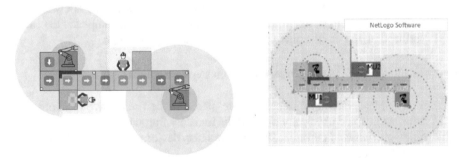

Fig. 5. Safety bubble design results

For more details about the safety bubble design results, the interested reader can refer to [3, 4]. The next section presents a formal verification model of the safety bubble.

3 Modelling and Formal Verification of the Safety Bubble

The objective of this section is to verify that there is no flaw in the designed safety bubble (Fig. 5), i.e., a possible violation of the safety rules. The failure of safety bubble design may lead to catastrophic effects, including human injury. Moreover, the behaviour of such systems is mainly characterized by complex phenomena such as concurrency and conflicts [11]. The diversity and complexity of these phenomena make the study of these systems more difficult and require the use of adequate tools and formalism. Among these tools and formalism, we underline the power and efficiency of Petri nets (PN), automata, multicriteria analysis, dioid algebra, model checking, to name a few.

In this paper, a formal model based on Petri nets is developed to prove and guarantee the design correctness of the safety bubble. Petri nets' accurate graphical and mathematical representation makes it possible to represent inherently logical interactions among parts or activities in a system [11]. Several situations can be modelled by PN such as synchronization, sequentially, concurrency, and conflict. A large body of theoretical results and practical tools have been developed around Petri nets [11, 12].

Before presenting the verification model, a brief introduction to Petri nets is given.

3.1 Petri Nets

Petri nets constitute a well-adapted formalism to the modelling and verification of complex systems [10]. Petri net is a bipartite graph composed of places and transitions that are connected by weighted arcs. Every place has a capacity and can hold a nonnegative number of tokens.

In this study, an extension of Petri nets is used - coloured Petri nets (CPN). CPN preserve useful properties of Petri nets and at the same time extend the initial formalism to allow the distinction between tokens. In this case, tokens may have "colours," i.e., data attached to them [11]. The arcs between transitions/places have expressions that describe the behaviour of the net. Thus, transitions describe actions and tokens carry values [12]. The CPN model permits hierarchical constructions and a strong mathematical theory has been built up around it [12]. In the following, CPN are used to model and formally verify the safety bubble approach.

3.2 Safety Bubble Model Description

In this section, we model the safety bubble using CPN to prove its correctness. The idea is to model the operators by tokens and test them with the safety bubble before the "on-line" phase on real human operators (Fig. 6). If a token manages to enter the robot's area without being detected, a flaw in the designed safety bubble is declared. This flaw is hence addressed to the safety manager to take major decisions: either to redesign the safety bubble from scratch or only add some safety devices to eliminate unsafe states.

The first PN model (Fig. 7) describes the movements of each token in RAS cell. The model consists of two elements: places and transitions. Place P represents a square (grey squares in Fig. 6) of the grid, transitions $T_{i,j}$, $1 \leq j \leq 8$ represent the possible movements from one square i to one of its neighbours, and finally places $PT_{i,j}$, $1 \leq j \leq 8$ represent

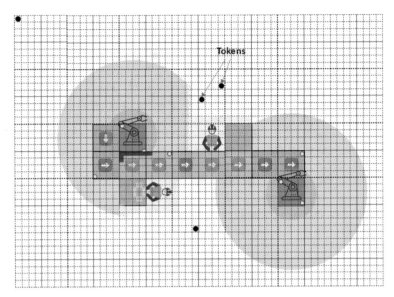

Fig. 6. Case study grid

movement authorization. If the movement from one square to another is feasible, the PT (representing this movement) contains a token.

Two cases are presented:

- *Case 1:* All directions are available (modelled by the existence of tokens in $PT_{i,j}$). In this case, the transitions $T_{i,j}$ are validated and only one will be fired. The one that will be fired is chosen randomly for possible movement.
- *Case 2:* Some directions are not available, given the existence of an obstacle (e.g., barrier, RAS units). For example, in Fig. 7, the token cannot move to the left positions given the presence of a barrier. As a result, no tokens are presented in $PT_{i,1}$, $PT_{i,7}$, and $PT_{i,8}$.

When a place P_i (located in the robot area) contains a token, a major flaw of the safety bubble is declared. Given the size of the Petri net for each square (Fig. 7), we have developed another more compact and complete model using coloured Petri nets (Fig. 8). The functioning of both models is the same, the only difference is that CPN is compact and include the concept of token colour.

The CPN model represents all the RAS cell units, the developed safety bubble, the movement of operators (transition Dpl), and flaws detection (place P_2). Furthermore, the developed model is modular, flexible, and highly adaptable to any RAS cell with several units and safety devices.

The transitions U1, U2, and U3 are fired one time. When U1 is fired the function FS puts n tokens representing the safety bubble squares (i.e., yellow squares in Fig. 5). Moreover, the function FI represents the injection function, i.e., the tokens that will be injected to verify the safety bubble design. Only one token is enough but we could inject several tokens. The transition Dpl models token movement from one square to one of its neighbours. This movement is done when there is no obstacle (presented by FE) such as a barrier, or RAS units. If, at a certain moment we fire the transition X (and eventually Y), then a major flaw exists. Otherwise, the safety bubble presents no flaws (Fig. 9). In fact, Fig. 9 shows CPN transitions (e.g., X et Y) firing number, i.e., if the firing number of the transitions X and Y is non-zero, then the RAS is completely unsafe.

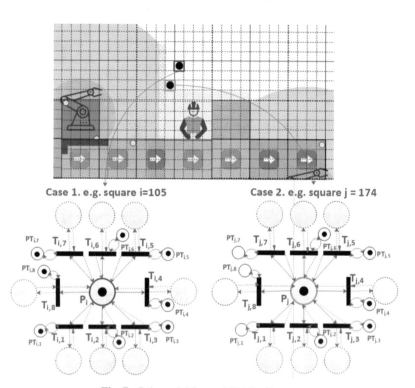

Fig. 7. Sub-model for each RAS cell square

This will assist the safety manager to take major decisions: redesign the safety bubble, add some safety devices to eliminate the unsafe states or move on to the next phase.

Some properties (liveness, deadlock-free) of the developed model are verified using CPN software tools. Moreover, the reachability graph is performed to detect flaws in the safety bubble developed in Sect. 2. The reachability graph determines all possible paths an operator can do/choose.

The developed model presents a limitation: combinatorial explosion. Due to the high complexity of the model (RAS with many cells and units), the computational time increases exponentially to determine the reachability graph. As a result, an optimization algorithm will be addressed in future work to overcome this limitation.

After validation of the RAS layout by the safety manager, the "on-line" phase can begin. This phase is out of this paper's scope.

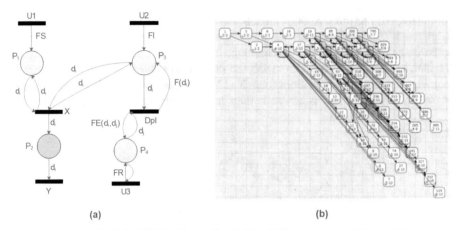

(a)   (b)

Fig. 8. (a) Global CPN of the studied RAS cell (b) state space of the model

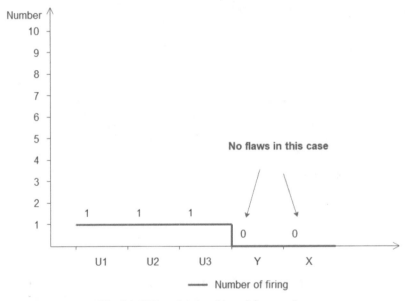

Fig. 9. CPN model transitions firing number

4   Conclusion

This article presents an extended version of the safety bubble concept to face the inadequacy of traditional approaches dealing with safety issues in an open RAS. Furthermore, some classical safety devices, e.g., barriers and rigid domes are so difficult to move and take time to reconfigure. As a result, more flexible and easy-to-deploy devices (e.g., safety plans, safety laser scanners, removable barriers) are used to increase flexibility and reconfiguration. Moreover, a formal model based on Petri nets is developed to prove and guarantee the design correctness of the safety bubble.

The effectiveness of the proposed approach has been validated for the "off-line" phase. In a future study, specific more user-friendly software must be developed to assist the safety manager. Indeed, the latter is not a Petri nets modelling expert and the usage of this formal model must be rendered easier for them.

References

1. Mehrabi, M.G., Ulsoy, A.G., Koren, Y.: Reconfigurable manufacturing systems: key to future manufacturing. J. Intell. Manuf. **11**(4), 403–419 (2000)
2. Koren, Y., et al.: Reconfigurable manufacturing systems. CIRP Ann. Manuf. Technol. **48**(2), 527–540 (1999)
3. Sallez, Y., Berger, T., Bonte, T.: The concept of "safety bubble" for reconfigurable assembly systems. Manufacturing Letters **24**, 77–81 (2020)
4. Sallez, Y., Berger, T.: How to build a cooperative safety bubble for a reconfigurable assembly system? In: Trentesaux, B., Cardin, T. (eds.) Service Orientation in Holonic and Multi-Agent Manufacturing, Studies in Computational Intelligence, vol. 762, pp. 187–197. Springer (2018)
5. ISO 10218-1: Robots and robotic devices–Safety requirements for industrial robots-Part 1: Robots, Geneva (2011)
6. ISO 12100: Safety of machinery – General principles for design – Risk assessment and risk reduction (2011)
7. Koo, C.H., et al.: Challenges and requirements for the safety compliant operation of reconfigurable manufacturing systems. Procedia CIRP **72**, 1100–1105 (2018)
8. ISO 13855: Safety of machinery-Positioning of safeguards with respect to the approach speeds of parts of the human body, https://www.iso.org/standard/42845.html
9. Benyoucef, L. (ed.): Reconfigurable Manufacturing Systems: From Design to Implementation. SSAM, Springer, Cham (2020). https://doi.org/10.1007/978-3-030-28782-5
10. Silva, M., Valette, R.: Petri nets and flexible manufacturing. In: Rozenberg, G. (ed.) APN 1988. LNCS, vol. 424, pp. 374–417. Springer, Heidelberg (1990). https://doi.org/10.1007/3-540-52494-0_38
11. Peterson, J.L.: A note on colored Petri nets. Inf. Process. Lett. **11**(1), 40–43 (1980)
12. Cortés, L.A.: A Petri net based modeling and verification technique for real-time embedded systems. Thesis, School of Engineering at Linköping University, Sweden, Diss (2001)
13. Tisue, S., Wilensky, U.: NetLogo: a Simple Environment for Modeling Complexity, Center for Connected Learning and Computer-Based Modeling Northwestern University, Evanston, Illinois, University (1999) http://ccl.northwestern.edu/netlogo/

Designing a Multi-agent Control System for a Reconfigurable Manufacturing System

Alexandru Matei[1]([⊠]) [ID], Bogdan Constantin Pirvu[1] [ID], Radu Emanuil Petruse[1] [ID], Ciprian Candea[2] [ID], and Bala Constantin Zamfirescu[1] [ID]

[1] Lucian Blaga University of Sibiu, Victoriei Blvd. 10, 550024 Sibiu, Romania
{alex.matei,constantin.zamfirescu}@ulbsibiu.ro
[2] Ropardo, Reconstructiei 2A, 550129 Sibiu, Romania
ciprian.candea@ropardo.ro

Abstract. This paper introduces a multi-agent control system for a reconfigurable manufacturing system designed to provide, in the context of Industry 4.0, test-before-invest services by the FIT EDIH. The product assembled is flexible, with multiple possible assembly sequences. The manufacturing system is composed of multiple interchangeable manufacturing cells that allow any layout configuration with autonomous transporter units that move intermediate products from one cell to another. The reference architecture used to instantiate the developed prototype and its multi-agent control system with the required agents, concept and predicate ontology are subsequently presented. The hardware and software implementation are also detailed. The multi-agent system is implemented using SPADE framework. Each manufacturing cell is controlled using 4diac framework while transporters use Robot Operating System. Manufacturing orders are placed by a client using a web interface.

Keywords: Industry 4.0 · Multi-agent system · Intelligent manufacturing system · Reconfigurable manufacturing system · Cyber-physical system

1 Introduction

The global competition and mass customization are pushing companies to produce better products at a faster pace at an affordable price. For manufacturing systems this implies the capacity to cope effectively with lot size in production. Thus, subsystems and/or workstations of production systems have more processing and communication capabilities to improve resource allocation (i.e., who goes where does what) or their composition (i.e., configuration of a workstation/sub-system for a given product). This unprecedented interaction and collaboration within a network of artefacts is called Industry 4.0 and became possible by the progress in ICT and by decentralizing and/or distributing decision-making in more intelligent subsystems throughout the whole production system. Going a step forward, considering sustainability and social aspects as well, Industry 5.0 is the coined term in Europe [1] for human-centred production, where the Industry 4.0 technology is augmented with the flexibility, adaptivity and creativity of humans towards a man-machine symbiosis.

© The Author(s), under exclusive license to Springer Nature Switzerland AG 2023
T. Borangiu et al. (Eds.): SOHOMA 2022, SCI 1083, pp. 434–445, 2023.
https://doi.org/10.1007/978-3-031-24291-5_34

Developing and utilizing such manufacturing systems is not trivial and not cheap, especially for SMEs. European Digital Innovation Hubs (EDIH) [2] are a key instrument within the European strategy to support the digital transformation throughout all economic sectors and regions by providing four main services: training, testing technologies before investing in them, augmentation of the innovation ecosystem and support to find investments. Adequate testbeds to showcase, experiment and train are required to understand how intelligent manufacturing systems work and what their impact can be on an SME.

In the following sections the collaborative manufacturing system within our FIT EDIH is presented. Section 2 provides the motivation for this work. Section 3 briefly describes the reference architecture together with the hardware description of the system prototype. Section 4 presents the design of the multi-agent control system (MAS) for the reconfigurable manufacturing system (RMS) together with software implementation details regarding the MAS. The last section concludes the paper together with future developments within the FIT EDIH ecosystem.

2 Rationale

As a result of modern manufacturing processes requirements which face an increase of product complexity and a market demand for customization, one can observe during the last decade an increased interest for Industry 4.0-related topics demonstrated by a surge of scientific publications [3]. Additionally, industrial disruptors (e.g., impact of electric vehicles on the automotive ecosystem [4]) contribute to the evolution of manufacturing processes, as demand prediction is becoming increasingly challenging. Improving production systems was always a top priority, one of the pioneering researchers being Martin K. Starr who introduced in 1965 the modular production concept [5] suggesting that production modularity can be a solution to avoid manufacturing offshoring. Further, the author reanalysed his early publication in 2010 [6] where he emphasized the relevancy of the initial concept considering current production challenges.

Nowadays, modular production systems, a backbone for Industry 4.0, have transitioned from the conceptual phase towards commercial applications achieving a superior technological maturity. In literature, solutions for solving the optimization problems of RMS include [7, 8]: simulated annealing, genetic algorithms, multiple objective particle swarm optimization, decision trees together with Markov analysis, etc. Moreover, recent supply chains disruptions caused by COVID-19 accelerated the production systems innovation. In a recent literature review [7], a classification of RMSs is made based on the configuration level: both system and machine level, system level with or without layout design, and machine configuration.

Looking at the latest developed modular RMSs in Table 1, we can characterize them by the following features: mass customization (MC) or product modularity (PM) [9]; human-centred production (HCP); commercial (C) or testbed (TB). The RMS in this paper is an HCP testbed that allows both product modularity and mass customization.

Besides having a system and machine level configuration, the designed manufacturing prototype was motivated by the real need to have full accessibility to all hardware and software components without requiring prohibitive third-party interventions for the

Table 1. Selection of available industry or research modular manufacturing systems

Source	MC/PM	HCP	C/TB
UINST Testbed [10]	Both	No	TB
SMART [11]	PM	No	TB
SmartFactory^{KL} [12, 13]	PM	No	TB
Bosch Manufacturing Solutions [14]	Both	Yes	C
ScalABLE4.0 [15]	MC	No	TB
EID Robotics [16]	MC	No	C
Huawei I4.0 Testbed [17]	MC	No	TB
Testbed Prague [18]	MC	Yes	TB
Innovation Lab Testbed [19]	Both	Yes	TB
Industry 4.0 Testlab [20]	PM	Yes	TB
This paper	*Both*	*Yes*	*TB*

specific requirements of the regional SMEs supported by FIT EDIH. Therefore, all the hardware components are standard in industrial automation, while for the software ones open-source frameworks and platforms were chosen. The prototype required function-alities were developed in collaboration with the regional companies that are partners of Smart Factory Romania [21].

3   System Design

3.1   Reference Architecture

This section synthetizes the reference architecture used to develop the RMS prototype. A detailed description of the reference architecture was given in [22]; it is a synthesis of some high-level architectures: OSMOSE [23], IoT-A [24], and BEinCPPS [25]. This architecture has BEinCPPS structural perspective as a starting point, a middle domain is added in a similar way to OSMOSE philosophy, while IoT-A is used as a guidance for the underlying architectural reference model.

Figure 1 presents the functional perspective of the reference architecture, called SoRA (Socio-centric Reference Architecture). From the structural viewpoint, it has three domains: *Design, Socio Cyber-Physical* and *Execution.* The *Design* domain is dedicated to cyber-physical-systems (CPS) design with a focus on the human and social factors. The *Socio Cyber-Physical* domain manages processes, actions, and system through which the human factor is prepared and trained to work in the cyber-physical environment. At the same time, the cyber-physical environment is adapted to the characteristics of the human factor as described in the *Design* domain. Briefly, it helps in achieving the required balance between social and cyber-physical factors. The *Execution* domain consists of all the necessary resources to achieve the system functionality.

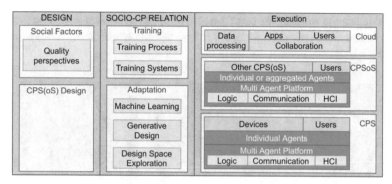

Fig. 1. Functional diagram of SoRA (adapted from [22])

Social factors define relevant human factor aspects like quality standards, security and safety standards, ergonomics, etc. *Training* defines the processes, documentation, and training systems that an operator can follow and access to familiarize with Cyber-Physical System of Systems (CPSoS). *Adaption* consists of several system capabilities that model the operator, work environment and CPS to improve and adapt the manufacturing processes. *Cloud* contains data processing functions and user collaboration applications. *CPSoS* represents the factory level, aggregating multiple CPS from the bottom layer together with external systems. *CPS* represents the device level of the physical equipment, and the human operators present on the manufacturing floor.

When instantiating this reference architecture, in the *Design* phase we decided that some the manufacturing cells will use cobots that allow human-robot collaboration, if needed for some advanced assembly processes. In this case, the human operator presence is not a must, but the human can always participate actively in the manufacturing process. The *Socio Cyber-Physical* domain is tackled within a training station for manual operation detailed in [26], while the CPS *adaptation* is explored in some previous works [27, 28]. The *Cloud* is represented by a database and a web application that can access the MAS of the manufacturing system. The *CPSoS* is the entire physical RMS that is composed of multiple *CPS*, each *CPS* being a manufacturing cell.

3.2 Hardware Description

The developed RMS is depicted in Fig. 2. The system is composed of different manufacturing cells that can be interconnected on any of the four sides using special connectors together with an autonomous AGV that transports the intermediate products from one cell to another.

The system cells have an aluminium frame, a standard square footprint, and a Han-Modular connector on each side for easy connection to another cell. Each cell is equipped with a programmable logic controller (PLC) that controls low level devices (actuators, valves, motors, etc.), a network router that manages the Ethernet connection to other manufacturing cells and a small factory PC with integrated screen for interfacing with human operators and other advanced devices or systems (cobots, CNC, MAS, data storage, etc.). Each cell is also equipped with sliding windows that, when lowered, fence the AGV inside the working area and can also be raised to allow the AGV to move from one cell to another. A possible configuration of several manufacturing cells can be seen in Fig. 2.

Fig. 2. Reconfigurable manufacturing system (Left). Assembled product (Right) with all slots personalized (top), intermediate assembly (centre), modules (bottom)

Next, we detail each manufacturing cell's responsibilities.

- **Warehouse** cell: contains a place to store product parts together with a cobot that can load the product parts into the storage area or to unload product parts onto the AGV for transportation.
- **Assembly** cell: contains a pneumatic mechanism to unload parts carried by the AGV and assemble them.
- **Customization** cell: contains tools to personalize the product by engraving an image using a CNC.
- **Testing** cell: contains a verification device – industrial camera – that checks if the product is assembled and personalized correctly.
- **Charging** cell: acts as a parking and charging station for AGVs.
- **Packaging** cell: contains a cobot that unloads the final product from the AGV and places it inside a packaging box. This cell can also be accompanied by a human operator that can work collaboratively with the robot to further customize the product and order by placing stickers, paint, smooth the rough edges if any, or do other special requests from the client for example.
- **Barebones:** is an empty module that can be used as parking space for AGVs. it represents the starting point for creating new manufacturing cells.

The product that is assembled on this RMS is a modular tablet - see Fig. 2 - composed of a main screen, a bus with six slots at which three types of modules can be connected: battery module, speaker module and flashlight module. The modules are available in different colours and can be further customized by engraving or marking with stickers. Each module can be connected to any of the six available slots, increasing the number of possible final tablet configurations that the final user can order.

Figure 3 describes the possible assembly sequences depending on the product customization. The technological process is flexible enough to allow multiple paths in assembling a customized product for the end user. The system can start with any of the following three operations: customize the modules, assemble the bus with modules or assemble the bus with the tablet. Because they don't impose a fixed assembly sequence, the manufacturing cells can be assigned a more balanced workload.

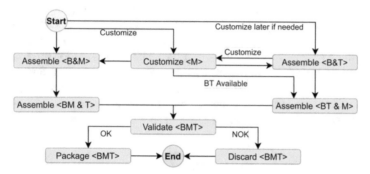

Fig. 3. Technological process for product assembly. B – Bus; M – Modules; T – Tablet Screen

4 Prototype Implementation

4.1 Multi-agent Control System

To achieve the information model specific for the reference architecture associated with the manufacturing line, a multi-agent simulation was performed. The MAS presented follows the IEEE standard on Industrial Agents [29], a hybrid loosely coupled interaction mode between high level control (agents) and low-level control (real-time hardware) devices. For the design of the simulation and control layers of the system, the classic method of developing a MAS was used, which involves the main phases: identifying the agents, describing the interaction between agents and specific behaviours, defining the ontology, implementing, and testing.

In the first step, we identified the following classes of agents: *Client*, *Order*, *Order Management* (OMA), *Knowledge Management* (KMA), and *Resource* agents for every manufacturing cell and transporter units. The *Client* agent represents the human client that places orders in the system using an intuitive user interface. It can create, view, update or delete orders by interacting with the OMA, fetch possible product configurations and customization from KMA and monitor the ongoing orders by interrogating the *Order* agent assigned to the order placed by the client. The *Order Management* agent manages current orders and processes order requests from the *Client* agent. It also has the responsibility to instantiate or terminate new order agents when necessary. An *Order* agent is created for each new order placed in the system. It manages a single order, and has the responsibility to plan and negotiate with other resource agents the execution of the order. It queries the KMA regarding its personalized order to get a complete recipe, task and activities needed to be followed to execute the order. It communicates with the order agents and negotiates, assigns, and monitors the tasks. KMA contains information about the RMS and the products that can be manufactured using the current layout, infrastructure, and parts available. It responds to client agent requests regarding the available products and possible customizations to be made or to order agents regarding assembly recipes and tasks specific to its current order.

Each of the *Resource* agents - *Warehouse, Assembly, Customization, Testing, Charging, Packaging* and *AGV* (Autonomous Guided Vehicle) are responsible to advertise their manufacturing or transport services, negotiate task execution and schedule accepted tasks. Negotiating tasks involves finding an available time window based on the request from an *Order* agent that meets requirements like maximum execution time, earliest start time or execution deadline. In response to the order agent's request, they can accept, decline or propose a new task execution window that meets the order requirements. Figure 4 illustrates a snippet of the interaction sequence after the task negotiation phase between an *AGV* agent, a *Warehouse* agent and an *Assembly* agent.

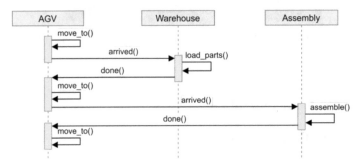

Fig. 4. Interaction sequence diagram between agents

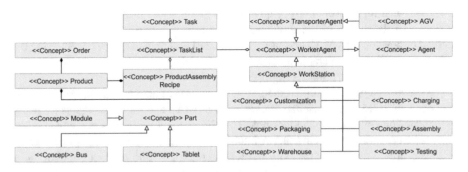

Fig. 5. Concept ontology

Figures 5 and 6 present relevant parts of the concept ontology and predicate ontology respectively, both created to define the RMS's MAS. The concept ontology describes in detail each concept, and highlights the relationships between concepts as some more advanced concepts inherit from basic concepts. This ontology contains concepts for

Parts, Product, Agents, Tasks and Recipes. In an ontology, predicates play the role of connecting elements between different concepts, describing actions and states. The ontology of predicates is graphically highlighted with green in Fig. 6, together with their relationship to other concepts. It contains the necessary predicates that describe transport actions or capabilities, task dependencies, part reservation or the specification of the parts that compose a certain product.

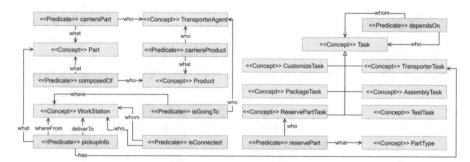

Fig. 6. Predicate ontology

Beside defining the system on an abstract semantic level, the concepts and predicates defined in the ontology contain the domain knowledge and are used by the agents of the MAS to communicate in a structured yet flexible way. The ontologies define the information model efficiently, allowing easy data and information retrieval. Ontologies are also meant to facilitate collaboration and prototype usage by dispersed team members belonging to different SMEs.

4.2   Software Implementation Details

For the client user interface, a web portal was developed (see Fig. 7) using Node.js and the Quasar framework connected to the MAS through an API that allows configuring the manufacturing line and placing manufacturing orders to the system.

The advantages of this framework include that it is open source, its compatibility with different browsers and auto adaptation to screen size, that make it easy to be accessed from a mobile device without much code changes. The web portal has features like creating an account for clients, placing personalized order for logged in clients or monitoring order status. An administrator user has features such as the possibility to configure product parts, to configure available product personalization modes, to enable manufacturing cells. It is possible to visualize and monitor all orders and their status; an overview of the entire MAS or for each order is achievable through a message sequence together with message details and timestamps. For archiving purposes, an SQL database managed by PostgreSQL can be used.

For the selection of the MAS development framework, the communication capabilities of the platform, the scalability, transparent integration of people and agents and support for integration with IoT systems were considered. For the MAS implementation, we used the SPADE (Smart Python Agent Development Environment) [30] framework,

Fig. 7. Monitoring interface. Right; overview of MAS. Left: single order message sequence

an open-source MAS platform available on [31]. It is written in Python and uses the XMPP (extensible messaging and presence protocol) instant messaging protocol for communication between agents; it supports FIPA metadata and has a web-based interface available. SPADE is also capable to integrate with other FIPA compliant MAS platforms due to the flexibility of XMPP [30]. The programming model for SPADE agents is based on behaviours including not only classical behaviors such as: Cyclic, One-Shot, Periodic, Time-Out and Finite State Machine, but also BDI (belief desire intention) behaviour [32]. It allows mixing of procedural, object-oriented and logic programming in the same agent, as well as creating personalized, complex behaviours. All agents execute inside an Agent Logic Container (ALC). Communication with the *Web Interface* component is done through the API exposed by ALC.

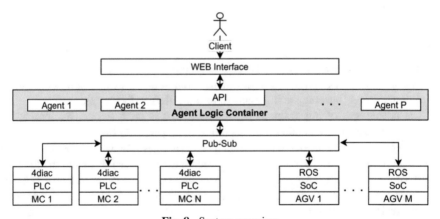

Fig. 8. System overview

AGVs are controlled on low level by a microcontroller and on high level by an SoC that runs Robot Operating System (ROS) on top of a Linux distribution. Low level control manages motors, drivers, battery, sensors, and wireless charging. High level control manages AGV functions like environment mapping, positioning inside the map, path planning and obstacle avoidance. Each manufacturing cell PLC is programmed using 4diac, an open source framework for distributed control of industrial processes based on IEC61499 industrial standard.

Figure 8 offers an overview of the information flow and the main systems involved, from high level control systems and human interfaces, down to the hardware devices level.

The communication between hardware level control – 4diac, ROS – and MAS is realized through MQTT pub-sub protocol for its lightweight, IoT oriented capabilities. Each manufacturing cell has its own id that is used to define its topic filter. The JSON formatted messages received from MAS are translated by 4diac or ROS to the hardware system. Table 2 contains a part of the low-level interface for the assembly cell: the topics preceded by the station id and message required fields.

Table 2. Assembly Station low level interface

Subscribing Topics	Publishing Topics
Init Check Topic: *id*/Initialization	Returns initialization status Topic: *id*/InitializationOut RETURN {"status": X, ["failed": "debug message"]}
Check Status Topic: *id*/CheckStatus	Returns station status [*free/busy/fault*] Topic: *id*/CheckStatusOut RETURN {"status":X}
Extract Topic: *id*/Extract Body: {"Position1":True, "Position2": False, …}	Returns extraction status [done/fault] Topic: *id*/ExtractOut RETURN {"status": X, ["failed": "debug message"]}
Deploy Topic: *id*/Deploy Body: {"Position1":True, "Position2": False, …}	Returns deploy status [done/fault] Topic: *id*/DeployOut RETURN {"status": X, ["failed": "debug message"]}

5   Discussion and Further Development

The paper presents a prototype of an RMS that is controlled by a MAS developed using SPADE framework. It provides basic functionalities for real-time layout reconfiguration, mass customization, and product modularity. The prototype will be exploited as a testbench within the FIT EDIH by interested SMEs that will have the opportunity in the next three years to test-before-invest in different scenarios based on their specific use cases. This will be done in the recently won project FIT EDIH in the call DIGITAL-2021-EDIH-01 funded by the European Commission.

Therefore, its further development is market-driven and depends on the use-cases required by regional SMEs. Apart from the AGV, the implementation was straightforward and did not pose any specific challenge. Requiring a suitable small size AGV, that was unavailable on the market, we had to build it inhouse. While developing the AGV, we encountered hardware issues in integrating all the necessary components in a small factor encasing, or software issues in localization and mapping due to sensor being partially obstructed that needed AI augmentation [26]. Therefore, the control systems have not been yet tested with multiple AGVs in human-robot collaboration scenarios.

Acknowledgements. This work is supported by the DiFiCIL project (contract no. 69/08.09.2016, ID P_37_771, web: http://dificil.grants.ulbsibiu.ro), co-funded by ERDF through the Competitiveness Operational Programme 2014–2020.

References

1. Renda, A.. et al.: Industry 5.0, a Transformative Vision for Europe: Governing Systemic Transformations towards a Sustainable Industry, European Commission. Directorate-General for Research and Innovation (2022). https://doi.org/10.2777/17322
2. European Digital Innovation Hubs | Shaping Europe's digital future, https://digital-strategy. ec.europa.eu/en/activities/edihs, Accessed July 2022
3. Miqueo, A., Torralba, M., Yagüe-Fabra, J.A.: Lean manual assembly 4.0: a systematic review. Appl. Sci. **10**, 8555 (2020). https://doi.org/10.3390/app10238555
4. Arden, W.: The EV disruption (2019). https://www.fia.org/marketvoice/articles/ev-disruption, Accessed July 2022
5. Starr, M.K.: Modular production - A new concept. Harv. Bus. Rev. **43**(6), 131–142 (1965)
6. Starr, M.K.: Modular production – a 45-year-old concept. Int. J. Oper. Prod. Manag. **30**(1), 7–19 (2010). https://doi.org/10.1108/01443571011012352
7. Sabioni, R.C., Daaboul, J., Le Duigou, J.: Optimization of reconfigurable manufacturing systems configuration: a literature review. In: Roucoules, L., Paredes, M., Eynard, B., Morer, P., Rizzi, C. (eds.) JCM 2020. LNME, pp. 426–435. Springer, Cham (2021). https://doi.org/10.1007/978-3-030-70566-4_67
8. Sabioni, R.C., Daaboul, J., Le Duigou, J.: Concurrent optimisation of modular product and Reconfigurable Manufacturing System configuration: a customer-oriented offer for mass customisation. Int. J. Prod. Res. **60**(7), 2275–2291 (2022). https://doi.org/10.1080/00207543.2021.1886369
9. Doran, D., Hill, A.: A review of modular strategies and architecture within manufacturing operations. Proc. Institute. Mech. Eng. Part D: J. Automobile Eng. **223**(1), 65–75 (2009). https://doi.org/10.1243/09544070JAUTO822
10. Kim, D.-Y., et al.: A modular factory testbed for the rapid reconfiguration of manufacturing systems. J. Intell. Manuf. **31**(3), 661–680 (2019). https://doi.org/10.1007/s10845-019-01471-2
11. Kovalenko, I., Saez, M., Barton, K., Tilbury, D.: SMART: A system-level manufacturing and automation research testbed. Smart Sustain. Manufact. Syst. **1**(1), 232–261 (2017). https://doi.org/10.1520/SSMS20170006
12. Gorecky, D., Weyer, S., Hennecke, A., Zühlke, D.: Design and instantiation of a modular system architecture for smart factories. IFAC-PapersOnLine **49**(31), 79–84 (2016). https://doi.org/10.1016/j.ifacol.2016.12.165
13. First skill-based application in operation in Kaiserslautern. https://smartfactory.de/en/first-skill-based-application-in-operation-in-kaiserslautern/, Accessed July 2022
14. BOSCH. How future production looks like – The Modular Production System. https://www.boschmanufacturingsolutions.com/news-and-highlights/modular-production-system/, Accessed July 2022

15. ScalABLE4.0. https://www.inesctec.pt/en/projects/scalable4-0, Accessed July 2022
16. EID Robotics. Building an Automated Assembly Line. https://eidrobotics.com/building-an-automated-assembly-line-how-to-minimize-costs-and-ramp-up-time/, Accessed: July 2022
17. Huawei. Introducing Huawei Industry 4.0 Test Bed. https://e.huawei.com/en/material/local/ccb33d2eeab84e68a70938335d1a6e83, Accessed July 2022
18. Testbed Prague. https://ricaip.eu/testbed-prague/, Accessed July 2022
19. Watson, L.: Use a testbed to enable Industry 4.0 manufacturing. Manufacturing AUTOMA-TION (2021). https://www.automationmag.com/use-a-testbed-to-enable-industry-4-0-manufacturing/, Accessed May 2021
20. Industry 4.0 Testlab. https://techlab.uts.edu.au/lab/industry-4-0-testlab/, Accessed July 2022
21. Smart Factory Romania. https://smartfactoryromania.ro/en/, Accessed June 2022
22. Ghetiu, T., Pirvu, B.: Insights into SoRa: a reference architecture for cyber-physical social systems in the industry 4.0 Era. In: Proceedings of the International Conference on Innovative Intelligent Industrial Production and Logistics - IN4PL, ISBN 978–989–758–476–3, pp. 47–52 (2020). doi: https://doi.org/10.5220/0009982300470052
23. Gusmeroli, S., et al.: OSMOSE: a paradigm for the liquid-sensing enterprise. In: IWEI Workshops (2015)
24. Bauer, M., et al.: Internet of Things – Architecture IoT-A Deliverable D1.5 – Final architectural reference model for the IoT v3.0. IoT-A (257521) (2013)
25. Isaja, M., Fischer, K., Rotondi, D., Coscia, E., Rooker, M.: D2.2 - BEinCPPS Architecture & Business Processes, Ref. Ares(2017)3817966 - 30/07/2017 (2017)
26. Matei, A., Țocu, N.-A., Zamfirescu, C.-B., Gellert, A., Neghină, M.: Engineering a digital twin for manual assembling. In: Margaria, T., Steffen, B. (eds.) ISoLA. LNCS, vol. 12479, pp. 140–152. Springer, Cham (2021). https://doi.org/10.1007/978-3-030-83723-5_10
27. Gellert, A., Sorostinean, R., Pirvu, B.C.: Robust assembly assistance using informed tree search with Markov chains. Sensors. 22(2), 495 (2022). https://doi.org/10.3390/s22020495
28. Gellert, A., Sarbu, D., Precup, S.-A., Matei, A., Circa, D., Zamfirescu, C.B.: Estimation of missing LiDAR data for accurate AGV localization. IEEE Access 10, 68416–68428 (2022). https://doi.org/10.1109/ACCESS.2022.3185763
29. IEEE Standards Committee: IEEE Recommended Practice for Industrial Agents: Integration of Software Agents and Low-Level Automation Functions. IEEE Std 2660(1–2020), 1–43 (2021). https://doi.org/10.1109/IEEESTD.2021.9340089
30. Palanca, J., Terrasa, A., Julian, V., Carrascosa, C.: SPADE 3: supporting the new generation of multi-agent systems. IEEE Access 8, 182537–182549 (2020). https://doi.org/10.1109/ACCESS.2020.3027357
31. Palanca, J.: SPADE Documentation. https://spade-mas.readthedocs.io/en/latest/index.html, Accessed June 2022
32. Palanca, J., Rincon, J., Julian, V., Carrascosa, C., Terrasa, A.: Developing IoT artifacts in a MAS platform. Electronics 11, 655 (2022). https://doi.org/10.3390/electronics11040655

Author Index

© The Editor(s) (if applicable) and The Author(s), under exclusive license
to Springer Nature Switzerland AG 2023
T. Borangiu et al. (Eds.): SOHOMA 2022, SCI 1083, pp. 447–448, 2023.
https://doi.org/10.1007/978-3-031-24291-5

Printed in the United States
by Baker & Taylor Publisher Services